MW01502901

Carbon Nanoforms and Applications

Maheshwar Sharon

Madhuri Sharon

New York   Chicago   San Francisco
Lisbon   London   Madrid   Mexico City
Milan   New Delhi   San Juan
Seoul   Singapore   Sydney   Toronto

The *McGraw·Hill* Companies

Library of Congress Cataloging-in-Publication Data

Carbon nanoforms and applications / [edited by] Maheshwar Sharon, Madhuri Sharon.
    p. ; cm.
  Includes bibliographical references and index.
  ISBN 978-0-07-163960-6 (alk. paper)
  1. Nanostructured materials.   2. Carbon.   I. Sharon, Maheshwar.
II. Sharon, Madhuri.
  [DNLM: 1. Nanostructures.   2. Carbon.   3. Nanotechnology—methods.
QT 36.5 C264   2010]
  TA418.9.N35C337   2010
  620.1'93—dc22   2009030969

McGraw-Hill books are available at special quantity discounts to use as premiums
and sales promotions, or for use in corporate training programs. To contact a representative please e-mail us at bulksales@mcgraw-hill.com.

Carbon Nanoforms and Applications

1 2 3 4 5 6 7 8 9 0   DOC/DOC   0 1 4 3 2 1 0 9

ISBN 978-0-07-163960-6
MHID 0-07-163960-8

The pages within this book were printed on acid-free paper.

Sponsoring Editor
  Taisuke Soda

Acquisitions Coordinator
  Michael Mulcahy

Editorial Supervisor
  David E. Fogarty

Project Manager
  Smita Rajan

Copy Editor
  Danielle Shaw

Proofreader
  Carol Shields

Production Supervisor
  Richard C. Ruzycka

Composition
  Glyph International

Art Director, Cover
  Jeff Weeks

Dedicated to
Our beloved parents

About the Authors

Maheshwar Sharon is a retired professor of Chemistry from IIT Bombay and Professor Emeritus, CSIR and UGC. He is currently heading a Nanotechnology Research Center as an Adjunct Professor at Birla College, Kalyan, Maharashtra, India, with a team of 16 research scholars, working on various aspects of nanoscience and nano-technology. Dr. Sharon was the first professor to initiate and shape carbon nanotechnology at IIT Bombay. He is Fellow of the Royal Society of Chemistry (1968-2003), Fellow of the Electrochemical Society, Life Member of the National Academy of Sciences, Allahabad and the Carbon Society of India.

Madhuri Sharon is currently Adjunct Professor at the University of Mumbai, Principal at S.I.C.E.S College Mumbai, and Managing Director of Monad Nanotech Private Ltd., the first Indian company to commercially produce carbon nanomaterials and conducting glasses. She was Vice President at Gufic Biosciences and Director at Reliance Industries Ltd., of India. As Adjunct Professor of Mumbai University Dr. Sharon is guiding Ph.D. students working in different fields of nano-bio-technology.

Contents

Foreword

When macroscopic amounts of the molecule C_{60} and other fullerenes suddenly became available in 1990 and it was possible to probe the chemistry and physics of this unique material, one of the most fascinating and surprising spin-offs was the discovery of the nanotubes, which are the elongated cousins of the fullerenes. The fullerenes and nanotubes have given rise to vast new areas of carbon materials research. Not only did the discoveries shed totally new light on the structure of carbonaceous materials on a nanoscale, but they have also opened up vast new vistas of advanced materials behavior.

Although some 20 years have now passed since the original discoveries and vast amounts of new knowledge have been learned, there is still an infinite amount that is still not understood, and the fascination with these materials is almost as great now as at first. Indeed, the amazing promise of these novel materials in advanced tensile properties, as well as electrical and magnetic behavior, has still to be realized.

This book contains an excellent range of chapters covering much of the present state of our knowledge, and it will undoubtedly catalyze new studies, resulting in the development of further new understanding of pure carbon atoms, an amazing materials science. The book covers areas of fundamental science as well as analytical techniques and applications.

The fascinating promise for biologic advances is also covered. The chapters in this compendium provide an ideal introduction to the field and give a clear overall understanding of the strategically exciting field of carbon nanoscale science.

Sir Harold W. Kroto
Nobel Laureate
Florida State University
Department of Chemistry & Biochemistry
Tallahassee, Florida 32306-4390
Tel: 850 644 8274
Email: kroto@chem.fsu.edu
Web: www.kroto.info or www.vega.org.uk

Preface

Nanotechnology has become one of the essential topics in research and developing technologies for their applications. Many universities around the world are initiating postgraduate courses leading to MTech, MSc, and PhD degrees in various areas related to nanotechnology. Scientists have been developing and working on nanomaterials of various types of metals, such as tungsten, colloidal gold, and silver, but among these materials, carbon nanomaterials (CNMs) appear to take a lead in their applications.

The important areas are synthesis, characterization, and application of CNMs. The application of CNMs as an electron field emitter is almost at the verge of being developed as a technology for flat-screen displays used in computers and other devices. Storage of hydrogen and the development of fuel cells and supercapacitors using CNMs are also becoming very promising areas; very soon they will also become reality in their application in developing electric-driven vehicles. Because carbon is biologically compatible, CNMs are attracting attention for their application in biotechnologic systems such as drug delivery, treatment of malignancy, purification of drinking water, and so on.

Thus, considering the applications of CNMs in various disciplines of science, they are very likely to become a major part of nanotechnology. The syllabi of many nanoscience courses leading to MSc, MTech, and PhD programs include CNMs in two to three course papers. Keeping this in mind, it was thought appropriate to write a book called *Carbon Nanoforms and Applications*.

The selection of writers for the book was done very carefully. All of the authors who have contributed their chapters for the book are scientists working in their areas for several years. They have detailed their experiences in their chapters.

The book starts with a description of various forms of CNMs and deals with the necessity of developing a new classification of carbon, discussing why nanoforms of carbon are found to possess p-type character even though they are intrinsically pure and why they show high reactivity compared with macrosized particles.

Synthesis of CNMs is an important part of this technology; hence, several chapters are devoted to discussing various techniques used for the synthesis of CNMs. Techniques that are normally adopted to characterize these materials have also been included. A few chapters are devoted to discussing the physicochemical applications of CNMs, with special emphasis on fuel cells, supercapacitors, hydrogen storage, microwave absorptions, electron field emission, and lithium batteries. Other chapters are devoted to the application of CNMs in biosystems. For example, there are chapters on treating malignancy, delivering drugs, and antimicrobial activity of nanocarbon. Other optics covered are carbon nanomaterials as a scaffold for tissue culture and the possible role of nanocarbon in neurodegenerative diseases.

Prof. Maheshwar Sharon
Dr. Madhuri Sharon
Mumbai, India

PART I

Nanoforms of Carbon

CHAPTER 1

Introduction to the Nanoworld of Carbon

Maheshwar Sharon

Nanotechnology Research Centre
Birla College, Kalyan, India

Madhuri Sharon

MONAD Nanotech Pvt. Ltd.
Powai, Mumbai, India

Nanotechnology is the act of purposefully manipulating matter at the atomic scale, otherwise known as the "nanoscale."

1.1   Introduction

The fascination and interest humans have in smaller things and the idea that smaller things have greater potential have prompted us to miniaturize things, resulting in the creation of laptop computers, palmtops in place of desktop computers, and microchips that can control such machines. Now the world is demanding still smaller things that will be more efficient than today's things. So scientists have developed a technology called *nanotechnology*.

Nanotechnology is derived from the Greek words *nanos*, which means dwarf, and *technologia*, which means *systematic treatment of an art*. The term *nanotechnology* was coined in 1976 by Norio Taniguchi and was made popular by K. Eric Drexler in *Engines of Creation* (1986).

This technology of the future deals with nanosize structures and their manipulation at the atomic or molecular level. The prefix *nano-* represents 10^{-9} (0.000000001) or one billionth of a meter. This technology is based on the manipulation of individual atoms and molecules to build nanosize structures having complex atomic specifications.

Nanotechnology may be used to fabricate smaller, faster computer chips for more efficient computers, mobile phones, and navigation systems. It has led to the development of new lasers, such as the quantum dot laser, that enable faster communication and new powerful data storage systems. Nanotechnology does not only bring improvements in the area of semiconductor technology and microelectronics, but the mastery of materials and systems on the nanometer scale can also revolutionize traditional areas. Nanostructured metallic and ceramic materials are more buoyant, stronger, and more rugged.

Carbon that has existed in nature in various forms (e.g., coal, diamond, graphite, complex molecules) has attracted the interest of many nanoscientists. They realized that carbon is an extremely versatile molecule; it can form linear or zigzag chains, rings (benzene and other aromatic compounds), buckyballs (spherical molecules), sheets (graphite and buckytubes), or blocks (diamond). Chemists have been able to bond organic molecules to each of these forms of carbon, creating carbon nanotubes and many other nanoforms of carbon. Carbon nanotubes are more elastic and robust than steel. Polymers mechanically strengthened with carbon nanotubes are used in areas from medicine to aviation. Nanostructured surfaces behave as efficient catalysts.

However, it is important to realize that long before the term *nanotechnology* came into effect, scientists realized that chemical properties of smaller particles are different and better than large particles.

The chemistry of colloids is one such example; it was known for centuries that when smaller sized ions take the form of colloidal size, their chemistry becomes different from the chemistry of large-size particles.

Moreover, homeopathic medicine has been using the concept "the more the dilution of medicine, the more its power of action." If one calculates the number of atoms of the medicine present in a high dilution such as 200 M, the concept of Avogadro's number would appear to fail—its dilution is so high that there are no atoms of original medicine left in the 200-M dilution except the solvent. But this dilution is most powerful. Likewise, Ayurvedic medicine uses various herbs, and it is believed that "the more you grind an herb, the more is its power." Today we know that one of the simpler methods to make nanosize particles is by ball milling, in which grinding is done by a few solid balls kept in the rotor. Hence, it would be wrong to say that the concept of nanotechnology (i.e., the smaller the size, the more is its chemical activity) was only known recently. Ancient scientists did know about nanotechnology without being able to see the size of the particles. Today's scientists have been able to observe the size of these particles using instruments such as transmission electron microscopy (TEM).

Similarly, the concept of the chemical vapor deposition technique (CVD) used to prepare nanomaterial was known to Ayurvedic doctors. Medicines such as *Bhasma* are the product formed by the pyrolysis of materials in the absence of oxygen. *Bhasma* from herbs is prepared by keeping herbs in a completely covered clay ball, which is kept underground for several days in fire. The material that comes out after cooling is used as medicine. This process is almost the same as the CVD process.

Hence, we should study nanomaterials with an understanding that the concept of "smaller is better" was known in the past, but that ancient people had no idea about the dimensions of these materials, which we have now realized. As a result, current scientists are able to use nanomaterials for various applications.

This book concentrates mainly on nanocarbons and their application. Carbon is a vital constituent of all living organisms. Organic chemistry branches out with carbon chains in addition to many organometallic compounds, π-bonded complexes, and carbonyls; inorganic chemistry includes large quantities of carbon black, coke, graphite, carbon dioxide, calcium carbide, and so on. In the periodic table, carbon stands at the sixth place and is the first element of column IV. Out of its six electrons, four electrons in its valence shell play the significant role of forming three hybridizations (i.e., sp, sp^2, and sp^3), which has led to the formation of many stable forms of carbon in all dimensions. Figure 1-1 shows a pictorial view of different hybridizations seen in carbonaceous materials. Carbon's three-dimensional diamond and two-dimensional graphite structures are

FIGURE 1-1
Hybridization of
carbon.

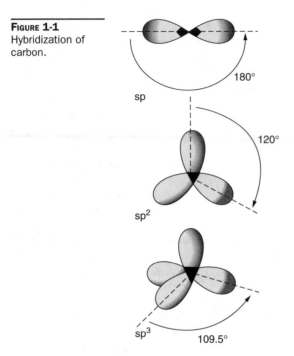

sp

180°

sp²

120°

sp³

109.5°

widely known. Carbon differs from the remaining elements of group
IV because of its smaller size, higher electronegativity, and the non-
availability of d-orbital.

Carbon is unique in forming π–π multiple bonds, such as C=C,
C≡C, C=O, C=S, and C≡N, which is not observed with any other element in this group. Carbon's strong tendency to form bonds with its
own atoms (the bond energy of a C–C bond is 348 kJ/mol) leads to
the formation of a long network of carbon structures. However, sp^3-
and sp^2-bonded carbon atoms are the main building blocks of all the
carbonaceous structures available so far. It is well known that
although both diamond (made from sp^3 carbon) and graphite (made
from sp^2 carbon) are constituted from the same basic element (i.e., carbon),
their properties are widely different, making the science of carbon all
the more interesting with the discovery of its new allotropic forms
(see Fig. 1-1).

1.2 Various Forms of Carbon

The structure and properties of various forms of carbon are distinctly
different. For example, diamond is shiny, and graphite is blackish;
diamond is the hardest material, and graphite is soft; and diamond
is an insulator, but graphite is semimetallic. The structure and some
properties of various forms of carbon are briefly discussed in this
section.

1.2.1   Diamond

When all of the four-valence electrons' orbitals (one s and three p) of carbon intermix with each other, they give four sp^3 hybrid orbitals of equal energy, which upon overlapping with neighboring carbon hybrid orbitals, create the isotropically strong diamond structure. Diamonds are usually colorless. However, some colored diamonds also occur in nature, such as the yellow diamond, which gets its color from the presence of traces of nitrogen, or the blue diamond, which owes its color to aluminum.

Although more than 90% of diamonds are cubic type, hexagonal and other "polytypes" of diamonds also exist, particularly as small clusters. The important physical and chemical properties of diamond come from its ultimate cross-link of crystalline carbon polymer, which has three-dimensional arrays of six-member carbon rings. In contrast to graphite, the layers of the rings are stacked in an ABCABC sequence along <111> directions, all the six-member rings are in chair conformation, and all the C–C bonds are staggered (Fig. 1-2A).

A single crystal of diamond is a cubelike structure with eight corners, each with one carbon atom. Each of these eight carbon atoms is shared by eight other cubes of the crystal. Additionally, six carbon atoms are situated on the six facets of the cube. These carbon atoms are also shared by the adjacent cube's facets. Thus, a single cube of diamond actually owns only three carbon atoms because half of each carbon atom is shared by the facets of the adjacent cube. Moreover, the center part of the cube contains four carbon atoms that are not shared by any other cube. The positioning of these four carbon atoms is such that they appear to be at the center of four quartets of the cube. Thus, a unit cell of a diamond contains eight atoms (see Fig. 1-2A).

1.2.2   Diamondlike Carbon

As the name suggests, diamondlike carbon (DLC) has great similarity to diamond. The existence of the DLC structure came to the front in the early 1970s after the development of different physical vapor deposition (PVD) and CVD processes. The mechanical and electronic

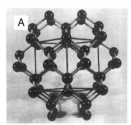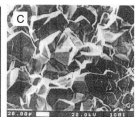

FIGURE 1-2   Different forms of diamond. (*A*) Channel-type diamond structure. (*B*) Single diamond from pyrolysed turpine. (*C*) Diamond thin film from pyrolysed turpine.

properties of DLC have been well studied, but the growth mechanism of these carbons has not been fully understood. Generally, DLC is amorphous or in a partly crystalline phase in the lattice of carbon structures. DLC is normally synthesized as thin film. Interestingly, the properties of DLC can be controlled by the deposition parameters. Thus, DLC has great importance in many areas of application; a thin film of DLC is used for coating tools to make it hard and rust resistant.

1.2.3   Graphite

By sharing three sp^2 electrons with three neighboring carbon atoms, carbon forms a layer of a honeycomb network of planar structure called *graphite*, and the fourth π-electron of carbon is delocalized over the whole plane of graphite. Because π-electrons are mobile, graphite conducts electricity in its plane. This conduction of electricity may occur only in this plane (i.e., not from one plane to another). This sp^2 carbon build layered structure (Fig. 1-3) has very strong in-plane bonding (covalent), but weak out-of-plane bonding (i.e., van der Waals forces). Because of this weak interplanar bond spaced at a distance of 3.35 Å, graphite sheets can slide upon each other. This is why graphite has been used as a lubricant and is considered a soft material. These planar honeycomb networks are arranged in two ways: ABABAB or ABCABCAB.

In the ABAB stacking arrangement (also called Bernal stacking after John D. Bernal, who proposed it in 1924), the in-plane carbon–carbon distance is 1.421 Å, the in-plane lattice constant a_0 is 2.456 Å,

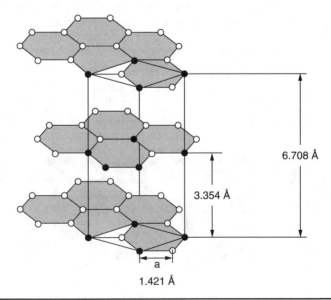

6.708 Å

3.354 Å

a

1.421 Å

Figure 1-3   Hexagonal graphite structure.

the c-axis lattice constant c_0 is 6.708 Å, and the interplanar distance is 3.354 Å (see Fig. 1-3).

The ABCABC (rhombohedral graphite) stacking arrangement of graphene sheets has a lattice constant a_0 = 2.456 Å and c_0 = 10.044 Å. Because the total energy for rhombohedral graphite differs slightly from that of the more stable Bernal ABAB stacking, ideal graphite has admixed with both forms. The unit cell of graphite contains four atoms, and the space group is $P6_3/mmc$ (D^4_{6h}). Graphite with defects or disorderliness with interplanar distance more than 3.345 Å is termed *turbostatic graphite*. Thermodynamically, graphite is the most stable form of solid carbon. Under high pressures and high temperatures, graphite changes to diamond, and after the pressure is released, diamond stays in a stable form under ambient conditions.

Single-crystal graphite is obtained in nature as flakes of smaller than 0.1 mm in thickness. These flakes usually contain defects and sometimes contain impurities such as iron and other transition metals. When single-crystal graphite is formed on molten iron, it is called *kish graphite*, which is used in many scientific investigations. When graphite is prepared by the pyrolysis of hydrocarbons at temperatures above 2000°C and subsequently heat treated to a high temperature in a pure, inert medium, a high-quality, highly oriented pyrolytic graphite (HOPG) is formed (see Fig. 1-3).

The third type of graphite (known as *graphite whisker*) is a cylindrical, hollow, rod-shaped carbon structure with a diameter in the micrometer range and a length in around one centimeter range produced by the arc discharge of two carbon electrodes. The diameter of the positive electrode is smaller than that of the negative electrode, and the gas pressure of the chamber is nearly 92 atm, which leads to formation of graphite whiskers on the cathode at the expense of anode. Carbon nanotubes can be synthesized in the same way by lowering the pressure.

1.2.4   Carbon Fibers

Carbon fibers are prepared synthetically. As per the precursor from which they are prepared, they are classified under three categories, polyacrylonitrile (PAN) based, pitch based, and vapor-grown carbon fibers (VGCF). Most of the carbon fibers are either PAN based or pitch based; they were developed between 1960 and 1970. On the basis of strength, they are categorized as high modulus, low modulus, and so on. The cost of PAN and pitch carbon fibers is always high because of the three-step synthetic process needed; production of VGCF has always been cheaper because of the single-step vapor grown technique used in its development. Figure 1-4 shows the SEM image of VGCFs formed from methane and hydrogen at 1000°C (Tibbetts, 1990). PAN and pitch carbon fibers are widely used in the defense and aerospace industries. Because they are ultimate in strength and

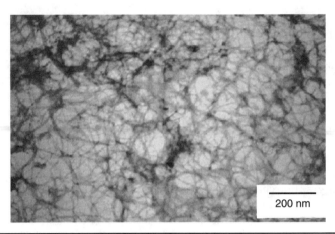

200 nm

Figure 1-4   Carbon nanofibers.

lightweight, they are popular in many sporting goods, musical instruments, brake discs for aircrafts, rocket nozzles, heating elements, and so on. VGCF can also be used for these applications, but further development and more studies are needed (see Fig. 1-4).

1.2.5   Carbon Black

Carbon blacks are spherical fine particles obtained from any synthetic technique used for preparation of carbon materials. Because of their smaller size, they are usually used as filler for improving the electrical, mechanical, and optical properties of different materials. Thermal black, charcoal black, acetylene black, plasma black, and furnace black are different types of carbon blacks obtained from the respective synthesis process or precursor. These carbon blacks are normally characterized by X-ray diffraction (XRD), High resolution transmission electron microscopy (HRTEM), and Raman spectroscopy to get information about their size, shape, crystallinity, and magnitude of disorderliness.

1.2.6   Activated Carbon

Activated carbon is an amorphous solid of varying shapes with a large internal surface area of specific pore size, which is generally controlled by a different activation process. Two types of carbon materials—granular or powder carbon and carbon fibers—are widely used for activation process. Activation can be done with a physical or chemical method. In physical activation, carbon is exposed to steam or CO_2 at a temperature around 1000°C. In the chemical method, carbon is mixed with any activating chemical, such as potassium hydroxide, phosphoric acid, or zinc chloride, and then pyrolyzed in a temperature range of 400° to 800°C. This process leads to the formation of complex arrays of mesopores or micropores on the carbon.

This porous carbon is one of the most important types of carbon materials widely used as an adsorbent (removal of SO_2/NO_x, volatile organic compounds from the air, toxic gases, water purification). These materials will be used in the future for adsorbed natural gas stored vehicles, super capacitors, medicines, secondary lithium batteries, refrigeration, and vapor sensors.

1.3   Nanoforms of Carbon

1.3.1   Buckminster Fullerene

In a very rare meeting, a group at Rice University along with Harry Kroto, the Nobel Laureate (Kroto 1985, 1988, 1992) from Sussex University, did some experiments and found an unusual symmetrical structure of carbon by laser vaporization of graphite. They gave it the name *Buckminster fullerene* (C_{60}) after the architect-engineer R. Buckminster Fuller, who designed geodesic domes. The discovery of the Buckminster fullerene later paved a path to a new branch of science called *nanotechnology*. At the same time, these researchers disproved the very truth of the fundamental physical properties of two allotropic forms of carbon, diamond and graphite.

Fullerene is the most stable cage cluster structure in which 12 pentagons are present (as required by Euler's theorem) and all the pentagons are separated by hexagons (the isolated pentagon rule). There are no other ways these pentagons and hexagons can arrange to form a closed cage structure of C_{60}. So, in a C_{60} molecule, carbon atoms are bonded in an icosahedral structure that is made up of 20 hexagons and 12 pentagons. The bonding between carbons atoms in fullerene are sp^2 type with some sp^3 character because of its high curvature.

After the discovery of Buckminster fullerene (C_{60}), other types of fullerene (e.g., C_{70}, C_{76}, C_{78}, C_{80}; Fig. 1-5), which are all constituted with different numbers of pentagonal, hexagonal, and heptagonal rings, were discovered. Fullerenes are related more closely to graphite because the bond distances between carbon atoms in fullerene are only 1 to 2% larger than those in graphite. C_{60} has been also

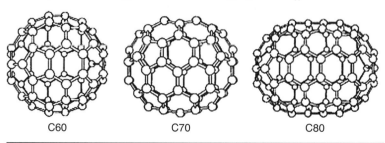

| C60 | C70 | C80 |

FIGURE 1-5   Structural model of fullerene.

synthesized for the first time from the natural plant-based precursor camphor by Sharon and colleagues (1994) (see Fig. 1-5).

1.3.2 Carbon Nano-onion

Discovery of fullerene opened many vistas for new types of carbon structures. Carbon onion is one such new type of fullerene product. It was named *nano-onion* by Ugarte in 1992 because it looks like an onion. Ugarte, a Brazilian electron microscopist working at Ecole Polytechnique Federale de Lausanne in Switzerland, found that when carbon nanoparticles and carbon nanotubes (from cathode soot obtained from arc vaporization of graphite) are irradiated by electrons of TEM operating under 300 kV, almost perfect spheres made up of concentric fullerenes are formed. Most of the nano-onion's diameter corresponds to C_{60} or C_{240}; the smallest has been C_{28} or C_{32}. Although fully formed nano-onions are perfect in structure with minimum defects and a diameter very close to that of fullerene, many nano-onions are found to be slightly faceted. The structural model of carbon onion is considered to consist of fullerenes of different number of carbon atoms and spacing of 3.4 Å between successive shells. But this model collides with theoretical results that larger fullerenes should be faceted rather than spheroid. More studies are needed to address the issue of the formation mechanism of carbon onions.

1.3.3 Carbon Nanocone

Nanocone is another form of carbon nanostructure obtained by pyrolysis of either camphor (Sharon, unpublished) at 900°C (Fig. 1-6) or by hydrocarbon in an industrial-grade carbon arc plasma generator with

Figure 1-6 Scanning Electron Micrograph (SEM) images of carbon nanocones obtained through pyrolysis of camphor.

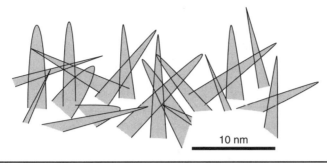

10 nm

FIGURE 1-7   Schematic of a single-walled carbon sheath called nanohorn.

a plasma temperature around 2000°C. Samples formed from this study included a large number of conical structures, disks, open nanotubes, and disordered material. Figure 1-6 shows the nanocones synthesized from camphor with different tip angles. The graphitic conical structure has one to five pentagonal rings depending on the angle of cones. An open disk does not have any pentagonal rings. However, these materials have not been studied extensively.

1.3.4   Carbon Nanohorn

Ablating graphite with a CO_2 laser in the absence of catalyst in an argon pressure of 760 torr resulted in formation of a new kind of aggregates of single-walled carbon sheath called nanohorns (Fig. 1-7).

1.3.5   Carbon Nanotubes

At the crossroads of traditional carbon fibers and fullerene, carbon nanotubes emerged as an important material in the nanoscale science. Carbon nanotubes are simply matchless to any other nanomaterials because of their unique properties, such as a smaller diameter (4 to 300 nm), long length (micrometer to centimeter), high mechanical strength, high thermal and chemical stability, and good heat conduction. The electrical properties of carbon nanotubes are even more fascinating.

When a single layer of graphene sheet (Fig. 1-8) is rolled to form a cylindrical shape, it is designated as single-walled carbon nanotubes (SWCNTs; Fig. 1-9A). If more than one graphene sheet is rolled to form a hollow, cylindrical shape, it is designated as multiwalled carbon nanotubes (MWCNTs; Figs. 1-9B and 1-9C). These graphene sheets are separated by a distance of 3.354 Å (see Fig. 1-3). They can be semiconducting or metallic depending on the diameter of the tube.

MWCNTs are much more complex because of the number of walls, interaction between the walls, disorderliness, and nonuniformity. In many ways, SWCNTs are superior to MWCNTs because they are free of defects, can be either semiconducting or metallic, and have better mechanical strength. MWNTs have very good mechanical character with a high Young modulus of around 1 terapascal (Fig. 1-10).

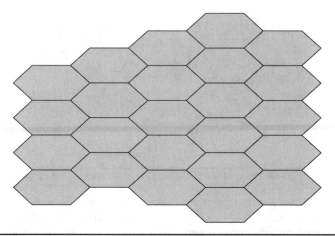

Figure 1-8   Schematic of a graphene sheet that, when rolled to form cylindrical shape, provides various types of nanotubes.

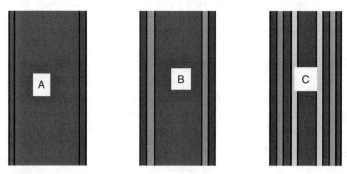

Figure 1-9   Schematic diagrams of carbon nanotubes. (*A*) Single walled. (*B*) Double walled. (*C*) Multiwalled.

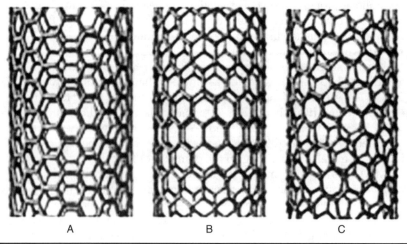

Figure 1-10   Different types of carbon nanotubes. (*A*) Armchair. (*B*) Zigzag. (*C*) Chiral. (*D*) Single-walled nanotube (one end is half of a fullerene).

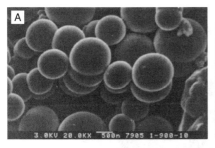
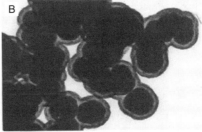

FIGURE 1-11   Carbon nanobeads synthesized from camphor with the chemical vapor deposition process. (*A*) Solid carbon nanobeads. (*B*) The inner portion is amorphous, and the circumference is made of graphitic carbon and is interconnected. (*C*) Carbon nanobeads with a graphitic surface and an inner mesh-type structure.

1.3.6   Carbon Nanobeads

Sharon and colleagues have been able to synthesize various types of carbon nanobeads (Fig. 1-11) through pyrolysis of camphor. Solid beads staggered together like a collection of soccer balls could be obtained by using cobalt nanosize powder as a catalyst during the pyrolysis of camphor (Fig. 1-11*A*). A special type of carbon beads (Fig. 1-11*B* and *C*) has been synthesized from camphor with the CVD method (Sharon, 1998). The circumference of nanobeads is made of broken graphene sheet about 80 nm in thickness (see Fig. 1-11*B*). The inner portions of the beads are made of amorphous carbon. These beads are interconnected (see Fig. 1-11*B*). Nanobeads shaped like the wheels of a motor bike with their spokes (see Fig. 1-11*C*) could be obtained by using an alloy of Ni/Co nanosize powder.

1.3.7   Carbon Nanofibers

Both carbon nanotubes and carbon nanofibers appear almost the same under the SEM. Both are similar to a hollow cylinder. Whereas the outer layer of carbon nanotubes is made up of unbroken graphene sheet, carbon nanofibers are made up of broken graphene sheet. Carbon nanofibers made from camphor with the CVD process is shown in Fig. 1-12.

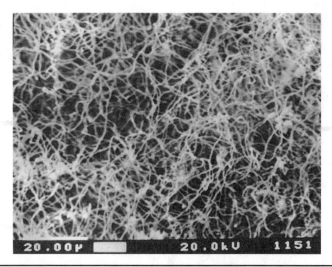

FIGURE 1-12   Carbon nanofibers of around 100 nm diameter synthesized from camphor.

1.3.8   Octopuslike Carbon Nanofibers

Depending on the raw material used for the synthesis of carbon nanomaterials (CNM) and the nanocatalyst used for the process, different types and shapes of carbon nanofibers have been reported. For example, carbon nanofibers grown by the CVD process from camphor using a nickel powder catalyst (Pradhan, 2003) that are 20 to 50 nm in size have resulted in formation of octopus-shaped carbon nanofibers (Fig. 1-13).

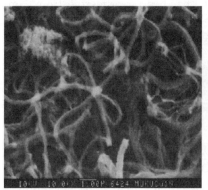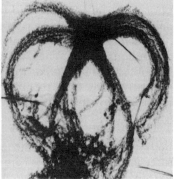

FIGURE 1-13   Octopus-type carbon nanofibers synthesized from camphor using nickel as a catalyst.

FIGURE 1-14   (A) Cactuslike carbon nanofibers. (B) Vertically growing cylinderlike carbon nanofibers.

1.3.9   Cactus-Type Carbon Nanofibers

Cactus-type carbon nanofibers (Fig. 1-14A) have been fabricated by pyrolyzing camphor using a nanosized nickel catalyst electroplated over copper plate (Pradhan et al., 2003). When large-size nickel particles were used as a catalyst, vertically growing cylindrical carbon nanofibers were formed (Fig. 1-14B).

1.3.10   Carbon Fibers from Different Types of Natural Precursors

When plant-derived precursors or kerosene are pyrolyzed at a temperature of 800°C in an inert atmosphere, various types of useful porous structures may be obtained. The advantage of using plant-derived precursors is that their basic structures are not destroyed during pyrolysis, resulting in some interesting channel types as well as porous structures. Otherwise, these structures would be difficult to synthesize. These carbon fibers may possess some suitable absorbing properties for some specific gases. Sharon and colleagues are engaged in exploring the applications of these fibers (Fig. 1-15).

1.3.11   Carbon Fiber Loaded with Nanosize Nickel Powder

Sometimes there is a need to load carbon nanomaterials with metal for special applications such as electrode materials for fuel cells. These materials can act like electrocatalysts for some specific oxidation/reduction reactions. Sharon and colleagues have been able to synthesize carbon fibers loaded with nanosize nickel powder by pyrolyzing turpine oil at 800°C in the presence of nanosize nickel powder (Fig. 1-16).

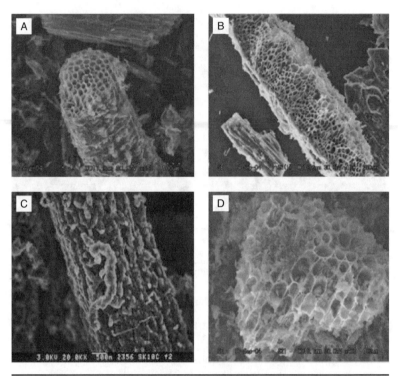

FIGURE 1-15   Carbon fibers synthesized from different plants. (*A*) Corn stem, giving a channel-type structure. (*B*) Anacardium seed, giving a porous fiber. (*C*) Pyrolysis of kerosene, giving a bitter gourdlike structure. (*D*) Bitter almond seed, giving a highly porous structure similar to a bee's hive.

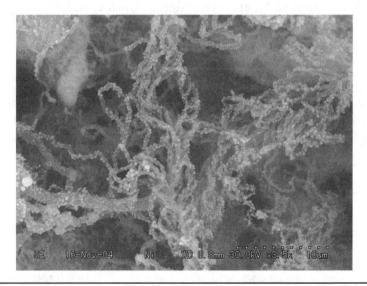

FIGURE 1-16   Carbon nanofibers loaded with nanosized nickel powder.

1.4   To Which Class of Allotrope Do These Other Forms of Carbon Belong?

The observations presented so far show that there are many forms of carbon. But as per conventional classification, there are only two allotropes of carbon: graphite and diamond. So the question arises as to which class of allotrope these other forms of carbon belong. To classify them in a way that all types of carbon could be covered, a classification based on the nature of carbon atoms present in the carbonaceous material may be more appropriate. Such classification may also open a Pandora's box for developing many new varieties of carbon than what we have been able to even conceive. For example, diamond has atoms with 100% sp^3 type of carbon, and graphite has 100% sp^2 carbon atoms. The former is an insulator because of its very large band gap (5.5 eV), and the latter is a conductor (band gap ≤0.25 eV). These two examples immediately give us a clue that there is a possibility of millions of permutations and combinations of sp^3 and sp^2, giving an innumerable variety of carbon with varied band gaps in the range of 0.25 eV to 5.5 eV. DLC and glassy carbon are two examples of such combinations. Neither can be classified as diamond or graphite. Therefore, a classification of carbon on the basis of ratio of sp^3 to sp^2 carbon atoms present in the material may be more appropriate than the conventional classification. We can thus classify carbon into three forms: diamond (100% sp^3), intermediate carbon (with different ratios of sp^3 to sp^2 carbon atoms), and graphite (100% sp^2).

All forms of carbon fall into one of these three categories. This classification also provides the opportunity to understand the reasons for carbon showing its novelty. Carbon nanobeads, carbon nanotubes, carbon nanofibers, active carbon, charcoal, and fullerenes all fall into the intermediate group of carbon because they contain different ratios of sp^3 to sp^2.

1.5   Why Naturally Occurring Diamond Is p-Type

It is interesting to observe that naturally occurring diamond in pure form (intrinsic diamond) shows a p-type character. Conversion of material into either an n- or p-type depends on the nature of the dopant added to the materials. If the dopant has a higher number of electrons than the host atom, the material becomes n-type; if the dopant has a lower number of valence electron than the host, the material becomes p type. This behavior is reasonably well understood, but why pure diamond exists in p form is not very well understood. Perhaps this strange behavior of diamond can be explained by considering the electronic configurations of carbon atoms bonded to form species such as diamond and graphite.

Carbon has four electrons in its outer shell with a configuration of $2s^2 2p^2$. The energy level of 2s and 2p are so close that these two orbitals get hybridized, forming various combinations such as sp, sp^2, and sp^3. This is the reason that carbon forms compounds with single (C–C, which can have sp^3 configuration), double (C=C, which can have sp^2 configuration), or triple bonds (sp configuration). Graphite has a hexagonal structure with an alternate double bond with a configuration of sp^2. Diamond has all single-bonded carbon with a configuration of sp^3.

When carbon forms a graphite structure with an sp^2 configuration, its p_x and p_y orbitals contain one electron each, and p_z has only one electron, which is shared with another carbon, forming a σ-bond. The electrons in p_x and p_y overlap with the p_x and p_y orbitals of other carbons. As a result of this overlapping, a π-bond is formed, which is filled with electrons (Fig. 1-17A). When a diamond-type structure is formed, it uses the sp^3 hybridization. As a result, all four bonds are σ type (Fig. 117B). The electrons of px, py, and pz are used to form σ-bonds; thus, neither px nor pz has any free electrons to overlap as they did with the sp^2 configuration. As a result, unlike with the sp^2 configuration, the π-orbital has no electron. The absence of an electron is considered as the presence of a hole. In other words, with the sp^3 configuration, the π-bond is populated with a hole, and with the sp^2 configuration, the π-bond is populated with an electron. In graphite, the π-bond is filled with an electron, so it should show n-type behavior. Removal of this π-charge by addition of a hole to graphite should lead to the structure of a diamond. If we assume the existence of some kind of equilibrium between the graphite and diamond (based on the charge density), then equilibrium between the sp^2 and sp^3 carbon atoms can be conceived (see Fig. 1-17).

If this concept is accepted, then diamond should intrinsically exist as p-type material, and graphite should exist as n-type material.

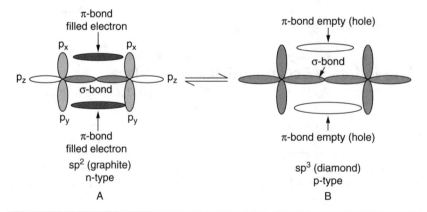

Figure 1-17 Schematic diagram showing overlapping of electrons of p-orbitals forms a π-bond in sp^2 carbon, making the material π type (A) and absence of electrons in the π-bond when sp^3-configured carbon atom forms make the material p-type (B).

This is what is normally observed with these materials. The conversion of graphite to diamond can therefore be viewed as substitution of an sp^3-configured carbon atom into sp^2-configured carbon atoms. Extending this argument, it can also be suggested that carbon material can show intrinsic properties by adjusting the ratio of sp^3 and sp^2 carbon atoms. Sharon (1998) has observed that if this ratio is around 0.33, then the carbon material shows intrinsic behavior.

1.6 Effect of Size on the Reactivity of Nanosize Particles

Why does nanosize material exhibit different chemical behavior? Experimentally, it has been observed that the absorption spectra of certain materials increase as the size of the particles increases. For example, as the size of PbS increases, its λ_{max} shifts toward a longer wavelength. Unfortunately, this result has been considered as the effect of the band gap on the size of the particle, which is incorrect because the band gap of any material depends on the crystal structure only. Unless the crystal habit changes with the size of the particle, there can be no effect of size on the band gap. Most reports have been made by taking the absorbance of solution, and because the λ_{max} changes with size, it has been assumed that the band gap also changes with size. This is because the absorbance changes due to the effect of scattering of particles of different sizes. There is a need to carry out some experiment with solid materials with different sizes to confirm the conclusion that λ_{max} is equivalent to the band gap of the material.

But why are particles of smaller size more reactive than particles of larger size? Such behavior is possible only if the electron transfer in nanosize particles follows a different mechanism than the electron transfer in bulk materials.

The reactivity of nanosize materials can be understood by probing the effect of formation of the depletion region between a semiconductor material and an electrolyte. It should be realized that because semiconductors possess a Fermi level representing the energy of their electrons, electrolyte also possesses a similar Fermi level as an electrode potential. For example, in the conversion of:

$$Fe^{3+} + e \rightarrow Fe^{2+} \ldots 0.7 \text{ V} \tag{1.1}$$

Fe^{3+} requires an electron of 0.7 V to reduce it to give Fe^{2+}. Hence, it could be said that Fermi energy of the electron present in Fe^{2+} is 0.7 V. The scale to measure the energy of electrons in electrolytes is measured with respect to its hydrogen reduction potential, but in solid material, it is referred with vacuum scale (i.e., the energy of electron in vacuum is 0, and as it gets bounded with materials, its energy reduces to negative value). This means that Fermi energy of solids always has a negative value. It is possible to get a conversion factor for energy level measured in terms of the hydrogen reduction potential (the potential of hydrogen reduction is considered as 0) and Fermi level measured with respect to vacuum by:

$$\text{Fermi energy (eV)} = \text{Hydrogen scale (V)} - 4.48 \qquad (1.2)$$

Thus, 0.7V of the Fe^{3+} reduction potential is equivalent to -3.78 eV on a vacuum scale.

With semiconductors, there are two levels, the conduction band and the valence level. The energy of electrons present in the valence band (for an n-type semiconductor) is referred as the Fermi level. These energy levels are represented in Fig. 1-18. In Fig. 1-18, for the purpose of explaining the formation of the depletion region, the Fermi energy of electrolyte is shown more negative than the Fermi level of electrons in an n-type semiconductor. In some cases, it might be other way around. But in such cases, no depletion region can be formed with an n-type semiconductor. It should also be remembered that the higher the negative value of energy, the more strongly it is held by the compound. In other words, it would be more difficult to take away electrons from material whose Fermi energy is more negative. Therefore, when two materials are in physical contact and if freedom is given to electrons to transfer from one material to other, there would be a spontaneous tendency for electrons to flow from the material possessing a Fermi level of less negative value to the material whose Fermi level is more negative. For example, a bucket of water kept on the roof of a building will have a spontaneous tendency to fall to the ground if a hole is made in the bucket. For bringing water back to the roof, there would be a need to supply potential energy.

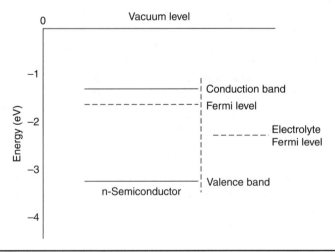

Figure 1-18 Schematic diagram representing various energy levels of semiconductors and electrolytes.

The electrons present in the semiconductor (Fig. 1-18) are more energetic than the electrons present in the electrolyte. Thus, if this semiconductor comes in close contact with the electrolyte, there would a spontaneous movement of electrons from the semiconductor to the electrolyte. If this happens, then there would be an accumulation of electrons from the semiconductor to the interface of the semiconductor and electrolyte. Electrons would be prevented from entering the domain of electrolytes to disallow the semiconductor from becoming positively charged or the electrolyte from becoming negatively charged. Because electrons have migrated from the inside of the semiconductor to the interface, all of the lattice sites from which the electrons have moved would become positively charged. To make matters simple, it is assumed that electrons have been transferred to the interface, creating positive charges such that they lie in one plane parallel to the interface at a distance of few hundreds of nm from the interface (Fig. 1-19). This transfer of electrons continues until equilibrium has been established between the two Fermi energies.

In Fig. 1-19*A*, various energy levels of an n-type semiconductor and the electrolyte are shown. When the semiconductor comes in contact with the electrolyte, the flow of electrons from the semiconductor to the interface occurs, and equilibrium is established. At equilibrium, a set number of negative charges are formed at the interface, and a similar amount of positive charge is formed inside the semiconductor at a distance *d* from the interface.

Assuming that the transfer of electrons to the interface and the creation of positive charge at $d = 100$ nm inside the material creates a potential of 1 volt, the electrostatic energy formed from this displacement of charges would be:

$$[1/(100 \times 10^{-7})] \, \text{V} \cdot \text{cm}^{-1} = 10^5 \, \text{V} \cdot \text{cm}^{-1} \qquad (1.3)$$

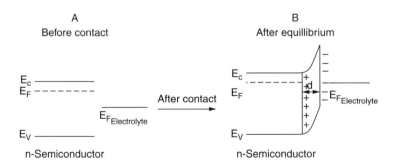

Notes: d = Depth of the depletion region; E_c = Conduction band;
      E_F = Fermi level; E_v = Valence band.

Figure 1-19 Schematic diagram showing the formation of the depletion region by making contact of an n-semiconductor with an electrolyte.

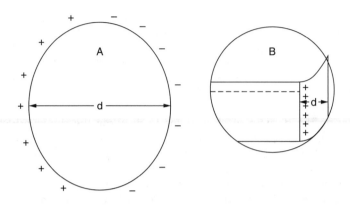

Figure 1-20 Schematic diagram showing charge distribution at the surface of a particle of size equal to the depletion region d (A) and particle size larger than the depletion region (B).

This electrostatic field is high enough to ionize all of the atoms present in this region (known as the depletion region). In other words, no extra charge (either positive or negative) can remain inside this depletion region.

Let us imagine that we have a particle (of size of depletion region d; Fig. 1-20A) that is in contact with electrolyte whose Fermi level is different, then the side of the particle that is in contact with the other materials will be charged negatively, and the other side (which is equivalent to side where positive charge is formed, as shown in Fig. 1-20) will contain a positive charge. Inside this particle, a very strong electric field exists that will not allow any charge to enter. If this particle is used for any charge transfer reaction, it will allow the transfer of electrons only through the surface in such a fashion that the total equilibrium of charges is not disturbed. On the other hand, if the size of the particle is larger than the depletion region (Fig. 1-20B), the transfer of charges will experience the resistance of the material because the electrons in this material have to be transferred through the material. In other words, electron transfer reactions with nanosize material do not depend on whether the material is a conductor or an insulator; rather, such transfer in the bulk material depends on the resistance of the material.

Hence, nanosize material shows better reactivity than large-size material. This discussion also suggests that although the band gap does not play any role in charge transfer process with nanosize material (because charge transfer takes place through the surface only), the band gap predominates with large size-particles (because the transfer of charge takes place through the bulk of the material). This is the main reason why nanosize particles show better chemical reactivity than large-size particle of the same materials.

1.7   Summary

Carbon nanomaterials are becoming popular because of their special properties and the possibilities of their unique applications. It has been puzzling as to why the size of the material should influence the properties. The same materials of larger size exhibit certain band gaps, but when their sizes are reduced to less than 100 nm, the band gap increases. Moreover, naturally occurring diamond and even synthesized undoped carbon nanomaterials exhibit p-type character when it is expected it to show intrinsic properties. Similarly, carbon is supposed to possess two types of allotropes—graphite and diamond—and with the discovery of fullerenes, a third form has been added. Does this mean that as we continue to discover new forms of carbon, the allotropes of carbon will continue to increase? Is it not possible to discover a classification method that could explain all possible type of allotropes of carbon, including those that have been discovered and others that might be discovered in the future? These are a few interesting questions that have been explored in this chapter. The chapter has also explained the various forms of carbon that have been discovered so far.

PART II

Synthesis of Carbon Nanomaterials

CHAPTER 2

Synthesis of Carbon Nanomaterials with the Arc Discharge Method

Yoshinori Ando

*Department of Materials Science and Engineering,
Meijo University, Nagoya, Japan*

2.1   Introduction

The arc technique is one of the earlier techniques used to make carbon nanotubes. The original method of arc plasma heating was reported in 1964 by Holmgren et al. Later, Krätchmer et al. (1990) developed mass production method of fullerenes by resistive

29

heating between contacting carbon rods. By applying DC arc discharge between the two carbon rods, the anode carbon is evaporated. The carbon deposited on the cathode was found to contain fullerenes (Saito et al., 1991, 1992). This technique led to the discovery of multiwalled carbon nanotubes (MWCNTs) by Iijima (1991) and Ando and Iijima (1993). The smallest innermost tube of MWCNTs was obtained by this method using pure hydrogen gas by Quin et al. (2000) and Zhao et al. (2004). Single-walled carbon nanotubes (SWCNTs) were also discovered (Iijima & Ichihashi, 1993) by the arc discharge of graphite using catalytic metals (Bethune et al., 1993). Mass production of SWCNTs is also possible with this technique (Journet, 1997; Zhao et al., 2003).

In an arc discharge unit, two graphite rods are used as the cathode and anode, between which arcing occurs when DC voltage is supplied (Fig. 2-1). A large quantity of electrons from the arc discharge moves to the anode and collides into the anodic rod. Carbon clusters from the anodic graphite rod caused by the collision are cooled to a low temperature and condensed on the surface of the cathodic graphite rod. The graphite deposits condensed on the cathode contain carbon nanotubes (CNTs), nanoparticles, and clusters. The graphite clusters synthesized in these initial experiments contained very small amount of CNTs, but later modifications to the procedure by Ebbesen and Ajayan (1992) have enabled a greatly improved yield. This apparatus must be connected both to a vacuum line with a diffusion pump and to a helium supply. The electrodes are two graphite rods, usually of high purity. Typically, the anode is a long rod approximately 6 mm in diameter, and the cathode is a much shorter rod 9 mm in diameter. Efficient cooling of the cathode has been shown to be essential in producing good-quality nanotubes. The position of the anode should be adjustable from outside the chamber so that a constant gap can be maintained during arc evaporation. A voltage-stabilized DC power supply is normally used, and discharge is typically carried out at 20 to 40 V and at a current in the range of 50 to 100 A. When a stable arc is achieved, the gap between the rods should be maintained at approximately 1 mm or less.

CNTs synthesized by arc discharge normally have multiwalled structures. After holes in the graphite rods are bored and filled with appropriately proportional composites of graphite powder and catalytic Co, Ni, Fe, and Y powder, SWNCTs can be synthesized by arc discharge on the cathode. The pressure in the evaporation chamber and the current are the most important factors in producing good yield of high-quality CNTs. An increase in the yield is evident as the pressure is increased, but there is a decrease in total yield at pressure that is too high. In addition, the current should be kept as low as possible, consistent with maintaining stable plasma. In practice, CNTs

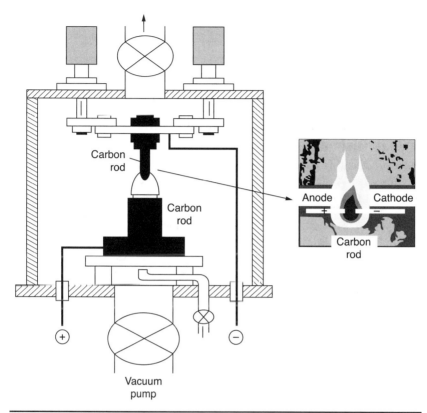

Figure 2-1   Schematic diagram of the DC arc discharge apparatus for preparing carbon nanotubes with a graphite rod.

are found in the core of the graphite deposits. MWCNs are covered with a large amount of carbon deposits or carbon particles.

2.2   Production of Fullerenes by Arc Discharge

The arc apparatus shown in Fig. 2-1 was used by Saito et al. (1992) by removing the Si block and using a graphite rod to achieve mass production of fullerene by arc discharge between two carbon electrodes (Saito et al., 1991, 1992). A similar set called a "contact arc" was used by Smalley's group (1995).

When the DC arc voltage is applied between the two carbon rods (see Fig. 2-1), the anode gets heated, and evaporated smoke of carbon produced in the atmospheric gas grows in the form of fullerenes. Some part of evaporated carbon was found to be deposited on the surface of the cathode; this deposit was found by Iijima (1991) to contain CNTs. Shinohara et al. (1992, 1993) tried to make metallofullerenes by using metal–graphite composite rods. By drilling a hole in the graphite anode, rare earth metals oxide, pitch, and graphite powder were packed and baked. When arc was generated with this electrode, metallofullerenes were produced.

Wang et al. (1996) were unsuccessful in producing fullerene by DC arc evaporation in the presence of hydrogen gas.

2.3   Production of Multiwalled Carbon Nanotubes by Arc Discharge

Mass production of MWCNTs by DC arc discharge was first achieved by Ebbesen and Ajayan (1992). A scanning electron micrograph (SEM) obtained from a section of a cathode deposit is shown in Fig. 2-2. In the micrograph, a number of fibrous MWNTs and nanoparticles can be observed. This cathode deposit was obtained by DC arc evaporation in He gas of 200 torr (Figs. 2-2 and 2-3).

To study the effect of gas, DC arc evaporation was done in He, Ar, and CH_4 gas at 100 torr (Ando, 1994). CH_4 gas was found to produce the best-quality MWCNTs. To understand the reasons for getting better results with methane gas, Ando et al. (1997) analyzed the gas present in the chamber and found that it contained hydrogen and C_2H_2. This suggested that methane during the arcing produced C_2H_2, which helped to get MWCNT. Hence, they proposed following thermal decomposition reaction:

$$2CH_4 \rightarrow 3H_2 + C_2H_2 \qquad (2.1)$$

FIGURE 2-2   Scanning electron micrograph of multiwalled carbon nanotubes and nanoparticles produced in He (200 torr).

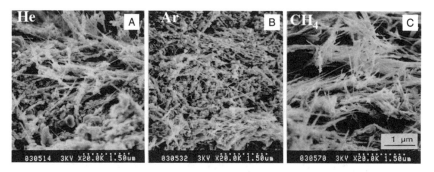

Figure 2-3 Scanning electron micrograph of multiwalled carbon nanotubes produced in different atmospheric gases at 100 torr. (*A*) He. (*B*) Ar. (*C*) CH_4.

As a consequence of this reaction, the chamber pressure increased twofold. To confirm whether the pressure, hydrogen, or C_2H_2 was main factor controlling the production of MWCNT, a separate experiment was done in the presence of C_2H_2 gas only. It was observed that C_2H_2 showed similar results as CH_4, but H_2 gas showed remarkably different results, as described in the following section.

2.3.1 Multiwalled Carbon Nanotubes: A Product of Hydrogen Arc Discharge

Zhao et al. (1997) and Wu et al. (2002) carried out arc discharge evaporation between two pure graphite rods in pure H_2 gas. A top view of the cathode deposit produced by hydrogen arc evaporation for 2 mm is shown in Fig. 2-4. The top surface is composed of three regions: a central black region A, a surrounding silver-gray region B, and a very

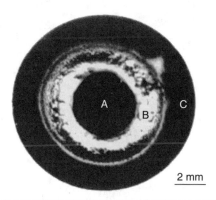

Figure 2-4 Top view of the cathode surface deposited by pure hydrogen DC arc discharge evaporation for 2 minutes. Regions A and B are 2 mm thick, and region C is very thin. Multiwalled carbon nanotubes exist in region A, and petal-like graphite sheets exist in region C.

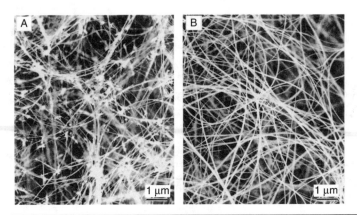

FIGURE 2-5   Scanning electron micrographs (SEMs) of the top surface of a cathode deposit. (*A*) Raw deposit produced by H_2 (60 torr) arc evaporation (the SEM micrograph was taken in region A, as shown in Fig. 2-4). (*B*) The same specimen after heat irradiation using infrared radiation at 500°C for 30 minutes.

thin outer region C. The thickness of the deposit is about 2 mm in regions A and B, but region C is thin and smaller than 0.2 mm.

SEM of carbon deposits from the region A showed it to be MWCNTs (Fig. 2-5), and deposits in the C region showed a roselike structure, called *carbon rose*. The carbon rose is aggregates of petal-like graphene sheets (Fig. 2-6) composed of several layers of parallel graphene sheets, the thinnest one consisting of only two graphene sheets. Similar structures were made by other methods, and they are now called *carbon nanowall* by other groups (Ando et al., 1998; Hiramatsu et al., 2004).

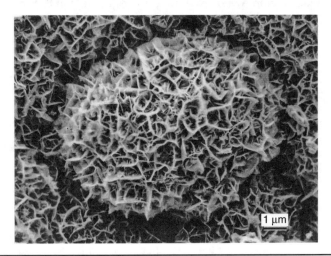

FIGURE 2-6   Scanning electron micrograph of petal-like graphene sheets observed in the C region of Fig. 2-4.

An example of a SEM of MWCNTs observed in region A of Fig. 2-4 is shown in Fig. 2-5A. In the micrograph, long and fine fibrous MWCNTs and a little coexisting carbon nanoparticles can be seen, which are fairly different from the MWCNTs produced in He arc discharge shown in Fig. 2-2. The purification of MWCNT was done by exposing to infrared radiation in air at 500°C for 30 minutes; the resulting SEM is shown in Fig. 2-5B. Almost all nanoparticles were removed, and very long MWCNTs and their bundles could be observed. A low-magnification SEM image of purified cathode deposit is shown in Fig. 2-5.

The part below the two arrows in Fig. 2-7 is a section of the purified cathode deposit, and the part above the arrow was collected from the top surface of the deposit. Randomly arranged MWCNTs can be seen on the top surface, and ones aligned to the thickness direction of the deposit can be seen in the section (Fig. 2-8).

The purified cathode deposit is similar to a sponge of 10 mm^2 surface area and 0.1 mm thickness and is easily handled by tweezers. High-resolution transmission electron micrographs (HRTEMs) of each MWNT produced by hydrogen arc of pure graphite rod were highly crystallized and had a very thin innermost tube (Ohkohchi, 1999; Zhao et al., 1999). In the case of the thinnest innermost tube, the diameter is only 0.4 nm, as shown in Fig. 2-8. It should be noted that the normal distance of MWNT layers is 0.34 nm (the same as the graphite's 002 spacing).

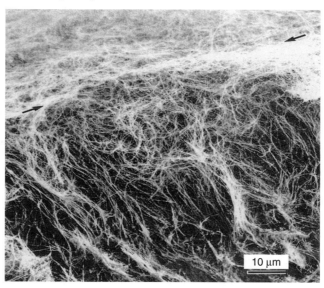

10 μm

Figure 2-7   Scanning electron micrograph of a section of purified multiwalled carbon nanotubes (MWCNT). The *line* connecting the two *arrows* is the upper boundary of the section, and above the line is the top surface of the cathode deposit.

FIGURE **2-8**
High-resolution transmission electron micrograph of multiwalled carbon nanotubes produced by hydrogen DC arc discharge. The diameter of the innermost tube is 0.4 nm.

2.4 Production of Single-Walled Carbon Nanotubes by Hydrogen Arc Discharge

Two years after the discovery of MWCNTs, SWCNTs were independently found by two groups (Bethune et al., 1993; Iijima and Ichihashi, 1993). Iijima and Ichihashi used a graphite rod, including a Fe catalyst, and Bethune et al. used a graphite rod, including a Co catalyst. By arc discharge evaporation of these graphite rods, including a magnetic metal catalyst, cotton soot–like SWCNTs were produced in the chamber rather than in the cathode deposit, as in the case of MWCNTs. Thus, the essential differences between SWCNTs and MWCNTs are that the former requires a catalyst, but the latter does not. Moreover, SWCNTs are produced in the whole chamber, but MWCNTs are confined to the cathode deposit only. From the latter difference, SWCNTs can also be produced by AC arc discharge (Ando et al., 2000).

Mass production of SWCNTs with arc discharge evaporation was first achieved by Journet et al. (1997) by using binary metals Ni and Y in the atmosphere of He. Ando et al. (2005) modified the method by rearranging the electrode configuration in a sharp angle of about 30 degrees and used it as the arc plasma jet (APJ) method. In the case of conventional arc method, more than half of the evaporated anode becomes cathode deposit, which does not include SWCNTs. In contrast, in the APJ method, more than 80% of evaporated anode becomes soot that includes SWCNTs. When soot was observed by HRTEM, bundles of SWCNTs; catalytic metal particles of Ni; and surrounding thick, amorphous carbon were seen. Removal of the amorphous carbon from SWNCTs with heat treatment is complicated because SWCNTs also get heavily damaged.

2.4.1   Purification of Single-Walled Carbon Nanotubes Produced by Arc Discharge Evaporation

For unproblematic purification of SWCNTs, the kind of catalyst and combination of atmospheric gas are important factors to be considered. For the reduction of amorphous carbon, hydrogen gas and a Fe catalyst were found to be effective by Zhao et al. (2003). By DC arc evaporation of 1.0% at Fe, including graphite rod in H_2–Ar mixture gas or H_2–N_2 mixture gas (Ohkohchi, 1993), large quantities of SWCNTs may be obtained. The transmission electron micrograph (TEM) and HRTEM of the as-grown SWCNTs is shown in Figs. 2-9A and 2-9B. The black particles seen in Fig. 2-9A are catalyst Fe metal, and their Fe particles are surrounded by thin, amorphous carbon,

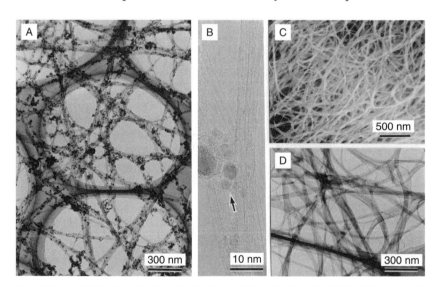

FIGURE 2-9   (A) A low-magnification transmission electron micrograph (TEM) of as-grown single-walled carbon nanotubes (SWCNTs). (B) A high-resolution TEM of as-grown SWCNTs. (C) A scanning electron micrograph of purified SWCNTs. (D) A TEM micrograph of purified SWCNTs.

which is smaller than 10 nm, as seen in Fig. 2-9B. Therefore, the removal of amorphous carbon by thermal heating at 400°C is straight-forward, and oxidized Fe particles may be removed by HCl. SEM and TEM micrographs of such purified SWCNTs are shown in Fig. 2-9C and 2-9D.

2.5 Summary

With the arc discharge evaporation of graphite rods, including rare earth metals, metallofullerenes can be produced. The same rare earth metals are helpful in increasing the production of MWCNTs. On the other hand, SWCNTs can be produced by using graphite rods, including different kinds of catalytic metals, such as magnetic metals Fe, Co, and Ni and their mixtures. The magnetic metals (metals which shows magnetic properties) are ineffective in producing metallofuller-enes, suggesting that the growth mechanism of metallofullerenes and SWCNTs should be quite different.

CHAPTER 3

Chemical Vapor Deposition and Synthesis of Carbon Nanomaterials

Mukul Kumar

Department of Materials Science,
Meijo University Nagoya, Japan

3.1   Introduction

Chemical vapor deposition (CVD) is a popular technique of growing thin films of various materials. A large variety of crystalline and amorphous materials can be grown with this method, both in the form of thin or thick films or in bulk. Whether it is a matter of insulating, semiconducting, conducting, or superconducting coating in electronics; a necessity of bulk structures in metallurgy; or a concern of hard coatings for wear resistance and corrosion protection of mechanical tools, CVD provides the foremost convenience and utmost control. That is why CVD is used from engineering college laboratories to industrial research laboratories on equal footing.

Although CVD is a well-established deposition technique since as early as 1950s, with the advent of revolutionary nanotechnology research in the past decade, it has attracted enormous interest toward the formation of nanomaterials. Because of the relative ease of creating materials of a wide range of accurately controllable stoichiometric composition, CVD has been gaining unprecedented success in different arena of modern technology. The aim of this chapter is to provide a broad outline of this versatile technique to graduate and postgraduate researchers. The chapter first acquaints readers with the fundamental aspects of CVD in general and then discusses the significance of CVD parameters for the synthesis of carbon nanotubes (CNTs) in particular.

3.2   Fundamental Aspects of Chemical Vapor Deposition

Here *deposition* means *transformation of vapor into a solid*, a process frequently used to grow solid film and powder materials. Basically, the two types of deposition process are physical vapor deposition and CVD. Whereas only a transformation of physical state is involved in the former (e.g., vacuum evaporation, sputtering, molecular beam epitaxy), essentially an irreversible chemical reaction takes place in the latter. By virtue of this, although CVD is a complex phenomenon, it is extremely versatile. Small changes in experimental parameters cause drastic changes in the results and hence lead to quite different branches of CVD, including thermal CVD, atmospheric pressure CVD, low-pressure CVD, plasma CVD, photo CVD, and laser CVD. CVD is so rich and versatile because of its adaptability to new technologies. It is virtually a homemade recipe that offers creativity and credibility.

Figure 3-1 shows a schematic diagram of a CVD setup in its simplest form. Briefly, in CVD, a mixture of gases passing over a hot surface undergoes chemical reactions that lead to a solid deposit on the surface.

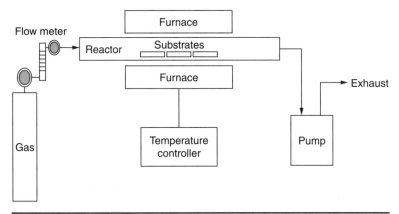

FIGURE 3-1   Schematic diagram of a typical chemical vapor deposition setup.

The key to controlling CVD to get a desired product of defined properties lies in understanding the underlying principles such as basic chemistry, thermodynamics, kinetics, and transport phenomena. A brief review of these fundamental aspects, from the viewpoint of CVD, will help readers understand the important parameters that govern the growth mechanism and essential steps involved in CVD.

3.2.1   Chemical Reactions in Chemical Vapor Deposition

As the name suggests, CVD is a chemical transformation of a vapor into solid deposits. The nature of the chemical reaction taking place in a CVD reactor greatly depends on the nature of the raw material that liberates the CVD precursor. Moreover, the CVD reaction is strongly affected by experimental parameters, such as reactor temperature, pressure, precursor composition and concentration, and flow rate. The reactor geometry is another important parameter.

3.2.2   General Steps in Chemical Vapor Deposition

A CVD reaction is almost always a heterogeneous process. Recall that in CVD, a mixture of gases passing over a hot surface undergoes chemical reactions that lead to a solid deposit on the surface. Such a process typically involves the following steps.

Vaporization and Transport of Precursor Molecules Into the Reactor
The chemical compounds that are fed to the reactor are first vaporized and fragmented. All fragments are not useful to CVD. The fragments that essentially take part in the CVD reaction are called *CVD precursors*. Thus, the formation of CVD precursor molecules is the first stage of an overall CVD reaction. This stage is both temperature and pressure dependent.

Diffusion of Precursor Molecules to the Surface

As-grown precursor molecules floating in the reactor encounter hot substrates and diffuse over the substrate (typically by surface diffusion). This stage primarily depends on the concentration gradient.

Adsorption of Precursor Molecules to the Surface

A fraction of diffused precursor molecules get adsorbed to the surface. Besides strong temperature dependence, this stage also depends on mutual interaction between the precursor molecule and the substrate material, which dramatically varies with materials and morphology.

Decomposition of Precursor Molecules and Incorporation Into Solid Films

This is the prime stage of deposition, the surface reaction. As soon as the adsorbed precursor molecules find enough energy to overcome the reaction barrier (bond energy), they get decomposed into atomic fragments and in turn get ready to make new compounds. Recall that bond breaking is a precondition to making new bonds and new compounds. Again, temperature plays a significant role in this stage.

Recombination of Molecule Byproducts and Desorption into the Gas Phase

While the desired species of the decomposed precursor lay down a solid deposit on the substrate, the rest of the species often recombine into gas-phase molecular byproducts and leave the surface via diffusion. This has mild temperature dependence.

These steps are sequential, and the slowest one is supposed to be the rate-determining step. Although this reaction mechanism sounds simple and convincing at a glance, it is often complex in reality. Because the process contains so many sequential steps, the overall reaction kinetics is often difficult to determine. Although it is certain that the slowest step would determine the overall deposition rate, the relative speeds of those steps vary from material to material and are not consistent in a wide range of temperatures and pressures.

3.2.3   Rate of Reaction: Physical Significance of Experimental Parameters

It is necessary to understand the possible growth mechanism based on careful observations of the growth rate. To meet this challenging task, the effects of experimental parameters, such as the reactant's concentration, surface area, pressure, temperature, and catalysts on the rate of reaction, must be analyzed. Additionally, the physical significance of these factors in driving chemical reactions and manner in which these parameters affect the rate of reaction must be understood.

Recall again that CVD is a reaction between two or more reactants (at least one in vapor phase) to form a solid deposit and that it involves formation of some transition product(s) in intermediate reaction(s). In every reaction, the reacting species (precursor molecules, ions, or free radicals) have to collide with certain minimum energy (E_a) and with a specific orientation (headway, sideway, or the other way around, with respect to the bond to be broken). The greater the collision probability, the greater the reaction probability, and in turn, the higher the reaction rate. With this rule of thumb in mind (barring exceptions), let us look into the effects of external parameters on the rate of reaction.

Effect of Concentration
In any heterogeneous reaction, it is a common observation that an increase in the reactant concentration increases the rate of reaction. Actually, providing more moles in a given volume increases the frequency of collisions per second, increasing the rate of reaction.

Effect of Surface Area
In heterogeneous systems involving a solid, it is a matter of general experience that the finer the particle, the faster the reaction. It is easy to understand this in the case of solid reactants in which only the outer surface is exposed and available for reaction. When the same reactant is crushed into fine or ultrafine particles, the exposed surface area per gram increases manifold, which causes an enormous increase in the collision probability; consequently, the rate of reaction is increased drastically. This effect is of highest interest in nanotechnology.

Effect of Pressure
In reactions involving a gaseous reactant, the effect of pressure increase is exactly the same as that of the concentration increase. That is, the higher the pressure higher is the number of molecules per unit volume, higher the collision frequency higher is the reaction rate.

Effect of Temperature
When the temperature of a system is increased, the thermal energy of each particle increases, and so does the number of high-energy particles in the system. Recall the kinetic interpretation of temperature (T crms^2). At higher temperatures, the molecules move faster and undergo a higher number of collisions per second. This is a straightforward explanation of why reaction rates increase with increasing temperature. Moreover, the Maxwell-Boltzmann energy-distribution plot at different temperatures (Fig. 3-2) illustrates how the number of high-energy particles increases with the increasing temperature.

Effect of Catalyst
Right from the early stages of studying chemistry, we have known that a catalyst is a material that accelerates a reaction but does not get

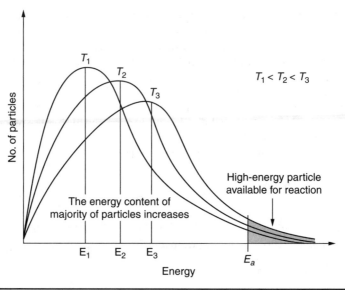

FIGURE 3-2   Effect of temperature on the energy distribution curve of gas molecules.

consumed at all. A catalyst artificially reduces the energy barrier of the reaction. How does this happen? Actually, catalysts (often solids) have certain lattice imperfections and crystal defects (point, line, and surface defects) that potentially attract liquid and gas molecules. These imperfections and defects offer a reactant to get adsorbed at defect sites and make temporary bonds with the surface atoms of the catalyst. In this process of atomic rearrangement (which happens in a series of fast steps within picoseconds), the newly formed bonds have significantly low bond energy, so they can be broken at a relatively low temperature. Thus, a catalyst provides an alternative route for the reaction with lower activation energy. Different catalyst materials offer different activation energies for the same reactions (Fig. 3-3).

A remark of caution: the above-mentioned explanation of the effects of experimental parameters is very superficial; nevertheless, it provides CVD beginners a track for pondering at the experiment site. Precise details of individual effect can be found in specialized books on CVD.

3.2.4   Rate of Film Growth

Having considered the effects of experimental parameters on the rate of chemical reactions, let us explore the rate of film growth, that is, the film thickness grown per unit time, which is generally measured in microns per minute. Actually, this is what we are ultimately interested in.

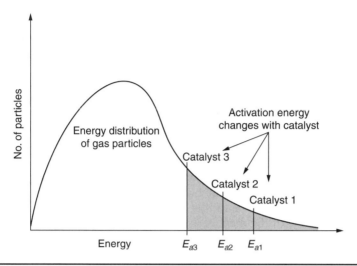

FIGURE 3-3   Effect of the catalyst on the value of activation energy.

From the kinetic theory of gas and Maxwell's law of velocity distribution to gas molecules, the approximate number of molecules striking per unit area of a plane surface is given by:

$$\frac{n}{4} c_{mean} \qquad (3.1)$$

where n = molar density (moles per unit volume) and c_{mean} is the mean velocity. From this relationship, it follows that the maximum possible deposition flux at a particular pressure and temperature can be:

$$J = 3.51 \times 10^{22} \frac{P}{\sqrt{MT}} \quad \text{molecules/cm}^2/\text{sec} \qquad (3.2)$$

This is very important relationship in CVD and is known as the *Knudsen equation*, where P is the partial pressure of precursor (in torr), M is the molar mass (in grams per mole), and T is the deposition temperature (in Kelvin). However, keep in mind that the actual growth rate that could be observed in experiments would be much lower than what is expected from the Knudsen equation. One main reason is that in the derivation of the Knudsen equation, the total flux of impinging precursors has been taken as the measure of deposition rate, which is practically untrue. All precursor molecules that strike the surface do not deposit there; the probability of getting bounced is much higher than that of getting stuck there. Moreover, the effective partial pressure at the surface is much lower than the macroscopic partial pressure of the precursor because the majority of the precursor molecules are away from the surface.

3.2.5 Transport Phenomena

In the previous section, we discussed that the CVD growth rate can be controlled by controlling the rate of surface reactions. However, chemical kinetics offers a much localized, surface-limited control, and a great deal of control is required right from the reactor inlet, where the reactants are fed into the outlet and where byproducts exit. This involves the transport phenomena, which is the study of heat, mass, and momentum transfer inside the reactor. The transport phenomena determine not only the deposition rate but also the uniformity of deposition, which is critical in practical applications. Proper understanding of transport phenomena is necessary for designing a suitable reactor for a specific type of CVD, although a discussion of this is beyond the scope of this book. The following sections briefly describe some of the most basic terms and equations of this critical phenomenon.

Mass Transport

Like in any industrial process, an efficient CVD technique demands maximum utilization of the input chemicals and attainment of the highest possible growth rate. This requires that sufficient amounts of the reactants be delivered to the growth surface. On the other hand, technical issues concerning practical applications demand a uniform composition everywhere, which requires that the reactants be delivered exactly in the same proportion at all points of deposition. To control these features, we need to understand mass transport (i.e., how molecules move from one point to another in the reactor).

Basically, the two methods of mass transport in fluids are convection and diffusion. Bulk flow of a fluid (constituent molecules moving together in a group) is called *convection*. This is driven down the pressure gradient. On the other hand, the macroscopic result of random motion of molecules is called *diffusion*. This drives down the concentration gradient. Let us review the essential features of the two modes of mass transport.

Convection

Convection is the bulk motion of fluid from the high- to the low-pressure side. In convective transport, almost all components of the fluid move in the same direction. Convective flow is measured as the mass (bulk or molar) moving across unit area in unit time, which is equivalent to the volume flowing per unit time. (The practical unit of mass flow is defined at 0°C and 1 atm, generally expressed in terms of standard cubic centimeters per minute or standard liters per minute.)

If n denotes concentration or density and u denotes velocity, the convection flux (flow per unit area per unit time) may be expressed as:

$$J_{conv} = n \times u = density \times velocity \qquad (3.3)$$

Note that various flux can be defined depending on whether one wishes to count mass, moles, or molecules.

$$\text{Mass flux} = \frac{kg}{m^3} \times \frac{m}{s} = kg/m^2/s \tag{3.4}$$

$$\text{Molar flux} = \frac{\text{no. of moles}}{m^3} \times \frac{m}{s} = moles/m^2/s \tag{3.5}$$

$$\text{Molecular flux} = \frac{\text{no. of molecules}}{m^3} \times \frac{m}{s} = molecules/m^2/s \tag{3.6}$$

The simplest CVD reactors are cylindrical tubes, typically 50 to 100 cm long and a few centimeters wide. In such tubes, the velocity of all molecules in the flow is not the same. The flow constitutes many layers; the central layer has the fastest moving molecules, and the molecules on the outermost layer (in contact of the tube wall) are almost at rest (Fig. 3-4). Thus, a velocity gradient is perpendicular to the flow direction. Nevertheless, as long as this gradient is invariant in time, the flow is said to be a laminar flow. This sort of flow can be described in terms of a *velocity field*, which is a vector trace of the flow path, describing the local average velocity and direction at any point. Such a contour of stream (with well-defined velocity at any flow point) is termed as *streamline*.

However, if the above-mentioned velocity gradient varies with time infinitesimal, the velocity vector cannot be well defined, and construction of flow contours is very difficult. The flow is then called *turbulent flow*. Obviously, such a flow is undesirable in CVD. So how could the convection in a CVD reactor be controlled? The first key is *Reynolds number*.

$$\text{Reynolds number: } Re = \frac{\rho u L}{\mu} \tag{3.7}$$

where, ρ = mass density, u = average velocity, L = relevant length (here, tube length), and μ = viscosity.

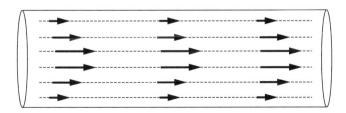

FIGURE 3-4 Streamlines of fluid flow in a tubular reactor. The lengths of the *arrows* indicate velocity of gas.

When Re is small (less than a few tens), the flow is smooth, and mass transport can be controlled, so this is recommended for CVD. For larger Re (a few hundreds), the flow becomes complex, and circulations and vortices appear, making the transport control difficult (but affordable). When Re is of the order of 1000, the flow is turbulent, flow contours are unpredictable, and control of mass transport is not possible. Fortunately, in commonly encountered CVD situations, Re is below 100, so the flow is well laminar. This is why calculation and manipulation of mass transport is feasible in CVD.

Diffusion

Diffusion is the result of the random (thermal) motion of individual molecules. Random velocity of gas molecules is much higher ($\sim 10^5$ cm/sec) than the convection velocities typically encountered in CVD processes. When the molecules are uniformly distributed (e.g., in a closed chamber), their random motions contribute no net motion in a specific direction. However, as soon as a concentration gradient is created (e.g., when some of the molecules are consumed in some reaction), the motion of molecules gives rise to a net macroscopic motion (although the motion of each individual molecule is still completely random) from higher to lower concentration.

By Fick's law, diffusion flux (J_{diff}) is proportional to the negative of the concentration gradient:

$$J_{diff} \propto -\frac{dn}{dx} \quad \text{or} \quad J_{diff} = -D\frac{dn}{dx} \qquad (3.8)$$

where n is concentration and D is the *diffusion constant* or *diffusivity*. Besides the nature of the material (typically, the mass [m] and diameter [a] of the molecule), D is a function of temperature (T) and pressure (P).

$$D \approx \sqrt{\frac{k^3}{\pi^3 m}\frac{T^{3/2}}{Pa^2}} \qquad (3.9)$$

where k is the Boltzmann constant.

Note that the applicable temperature range in typical CVDs is very limited ($\sim 100°$ to $1000°C$; the span is 1 order of magnitude wide), but the pressure range is much wider ($\sim 10^{-1}$–10^3 torr; the span is 4 orders of magnitude wide). Diffusivity is much higher at low pressures. Typically:

$$D = 0.1 \text{ to } 1 \text{ cm}^2/\text{s at room temperature, 760 torr} \qquad (3.10)$$

$$D = 75 \text{ to } 750 \text{ cm}^2/\text{s at room temperature, 1 torr} \qquad (3.11)$$

Fick's law of diffusion can be extended to formulate a differential equation describing the relationship between the spatial and temporal variations of concentration:

$$\frac{\partial n}{\partial t} = D\nabla^2 n = D\left(\frac{\partial^2 n}{\partial x^2} + \frac{\partial^2 n}{\partial y^2} + \frac{\partial^2 n}{\partial z^2}\right) \tag{3.12}$$

This important relationship is known as the *diffusion equation*. If we know the initial concentration and the boundary conditions, the diffusion equation allows us to compute the concentration at any later time at any point of the system.

The solution to this equation is not unique, but a unique solution can be obtained by imposing relevant boundary conditions, such as concentrations at the inlet and outlet ends of the reactor. Detailed analysis of solution finding is beyond the scope of this chapter. However, it is worth mentioning that any solution of the diffusion equation essentially consists of a particular combination of diffusivity D and time t in the pre-exponential term, typically $(\alpha D t)^{1/2}$, where α is a dimensionless constant. This quantity has the dimensions of length and is known as *diffusion length* (L_d). The diffusion length is the characteristic length scale for diffusion problems.

$$\text{Diffusion length:} \quad L_d = \sqrt{\alpha D t} \tag{3.13}$$

In CVD, the relevant time to calculate diffusion length is the residence time of gases in the reactor. (Roughly, $t_{resi} = V/F$ where V = reactor volume and F = volumetric flow/sec.) For one-dimensional reactors (where tube length > diameter, which is often the case), if the diffusion length is much longer than the reactor length, the concentration drop along the length is linear and is almost independent of time. If the diffusion length is much shorter than the reactor length, the concentration gradient is large and is highly time dependent. Thus, the ratio of the reactor length to the diffusion length is a parameter of practical importance. This is known as P_{eclet} number (P_e).

$$P_{eclet} \text{ number:} \quad P_e = \frac{L}{L_d} \tag{3.14}$$

General Case: Mixed Transport

In any flow, generally both the modes of mass transport are present simultaneously; the convection flux J_{conv} is in the direction of velocity vector u, and the diffusion flux J_{diff} is at right angles to that. However, because CVD substrates are impermeable, the streamlines do not penetrate the substrate but change their direction parallel to or away from the substrate (Fig. 3-5). That is, the convection flux to the substrate is 0. Hence, the only mode of mass transport to the substrate is diffusion

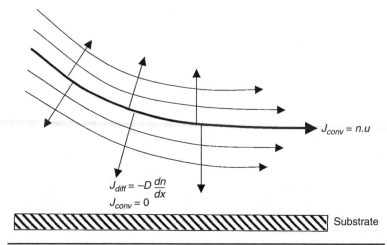

$$J_{conv} = n.u$$

$$J_{diff} = -D \frac{dn}{dx}$$
$$J_{conv} = 0$$

Substrate

FIGURE 3-5 Flow flux near the substrate. Convection flux to the substrate is 0 so that the deposition is solely contributed by diffusion flux.

that contributes to the deposition. However, as soon as deposition takes place, a temporary loss of species occurs near the substrate. Hence, to get a uniform deposition, it is necessary to achieve a balance between the species transport by diffusion and species loss by deposition. This is governed by *Thiele's law of diffusion*. Details of Thiele's diffusion can be found in standard technical books on CVD, where readers can learn how skillful calculations enable researchers to manipulate the diffusion flux to the substrate and to control the deposition composition as well.

Heat Transport

So far this chapter has discussed the motion of reactants in a CVD reactor while overlooking the temperature aspect. However, recall that CVD is almost always a high-temperature phenomenon. The concerns of temperature control are often encountered in the reactor because the entire reactor is not the reaction zone. Generally, only a limited portion is the deposition zone that is required to be heated to a certain high temperature while the rest is to be maintained at sufficiently low temperatures, so the reactants do not get spoiled before reaching the desired deposition zone. For the sake of such controls in a CVD reactor, it is imperative to understand the general laws of heat flow and temperature distribution inside the reactor. Fortunately, with the advent of high-tech electronic systems and the availability of sophisticated furnaces, temperature control is not a big problem, and researchers do not have to bother understanding it. Nevertheless, the following sections briefly review the essential concepts of heat flow so readers can visualize (or guess, at least) what the thermal trace of a typical reactor interior might look like.

Let us start with energy sources in CVD. The most commonly used sources are thermal ones, such as tube furnaces (using resistive heating), radiofrequency heaters (using inductive heating), and high-intensity radiation lamps (using radiative heating). Other popular energy sources include lasers and plasma, although they lead to two independent branches of CVD, known as *laser CVD* and *plasma CVD*, respectively. The discussion here will be restricted to thermal CVD.

When tube furnaces are used, the entire reactor gets heated, and heat flux is directed radially inward. However, when radio frequency induction coils or radiation lamps are used, only the heat susceptor kept inside the reactor gets heated and generates heat flux perpendicularly upward, although the reactor walls remain significantly cold. CVD reactors are thus broadly classified into two groups, hot- and cold-wall reactors. Schematic diagrams of hot- and cold-wall reactors, in their simplest forms, are shown in Fig. 3-6.

Heat Flow in a Hot-Wall Reactor

As mentioned earlier, a hot-wall reactor is a case with tube furnaces. Resistance elements are wound around a ceramic or silica tube, inside which the CVD reaction tube (often quartz tube) is inserted. By virtue of tubular geometry, thermal field in the reactor is quite symmetric radially inward. For small-diameter tubes, the temperature in the

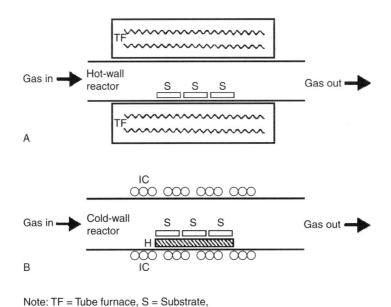

Note: TF = Tube furnace, S = Substrate,
      IC = Induction coil, H = Heat-susceptor

FIGURE 3-6   Schematic diagrams of hot-wall (*A*) and cold-wall (*B*) reactors in their simplest forms.

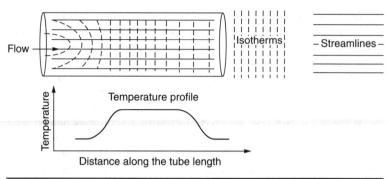

FIGURE 3-7   Schematic isotherms and the temperature profile in a hot-wall reactor.

central hot zone is fairly constant. Unless the gas flow is exceptionally high, the isotherms in the deposition zone are laminar and sufficiently steady, which makes the case quite straightforward. The situation may graphically be expressed as shown in Fig. 3-7. Practically, however, the existence of some flow-disturbance fixtures, potential entry-length problems, may interfere, which needs to be taken into account.

Hot-wall reactors are predominantly used in systems in which the deposition reaction is exothermic in nature because the high wall temperature minimizes or even prevents undesirable deposition on the reactor walls. In the case of multicomponent systems, the combination of several resistance elements in the tube furnace makes it possible to impose desired temperature gradients along the tube length to control CVD processes occurring in various sections of the reactor.

Heat Flow in a Cold-Wall Reactor

In cold-wall reactors, the situation is somewhat more complex. Here, instead of the entire reactor, only the susceptor (generally SiC-coated graphite) is heated by radiofrequency induction coils. The susceptor temperature may go up to 1000° to 1200°C while the reactor walls remain at sufficiently low temperature. This creates sharp temperature gradients away from the susceptor, which disturbs the streamlines of fluid flow. Consequently, the thermal fields in the reactor are not well defined and it is difficult to choose an appropriate model for transport calculations. For this reason, a lot of experimental studies have been conducted to define and manipulate the thermal fields in cold-wall reactors. An interesting way to visualize the flow patterns inside the reactor is to release smoke into the reactor. Flow visualization experiments suggest that there exists a boundary layer adjacent to the susceptor and that steep temperature gradients perpendicular to the flow affect the nature of the flow. Thus, we get a qualitative picture of heat flow in a cold-wall reactor. For a quantitative picture, however, it is necessary to measure the spatial variation of

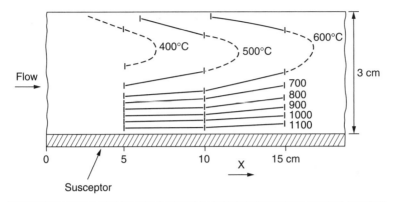

FIGURE 3-8 Experimental plot of isotherms in a cold-wall reactor ambience with helium (flow velocity, 50 cm/sec; susceptor temperature, 1200°C) (Ban, 1978).

temperature in situ at as many points in the reactor as possible. Such an experimental plot illustrating the isotherms in a cold-wall CVD reactor is shown in Fig. 3-8. From the figure, it follows that (1) there is a steep temperature gradient in the first 1.5 cm above the susceptor and a shallower gradient in the space above that, and (2) the average temperature of the gas increases along the susceptor. Moreover, both the temperature and the temperature gradient vary with the nature of the gas in the reactor.

Cold-wall reactors are suitable for CVD involving endothermic reactions that tend to proceed on the hottest surfaces of the system. A uniformly heated susceptor and substrates lead to technologically demanded growth uniformity. Moreover, the low temperature in the reactor space (beyond the susceptor and substrates) prevents undesirable gas-phase nucleation, parasitic reactions, and autodoping.

3.3   Synthesis of Nanomaterials

We have reviewed the fundamental principles and general steps of CVD, so it is pertinent to proceed to synthesis of nanomaterials. What is a nanomaterial? A simplistic answer is "ultrafine materials in a size range of a few nanometers to a few hundreds of nanometers." But *nanomaterials* is still a vague term encompassing a wide range of entirely different tiny materials, including nanoparticle, nanofiber, nanowire, nanotubes, nanodiamond, and nanowall, of any element, compound, or mixture. Actually, these are the key components of the revolutionary field of nanotechnology, which is still in its very beginning stage. Hence, the synthesis, characterization, and application of all these so-called nanomaterials are still in the test phase. It is more important to realize that, although CVD has great potential for synthesizing most of the above-mentioned examples of nanomaterials, each example belongs to a different class of expertise, requiring different materials, methods, processes, and parameters. Therefore, instead of presenting a vague outline of common nanomaterials synthesis, it is advisable to present a clear picture of a specific nanomaterial with relevant synthesis details. Hence, let us pick just one example—CNTs—as the subject matter of this section. CNTs are the most widely investigated nanomaterial in the world so far because of their amazing application potential in multidisciplinary applications.

3.3.1   Carbon Nanotubes

A CNT may be considered a graphene sheet (a single layer of graphite, meaning a hexagonal arrangement of carbon atoms in one plane) rolled to form a seamless cylinder a few nanometers in diameter and a few hundreds of microns in length. This is what is known as a *single-walled CNT* (SWCNT). However, when a multilayer graphite exists in the form of a hollow cylinder (in which all graphene layers are coaxially cylindrical), it is known as *multiwalled CNT* (MWCNT). The electronic properties of the resulting CNT depend on the direction in which the sheet is rolled up; that is, the electrical conductivity of a CNT depends on the orientation of its hexagonal network, called the *tube chirality*. A CNT may be a conductor or semiconductor just by virtue of different possible arrangements of carbon atoms constituting the tube. The electronic properties of a CNT can thus be tailored, which is of great significance in nanoelectronics. By virtue of ballistic electron transport in an ideal metallic CNT, its electrical conductivity may be a million times higher than that of copper.

CNTs also exhibit extraordinary mechanical properties. They are the strongest and the most flexible molecular material because of C–C covalent bonding and seamless hexagonal network architecture. Their Young's modulus is of the order of tera pascal (10^{12} Pa), that is, they are as stiff as diamond, the hardest material known. And they have a tensile strength about 200 giga pascal, that is, about 100 times stronger than steel but one sixth its weight. Hence, worldwide efforts are going on for growing CNTs in large quantities, and CVD has great potential for achieving this. Compared with the arc discharge and laser ablation methods, CVD is a simple and economic technique for synthesizing CNTs at low temperature and ambient economical pressure.

3.3.2   Schematics of Carbon Nanotube Growth

Figure 3-9 shows a schematic diagram of the experimental setup used for CNT growth using the CVD method in its simplest form. The process involves passing a hydrocarbon vapor (typically for 15 to 60 minutes) through a tube furnace in which a catalyst material is present at sufficiently high temperature (600° to 1200°C) to decompose the hydrocarbon. CNTs grow over the catalyst in the furnace, and the

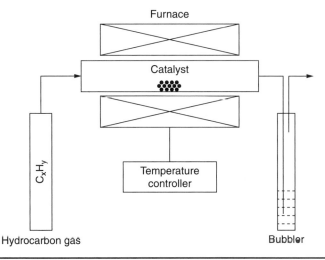

Furnace

Catalyst

Temperature
controller

C_xH_y

Hydrocarbon gas

Bubbler

FIGURE 3-9   A simple setup for growing carbon nanotubes.

CNTs are collected upon cooling the system to room temperature. In
the case of a liquid hydrocarbon (e.g., benzene, alcohol), the liquid is
heated in a flask, and an inert gas purged through it carries its vapor
into the reaction furnace. Vaporization of a solid hydrocarbon (e.g.,
camphor, naphthalene) is conveniently achieved in another furnace
(low temperature) before it reaches the main reaction furnace (high
temperature), as shown in Fig. 3-9. Catalyst material (solid, liquid, or
gas) is suitably placed inside the furnace or fed from outside. Pyrolysis of the catalyst vapor at a suitable temperature liberates metal
nanoparticles in situ; this process is known as the *floating catalyst
method*.

Alternatively, catalyst-plated substrates may be placed in the hot
zone of the furnace to catalyze CNT growth. Catalytically decomposed carbon species of the hydrocarbon are assumed to dissolve in
the metal nanoparticle and, after reaching super saturation, precipi-
tate out in the form of a fullerene dome extending into a carbon cylinder (similar to an inverted test tube, as shown in Fig. 3-10) having
no dangling bond and hence bearing minimum energy. When the
substrate–catalyst interaction is strong, a CNT grows up with the
catalyst particle rooted on its base (commonly known as the *base-
growth model*). When the substrate–catalyst interaction is weak, the
catalyst particle is lifted up with the growing CNT and continues
conducting CNT growth sitting on its tip (the so-called *tip-growth
model*). Formation of SWCNTs or MWCNTs is governed by the size of
the catalyst particle. Broadly speaking, when the particle size is a few
nanometers, SWCNTs form; bigger particles (i.e. tens of nanometers)
favor MWCNT formation.

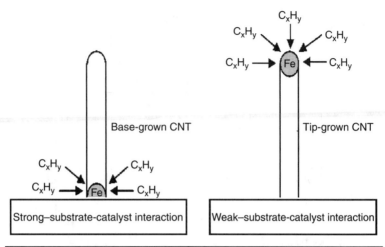

FIGURE 3-10   Two probable models for carbon nanotube (CNT) growth.

3.3.3   Parameters for Carbon Nanotube Growth

Three main parameters for CNT growth by CVD are hydrocarbon, catalyst, and growth temperature. Other important parameters are the carrier gas, gas flow, and CVD reaction time, which largely depend on the reactor geometry (length and diameter of the reaction tube). Hence, all parameters need to be carefully optimized on a particular experimental setup. With proper optimization of all these parameters, CNTs can be grown in a variety of forms, such as powder, thin or thick films, aligned or entangled, straight or coiled, or a desired architecture of nanotubes on predefined sites of a patterned substrate.

Hydrocarbons

The main raw material for CNT synthesis is a hydrocarbon, which is catalytically decomposed to liberate carbon atoms and constitute CNTs. Plenty of hydrocarbons (e.g., methane, ethylene, acetylene, benzene, xylene) have been used as a carbon source for growing CNTs. However, some other raw materials (e.g., carbon monoxide, ethanol, camphor, turpentine oil, kerosene, ferrocene) have also been successful at producing CNTs. Hence, it is believed that any carbon-containing material can produce CNTs, although optimum experimental conditions (e.g., temperature, pressure, catalyst) are quite different for different starting materials. Because CVD is a vapor-phase reaction, the vapor pressure of the raw material dictates the deposition conditions, which need to be carefully optimized.

In fact, CVD has been used for producing carbon filaments and fibers since 1959 (Walker et al.). Using the same technique, soon after the discovery of CNTs by Iijima (1991), Endo et al. (1993) reported CNT growth from pyrolysis of benzene at 1100°C. Yacaman et al. (1993) obtained clear helical MWCNTs at 700°C from acetylene instead. In these cases, iron nanoparticles were used as catalyst. Later, MWCNTs were also grown from ethylene (Hernadi et al., 1996), methane (Satishkumar et al., 1999), and many other hydrocarbons. On the other hand, SWCNTs were first produced by Dai et al. (1996) from disproportionation of carbon monoxide at 1200°C catalyzed by molybdenum particles. Later, SWCNTs were also produced from benzene (Cheng et al., 1998), acetylene (Satishkumar et al., 1998), ethylene (Hafner et al., 1998), methane (Kong et al., 1998), and so on by using various catalysts. Against the conventional use of these petrohydrocarbons, a green-plant product, camphor, was found to be an efficient material for producing high yields of MWCNTs (Kumar et al., 2001, 2002).

Because carbon nanomaterials will be in great demand in the future, there is a need to develop the process of synthesis using precursors that are not derived from sources such as petroleum products, so that in the event that these sources get depleted, the production of carbon nanomaterials is not adversely affected. Sharon and his research group therefore have been exploring the precursors that are derived from plants, the idea being that these precursors could be harvested as and when needed, and even in the event of scarcity of petroleum-derived precursors, production of carbon nanomaterials would not be adversely affected. Sharon and his research group are the first to have explored this possibility and have successfully synthesized various forms of carbon with the CVD process, such as C_{60} (Mukhopadhyay et al., 1994), glassy carbon (Sharon et al., 1996), diamondlike carbon film (Mukhopadhyay et al., 1997), semiconducting carbon (Sharon et al., 1999), carbon nanofibers (Pradhan et al., 2003), spongy carbon nanobeads (Sharon et al., 1998), and so on.

Temperature

General experience is that whereas low-temperature CVD (600° to 900°C) yields MWCNTs, higher temperature (900° to 1200°C) reactions favor SWCNT growth, indicating that SWCNTs have a higher energy of formation (presumably owing to small diameters, high curvature, and high strain energy). Perhaps this is why MWCNTs are easier to grow (than SWCNTs) from most of the hydrocarbons and why SWCNTs could be grown from selected hydrocarbons (e.g., carbon monoxide, methane) that have a reasonable stability in the temperature range of 900° to 1200°C. Commonly efficient precursors of MWCNTs (e.g., acetylene, benzene) are unstable at higher temperature and decompose before reaching the catalyst surface, which leads to the deposition of large amounts of carbonaceous compounds other than nanotubes. Within the temperature range of 600° to 900°C, diameter distribution of MWCNTs increases with increasing temperature. The lower is the CVD temperature, the narrower the diameter distribution. The ultimate goal should be to produce CNTs of specific diameter at the minimum temperature and hence at minimum energy cost.

Catalysts

Transition metals (Fe, Co, Ni) are commonly used catalysts for CNT growth because the phase diagram of carbon and metals suggests finite solubility of carbon in transition metals at high temperatures. This leads to the formation of CNTs under the growth mechanism outlined above (see Fig. 3-10). Solid organometallocenes (ferrocene, cobaltocene, nickelocene) are widely used catalysts because they liberate metal particles in situ, efficiently catalyzing CNT growth.

The catalyst particle size dictates the tube diameter. Hence, metal nanoparticles of controlled size, presynthesized by other reliable techniques, may be used to grow CNTs of controlled diameter (Ago et al., 2000). Thin films of catalyst coated on various substrates have also proven effective at obtaining uniform CNT deposits (Fan et al., 1999). Also, the material, morphology, and textural properties of the substrate greatly affect the yield and quality of the resulting CNTs. Zeolites supported with catalysts in their nanoropes have resulted in significantly high yields of CNTs with a narrow diameter distribution (Kumar and Ando, 2005). Alumina may be a better catalyst support than silica owing to the strong metal–support interaction in the former, which allows high metal dispersion and thus a high density of catalytic sites. Such interactions prevent metal species from aggregating and forming unwanted large clusters that lead to graphite particles or defective MWCNTs. The key factor to get high yield of pure CNTs is by achieving the hydrocarbon decomposition on catalyst sites only and avoiding spontaneous (uncatalyzed) pyrolysis. It is remarkable that transition metals have proven to be efficient catalysts not only in CVD, but also in arc discharge and laser vaporization methods. This indicates that these apparently different methods might inherit a common growth mechanism of CNT, which is not yet clear.

Some successful attempts at CNT synthesis have been made by using ferrocene (Rao, 1998) as a carbon-cum-catalyst precursor, although as-grown CNTs were mostly metal encapsulated. Recently, the use of ethanol has drawn much attention for synthesizing SWCNTs at relatively low temperature (800° to 900°C) on zeolite support impregnated with Fe–Co (Maruyama, 2002) as well as Mo–Co dip-coated quartz substrates (Murakami, 2003). Almost simultaneously, high-purity SWCNTs and MWCNTs were also produced from camphor using 1% ferrocene as a catalyst (Kumar, 2003a, 2003b), as shown in Fig. 3-11. Because of a very low catalyst requirement with camphor, as-grown MWCNTs are least contaminated with metal, but the oxygen atom present in camphor helps in oxidizing amorphous carbon in situ. Camphor's ring-based growth mechanism of CNTs was approved by in-situ mass spectroscopic study of benzene CVD (Tian et al., 2004).

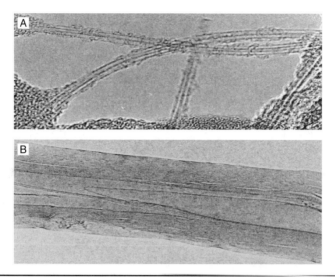

Figure 3-11 Typical images of single-walled carbon nanotubes (A) and multiwalled carbon nanotubes (B) grown from camphor.

The use of camphor as a CNT precursor is slowly spreading in Europe (Porro et al., 2006).

3.3.4 Producing Aligned Carbon Nanotubes

The CVD method is ideally suited for growing aligned CNTs on desired substrates for specific applications, which is not feasible with the arc or laser methods. Li et al. (1996) grew dense MWCNTs arrays on iron-impregnated mesoporous silica, and Pan et al. (1998) reported aligned CNTs larger than 2 mm in length over large-area mesoporous substrates from acetylene. However, Andrews et al. (1999) were the first to grow vertically aligned CNTs on a bare quartz substrate with simple pyrolysis of a xylene–ferrocene mixture. Similarly, CVD of camphor with 1% ferrocene resulted in very uniform carpets of aligned CNTs (Kumar, 2003b), as shown in Fig. 3-12. In such a process, although both the carbon source and the catalyst source are evaporated at the same temperature, the iron nanoparticles generated in situ are supposed to first disperse on the substrate kept in the high-temperature zone and then catalyze the CNT growth. The reason for vertical alignment of as-grown CNTs has been attributed to a uniform dispersion of iron particles everywhere on the entire substrate, causing the simultaneous growth of high-density CNTs from all particles. Camphor-grown vertically aligned CNTs showed good field-emission properties (Kumar, 2004). More organized assemblies of CNTs can be grown by patterning the substrate in a desired manner (Cao et al., 2004).

Figure 3-12   Carbon nanotube carpet grown from camphor on a quartz substrate.

3.3.5   Mass Production of Carbon Nanotubes

Because CVD is a well-known and established industrial process, CNT production by CVD is easy to scale up. MWCNTs of controlled-diameter distribution are being produced in large quantities (~100 g/day) from acetylene CVD using nanoporous materials as a catalyst support (Couteau, 2003). Wang et al. (2002) have developed a nanoagglomerate fluidized-bed reactor in which continuous decomposition of ethylene gas on an iron or alumina catalyst at 700°C produces a few kilograms of MWCNTs per hour. Dai's (1999) group has excellently scaled up SWCNT production from methane using a Fe–Mo bimetallic catalyst supported on sol-gel derived alumina-silica multicomponent material (Cassel et al., 1999). However, Smalley's laboratory at Rice University is still leading in mass production of SWCNTs with a high-pressure carbon monoxide technique (Nikolaev et al., 1999). In this technique, iron pentacarbonyl catalyst liberates iron particles in situ at high temperature, and the high pressure of carbon monoxide (~10 atm) enhances the carbon feedstock manifolds, which significantly speeds up the disproportionation of CO molecules into C atoms and thus accelerates the SWCNT growth. Unfortunately, however, the CNT purity decreases significantly (by 20 to 30%) when produced in large quantity (in bigger reactors). Hence, extensive research is still going on to produce high-purity CNTs with high yield.

Quite recently, camphor has shown an amazing CNT production efficiency of 50wt%. Thirty-minute CVD of 12 g of camphor at 650°C yields 6-g MWCNTs of fairly uniform diameter (~10 nm), and their as-grown purity is more than 88% (Kumar, 2005). But mass production of SWCNTs from camphor is still a challenge.

3.3.6 Producing Individual Carbon Nanotubes

Apart from large-scale production, CVD also offers to grow a single nanotube to be used as a probe tip in atomic-force microscopes (AFMs) or field emitter in electron microscopes. Such an applied growth is feasible in low-pressure CVD (~1 torr) so that the partial pressure of the hydrocarbon and hence the carbon concentration around an individual catalyst particle could be controlled precisely. Hafner et al. (1999a) grew single SWCNTs and MWCNTs (1 to 3 nm) rooted in the pores of silicon tips, which were quite suitable for AFM imaging. In another approach, single SWCNTs were directly grown on pyramids of silicon cantilever-tip assemblies (Hafner, 1999b). In this approach, an SWCNT growing on the silicon surface (controlled by the catalyst density on the surface) protrudes from the apex of the pyramid. As-grown nanotube tips are smaller than mechanically assembled nanotube tips by a factor of 3 and hence enable significantly improved imaging resolution.

Thus, CVD CNTs hold great promise toward fabricating sophisticated instruments and nanodevices. And when it comes to thinking of creating nanochips for the next generation of computers, CVD is the only choice to explore.

3.4 Summary

CVD is a process in which chemical species react in the vapor phase to form a solid product on some surface. CVDs are almost always surface-catalyzed reactions. This is the most popular and accepted technique for preparation of nanoforms of carbon. Various forms of nanosized carbon are obtained by this technique, depending on the catalyst, temperature, carrier gas, and raw material used. The nature of chemical reaction can be determined by in situ mass spectroscopic analysis and postdeposition elemental analysis. Thermodynamic calculations provide valuable guidelines for establishing the process conditions if the free energy values for reactants (vapor phase) and product (condensed phase) are known. Activation energy for the deposition reaction can be obtained from the slope of the experimental plot of the deposition rate versus the reciprocal temperature. Subsequent analysis of chemical kinetics leads to the reaction mechanism and growth model.

Study of transport phenomena is an essential element in designing efficient CVD reactors. Mass transfer occurs by both the convection and diffusion modes, although diffusion plays a dominant role in deposition. Control of heat transfer is crucial for getting high-quality depositions; the reaction should essentially be surface catalyzed, and self-pyrolysis, autonucleation must be avoided.

Thermal CVD is a simple and low-cost method of synthesizing CNTs in high yield and high purity with better control on the tube properties such as diameter length and orientation. The three prime parameters are the carbon source (hydrocarbon), catalyst, and CVD temperature. Others are the carrier gas, flow rate, catalyst support, and CVD duration. The key point for CNT growth is achieving hydrocarbon's decomposition at the catalyst surface (avoiding self-pyrolysis of hydrocarbon). Nanoparticles of transition metals serve as good catalysts for CNT growth. Whereas small metal particles (a few nanometers in size) favor SWCNT growth, bigger particles (a few tens of nanometers) result in MWCNTs. Whereas high temperature (900° to 1200°C) is required for SWCNT growth, MWCNTs can be grown at relatively low temperature (600° to 900°C). Hydrocarbons that withstand high temperature (900° to 1200°C) are preferable for SWCNT formation, but MWCNTs can be grown from common hydrocarbons or even from any carbon-containing compound (with proper optimization of the experimental parameters). For an industrially viable CNT growth technique, low-temperature and low-atmospheric pressure CVD processes (consuming low energy in turn) are advisable. For long-term, sustainable CNT growth technique, stress should be placed on exploring regenerative non–fossil fuels or plant-based carbon sources.

CHAPTER 4

Catalysts for Synthesis of Carbon Nanotubes

Bholanath Mukherjee

K.V. Pendharkar College, Maharashtra, India

Maheshwar Sharon

Nanotechnology Research Centre, Birla College, Kalyan, India

Madhuri Sharon

MONAD Nanotech Pvt Ltd, Powai, Mumbai, India

4.1 Introduction

The term *catalysis* was first introduced by Berzelius in 1836 to describe chemical substances that affect reactions, but do not take part in reactions. *Catalyst* cannot be defined in a way that is applicable to every

63

branch of catalysis; however, it may be described as a substance that alters the rate of a reaction and gets regenerated in due course of the reaction.

A catalyst usually increases the rate of reaction by decreasing the activation energy of the reaction. Catalysts are specific in their utility and may be added with promoters to increase their catalytic activity. The added promoter itself may or may not have catalytic activity. If the promoter also has catalytic properties, then the catalyst is called a *mixed catalyst*.

The entire surface of a catalyst (specially in solid state reaction) may not be catalytically active, but the sites (e.g. surface states) that are active are called *active centers*. The active centers may be atoms or groups of atoms which are deprived or posses more electrons than the atoms which are present in the bulk of material. These sites may behave like Lewis acid or base.

Certain substances present in the reaction system decrease the rate of reaction or stop it completely. Such substances are called *catalyst poisons*. Catalyst poisons may react with the active site or neutralize them. Catalysts have a definite life even in the absence of poisons. The deactivation is attributable to different reasons, such as surface's structural changes over a period of use or a decrease in surface area.

Nanosize catalysts can be prepared by different methods are called *nanocatalysts*. They can be in the form of nanocrystals, nano-films, nanowires, nanoclusters, and so on. In this chapter, different aspects of the use of metal nanocatalysts for synthesis of carbon nano-materials are discussed.

4.2   Nanometal Catalysts for Synthesis of Carbon Nanotubes

For the synthesis of carbon nanotubes (CNTs), a transition metal nanocatalyst is a necessity. The diameter of CNTs prepared depends on the size of the metal catalyst used. The growth rate, morphology, wall thickness, and structure of the CNT also depend on the size of the catalyst.

4.2.1   Required Properties of a Catalyst for Synthesis of Carbon Nanotubes

A good catalyst should have a strong catalyst metal–support interac-tion, a large surface area, and a large pore volume, even at high tem-peratures. Because of the strong catalyst metal–support interactions, the catalyst particles remain dispersed on the support and prevent their aggregation. This results in high-density catalytic sites favoring growth of single-walled CNTs (SWCNT) and preventing the growth of graphitic particles or defective multiwalled CNTs (MWCNTs) (Cassel et al., 1999).

By adding promoters such as S, Bi, or Pb to a Co nanocatalyst powder (Kiang et al., 1996), longer CNTs with larger diameter can be synthesized (Cheng et al., 1998). The mixture of inert carrier gases, such as He, Ar, or N_2, used during CNT synthesis have also been found to affect the CNT diameter. Different mixtures of carrier gases have different diffusion coefficients and thermal conductivities, which alter the rate of diffusion of carbon on the metal catalyst as well as cooling of the reaction mixture, which in turn affects the diameter of the CNTs (Ebbesen and Ajayan, 1992; Jung et al., 2003). However, this is not fully understood.

The most commonly used nanocatalysts are nanoparticles of Fe, Ni, Co, and rare earth elements such as Y. At high temperature, carbon has limited solubility in these metals (forming metal carbides), so it crystallizes easily. The bimetallic catalyst of Co and Fe supported on silica gel particles improves the quality, quantity, and uniformity of CNTs (Afre et al., 2006). Mixed or bimetallic catalysts, such as Fe–Ni and Co–Ni, have also been used with improved results. None of the metals Cr, Mn, Zn, Cd, Ti, Zr, La, Cu, V, and Gd are useful in the catalytic growth of CNTs (Deck and Vecchio, 2006).

4.2.2   Preparation of Metal Nanocatalysts

Preparation of Nanometal Oxides

Metal oxides obtained by various methods are used as sources of nanometal catalysts. Metal oxides are reduced by using hydrogen alone or mixed with argon at 750° to 950°C. Nanometal catalysts are susceptible to air oxidation even at slightly elevated temperatures. Hence, they are cooled to room temperature before they are exposed to air. Some of the methods used to prepare nanosize metal oxides are discussed below.

Hydrolysis   Many metal oxides of nanosize are prepared from aqueous solution by hydrolysis. Here *hydrolysis* mainly refers to decomposition of water molecules in the hydrated metal ions. It must be noted that not all hydrated metal ions can be hydrolyzed (e.g., alkali metals, alkaline earth metals, inner transition metal ions).

Solvolysis   Hydrolysis of metal ions using polar solvents other than water, such as propylene glycol (Souad Ammar et al., 2001) and surfactants such as dodecyl sulphonate (Moumen et al., 1996; Moumen and Pileni, 1996a, 1996b), have been used to prepare its analogs with iron, cobalt, and others. Their micellar solution at higher pH precipitates metal oxide of the size 2 to 5 nm. Sonication is also used to accelerate the rate of solvolysis and hydrolysis (Kumar et al., 2000).

Oxidation   Metal carbonyls such as $Fe(CO)_5$, when decomposed in solvents, form very fine metal particles. Likewise, alkoxy compounds of metal also decomposes in the presence of moisture giving nano size metal oxide powder. For example, nanosize TiO_2 is prepared by hydrolysing titanium alkoxide. These metal salts undergo oxidation very easily, even on slight exposure. Nanosize materials have a tendency to agglomerate. Decaline can be used as a solvent with oleic acid as stabilizing ligand (Bentzon et al., 1989).

Metathesis   In simple terms, a metathesis reaction is double-decomposition reaction in which two compounds, MA and NB, react to form MB and NA. Two different techniques are suggested by reacting metal halide with a metal alkoxide and metal halide with ether (Arnal et al., 1997). They can be expressed as:

$$MX_n + M(OR)_n \rightarrow 2\,MO_{n/2} + n\,RX \tag{4.1}$$

$$MX_n + n/2\,ROR \rightarrow MO_{n/2} + n\,RX \tag{4.2}$$

The removal of volatile alkyl halide makes the reaction possible.

Thermolysis   Precursors in which metal ions are bonded to ligands through an oxygen atom; upon decomposition, they give metal oxide. Cupferron complexes (Rockenberger et al., 1999) have been used to prepare oxides of transition metals by thermolysis. It has been observed that the size of the oxide particles depends on the temperature of the reaction. Thermolysis is a viable solution when hydrolysis fails.

Solvothermal Method   This method uses a solvent at a temperature higher than its boiling point in a closed vessel that can withstand high pressure. Many nanosize oxides are synthesized by this method (Rajamathi and Seshadri, 2002).

Solvothermolytic Method   This is a urea-based method used to prepare nanosize metal oxides using transition metal salts and molten urea. Thermolysis is usually carried out in a quartz boat at 650° to 750°C at ambient pressure.

Decomposing Metal Nitrate   Another method of preparation of single and binary mixed metal oxides, similar to the solvothermolytic method, is by decomposing metal nitrates with polyvinyl alcohol (PVA) or a mixture of PVA and polyacrylic acid (Pramanik, 1996).

Nanosize catalysts have also been prepared by soaking MgO powder with an alcoholic solution of $Fe(NO_3)_3$ or $Co(NO_3)_2$. The solution is then dried and reduced by hydrogen (Deshpande et al., 2004) to get nanosize metal powder.

Preparation of Nanometal

Some of the common methods for preparation of nanometals are discussed in the following sections.

Sonochemical Method Sonochemical synthesis is the result of acoustic cavitation. It involves the creation and growth of a bubble that ultimately collapses (i.e., the bubble implodes). According to Flannigan and Suslick (2005), the implosion creates a temperature of 20,000°K and a pressure of 2000 atm.

Sonication of an aqueous transition metal salt hydrazine (e.g., copper [II] hydrazine carboxylate) in aqueous solution has successfully resulted in nanoclusters (Gibson and Putzer, 1995) of nanosize copper particles in the presence of hydrogen and argon (Dhas et al., 1998). Nanoparticles of many metals, such as Au, Ag, Pt, Pd, and Rh (Okitsu et al., 1996), have also been prepared by sonication.

Microwave Heating Method Microwave heating in the frequency range of 900 MHz to 2.45 GHz is used to synthesize nanosize metal particles. Nanosize colloidal clusters of many metals, such as Pt, Au, Ir, Rh, and Pd, have been prepared by this method (Komarneni, 1992; Tu and Jia, 2000), and nanosize particles of Ag have been prepared by microwave irradiation of $AgNO_3$ in poly-(N-vinyl-2-pyrrolidone) without any other reducing agent (He et al., 2002).

CoMoCAT Method A unique catalyst system made up of Co–Mo is used to prepare selectively SWCNTs by disproportionation of CO. The CoMoCAT catalyst is a bimetallic catalyst of Co and Mo. It is usually supported on silica or magnesia. A silica-supported CoMoCAT is prepared by coimpregnation of an aqueous solution of cobalt nitrate, $Co(NO_3)_2.6H_2O$, and ammonium heptamolybdate salt, $(NH_4)_6 Mo_7O_{24}.4H_2O$, on silica. The impregnated silica is dried in a convection oven at 110°C for 12 hours and then calcinated at 500°C in hydrogen flow for 3 hours. To prepare a magnesia-supported catalyst, an alcoholic solution is used instead of aqueous solution because MgO is not soluble in water (Herrera et al., 2003; Lolli et al., 2006; Yongqiang and Resasco, 2005).

Dissociation of CO, in the absence of hydrogen takes place according to the reaction:

$$2CO \rightleftharpoons C + CO_2 \tag{4.3}$$

This reaction is known as *CO disproportionation* or a *Boudouard reaction* (Balakos and Chuang, 1993). It gives 80 to 90% SWCNTs (Maeda et al., 2000). By altering the reaction conditions, SWCNTs with varied diameters can be produced (Alvarez et al., 2002; Herrera et al., 2003). The presence of Mo in the catalyst prevents sintering of Co. Mo forms Mo_2C, which retards the growth of CNTs. Hence, the formations of other forms of carbon are prevented with an increase in selectivity (Jorio et al., 2005). Recently, Resasco and his team have prepared vertically aligned SWCNTs (VSWNTs) using the CO disproportionation technique on a Co–Mo bimetallic catalyst. This produced a VSWNT forest of almost 40 μm in length (Zhang et al., 2006).

Direct Precipitation   Finely dispersed metal particles can be synthesized in solution by direct precipitation from homogenous solution using proper reducing agents such as organic acids, alcohols, polyols, aldehydes, hydrazine, and $NaBH_4$ (Goia and Matijevic, 1998).

Hydrothermal Synthesis   Hydrothermal synthesis of nanopowders using urea (Li et al., 2006) is a recently developed technique. A urea-based homogenous precipitation process is used to prepare nanosize particles (Kundu et al., 2003; Taegyung amd Hwang, 2003).

Biosynthesis of Nanometal Catalyst   Use of microbes for synthesizing nanometals with a "bottom-up" approach has now become an accepted technology. *Desulfovibrio desulfuricans* and *Escherichia coli* have been used to prepare nanocatalysts. In the presence of hydrogen gas or organic formates, microbes take metal from their solution and deposit on the cell surface in the form of nanocrystals. Pd, Ag, and Pt nanocrystals have been prepared by this method.

4.3   Role of Catalysts in the Growth Mechanism of Carbon Nanotubes

The exact role of the catalyst in the mechanism of growth of CNT is still controversial. Catalysts have been shown to be necessary for the growth of SWCNTs, although growth of SWCNTs without catalyst has also been reported (Brabec et al., 1995; Charlier et al., 1997, 1999; Maiti et al., 1995; Robertson et al., 1992). One of the proposed explanations for the growth of SWCNTs during the carbon vapor deposition (CVD) method suggests that the adsorption and decomposition of the hydrocarbon source into carbon and hydrogen on the surface of the catalyst are responsible. The carbon atoms dissolve in the metal catalyst, forming a metal–carbon solution. When the solution becomes supersaturated with carbon, precipitation of crystalline tubular carbon solids in an sp^2 structure takes place (Amelinckx et al., 1994; Baker, 1989; Cassel et al., 1999; Tibbetts, 1984; Tibbetts et al., 1987).

The catalyst particles are believed to be spherical. Precipitation of carbon does not take place at the apex of the catalyst sphere, accounting for the hollowness of the tube. Because of the absence of dangling bonds and low energy, formation of a tubular structure is preferred to other forms of carbon such as graphitic sheets with open edges.

Three models have been proposed for elongation of CNTs, the base-growth, tip-growth, scooter models.

In the base-growth model, the catalyst particles remain attached to the support surface, and the CNTs grow upward from the catalyst surface with a closed end. This model is applicable if the catalyst metal and support have strong interactions.

In the tip-growth model, the catalyst particles get detached from the support surface and remain at the tip of the growing CNTs. The tip-growth model is applicable if the catalyst metal and support have weak interactions (Cassel et al., 1999).

The scooter model suggests that one or a few atoms of catalyst particles remain anchored at the open end of fullerene cluster (Thess et al., 1996) and maintain the uniformity of the diameter of the CNTs. Although the catalyst particles are strongly bound to the tip, they are found to be mobile at the tip (Lee et al., 1997). The catalyst particles prevent the formation of carbon pentagons by "scooting" around the open tip, preventing tube closure. The catalyst particles thus help in forming carbon hexagons and lengthen the tube. When the metal cluster at the tip becomes sufficiently large, it peels off and tube closure take place.

CNTs can also be grown on catalytic films. The diameter of the tube then depends on the thickness of the film. A thickness of about 13 nm gives MWCNTs 30 to 40 nm in diameter, and a thickness of 27 nm gives MWCNTs 100 to 200 nm in diameter (Park et al., 2002).

Double-walled CNTs have been prepared using a multilayered metal catalyst film of Al–Fe–Mo on silicon using the CVD technique at 600° to 1100°C (Cui et al., 2003).

Another mechanism states that C_2, is formed from the precursor which combines in the reaction vessel to form CNTs, is formed on the surface of the metal catalyst followed by formation of a rodlike carbon. Finally, graphitization of the wall takes place. This mechanism is based on in-situ transmission electron micrography observation (Yasuda et al., 2002) (Fig. 4-1).

In this figure mechanism of CNT growth has been shown. In Fig. 4.1*A* formation of carbon tube takes at the bottom in the presence of the catalyst. As the tube grows lengthwise, catalyst also move along the length (Fig. 4.1*B*)

Experimentally (especially in the CVD process) though catalysts are kept at one place in the furnace, but CNTs are found all over the tube. This can happen only if the precursor gets decomposed into some smaller fragments (initiated by the catalyst) and vapor of these fragments then depending upon the pyrolysis parameters of the of the CVD, gives CNTs of various shape and size. However, there is a need to understand the actual role of the catalysts in the formation of CNTs.

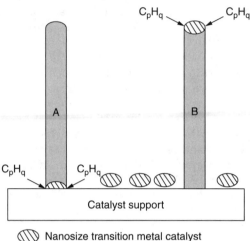

C_pH_q C_pH_q

A   B

C_pH_q C_pH_q

Catalyst support

⬭ Nanosize transition metal catalyst

FIGURE 4-1   (A) Base growth model. (B) Tip-growth model for synthesis of single-walled carbon nanotubes.

4.4   Purification of Synthesized Carbon Nanotubes

Purification of synthesized CNTs is needed because they often contain leftover catalyst (metal) particles. Moreover, during CNT synthesis, some amorphous carbon or diamond like carbon are also formed, which is not desired. Various chemical and physical procedures, such as oxidation, acid treatment, functionalization, chromatography techniques, annealing, microfiltration, ultrasonication (Guo, 1995; Lyu, 2004), centrifugation (Lee et al., 2004), and ferromagnetic separation, may be used for purification.

Refluxing of the as-grown SWCNTs in nitric acid solution not only dissolves the metal catalyst particles, but also oxidizes the amorphous carbon particles (Liu et al., 1998). However, the use of acid requires many considerations because it has been found to damage the morphology of CNTs.

Treatment with HCl is also done for purification, and similar to nitric acid, HCl also affects CNT morphology (Chiang et al., 2001a, 2001b; Goto et al., 2002; Moon et al., 2001).

Purification of SWCNTs from ferromagnetic catalyst particles has been done with sonication and magnetization (Thien-Nga et al., 2002).

4.5   Various Catalysts Used for Synthesis of Carbon Nanotubes

Table 4-1 provides a list of various catalysts used in the synthesis of CNTs.

Catalysts used	CNM formed	References
Ni, Co, Cu, Nb, Pt, Co–Ni, Co–Pt, Co–Cu, Ni/Pt	SWCNTs	Guo l., 1995
Ni–Co, Co–Y, Ni–Y	SWCNTs	Journet et al., 1997
Iron pentacarbonyl (ferrocene)	SWCNTs	Bhowmick et al., 2008 Bronikowski et al., 2001 Height et al., 2003 Liu et al., 2002 Nikolaev et al., 1999
Co, Y, Ni–Co, Ni–Y, Ni–Fe, Co–Y, Co–La	SWCNTs	Munoz et al., 2000
Co–Ni, La	SWCNTs	Guillard et al., 2000
Ni, Co, Y, Fe, Ni–Y, Ni–Co, Co–Y, Ni–La	SWCNTs	Maser et al., 2001
CoMoCAT	SWCNTs	Resasco et al., 2001
Mg	CNTs	Mottel et al., 2001
$CoCl_2$	SWCNTs	O'Loughlin et al., 2001
Fe	Aligned SWCNTs	Andrews et al., 2002
Iron nanocluster	SWCNT, sMWCNTs	Cheung et al., 2002
Iron (III) phthalocyanine	MWCNTs	Komatsu and Inoue, 2002
Iron acetate or silica	MWCNTs	Coquay et al., 2002
Polyamidoamine dendrimers, Fe–SiO_2	CNTs	Choi et al., 2002
MgO/iron nitrate	SWCNTs, MWCNTs	Mauron et al., 2003
Fe–Al_2O_3	MWCNTs	Corrias et al., 2003
Iron nitrate coated on aluminum	CNTs	Emmenegger et al., 2003
Fe–SiO_2	MWCNTs, CNFs	Perez-Cabero et al., 2003
Fe deposited on silica	CNTs	Nerushev et al., 2003
Fe–Co on zeolite	SWCNTs	Maruyama et al., 2003
Fe–Mo–MgO	DWCNTs	Lyu et al., 2003
Co deposited on anodic oxide	CNTs	Jeong et al., 2003

TABLE 4-1 Catalysts Used in the Synthesis of Different Carbon Nanotubes

Catalysts used	CNM formed	References
Fe–Al$_2$O$_3$	CNTs	Weizhong 2003
Co, Ni, Fe, Co sulphide	CNTs, CNFs	Hong et al., 2003
Fe(CO)$_5$	CNTs	Liu et al., 2003
NiCl$_2$	CNT bundles with mesoporous structures	Hu et al., 2003
Cobalt	MWCNTs	Li et al., 2004
Fe–Mo-MgO	SWCNTs	Lyu et al., 2004
Ferrocene, Fe, Fe–Pt	MWCNTs, nanofilaments	Lee et al., 2004
Fe	MWCNTn	Shao et al., 2004
Mo, Ni, or Pd–Fe on Al$_2$O$_3$	CNTs	Shah et al., 2004
Ferrocene and sulfur	Amorphous CNTs and Fe/C coaxial nanocables	Luo et al., 2006
CoMoCAT	VSWCNTs	Zhang et al., 2006
Multilayered Al/Fe/Mo catalyst on silicon substrates	Aligned SWCNTs and MWCNTs, DWCNTs	Cui et al., 2003
Fe/Mo bimetallic species supported on a novel silica–alumina	SWCNTs	Cassel et al.,1999
Bismuth, lead, and tungsten on cobalt	SWCNTs	Kiang et al., 1996
Bimetallic catalyst of Co and Fe supported on silica gel	CNTs	Afre et al., 2006
Fe, Ni, Co	Well-aligned mats of MWCNTs	Deck and Vecchio, 2006
Co–MO/SiO$_2$	SWCNTs	Alvarez et al., 2002 Herrera et al., 2003
CoMoCAT	SWCNTs	Jorio et al., 2005 Lolli et al., 2006 Maeda et al., 2000 Yongqiang and Resasco, 2005

TABLE 4-1   Catalysts Used in the Synthesis of Different Carbon Nanotubes (*Continued*)

Catalysts used	CNM formed	References
Anodic aluminum oxide (corona discharge plasma method)	MWCNTs	Li et al., 2006
Fe	Carbon filaments	Tibbetts et al., 1987
Fe, Co, Ni	Carbon filaments	Baker, 1989
Ni–Co	SWCNTs	Thess et al., 1996
Ni	SWCNTs	Lee et al., 1997
Fe	CNTs	Park et al., 2002

CNF = carbon nanofibres; CNTs = carbon nanotubes; DWCNT = Double walled carbon nanotube; MSWCNT = multiwalled carbon nanotubes; SWCNT = single-walled carbon nanotubes; VSWCNT = vertically aligned single-walled carbon nanotube.

TABLE 4-1 *(Continued)*

4.6 Summary

In this chapter, the importance of nanocatalysts in the synthesis of different carbon nanomaterials has been emphasized. Various methods for the synthesis of nanometals and nanometal oxides were discussed along with a brief introduction to the role of catalysts in the growth of CNTs. CNTs synthesized using metal catalysts need purifications to remove remnant catalysts in the product. Hence, various methods of purification were also discussed.

CHAPTER **5**

Preparation of Thin-Film of Carbon Nanomaterial by Pulsed-Laser Deposition

Golap Kalita

Department of Electrical Engineering, Chubu University, Aichi Prefecture, Japan

Maheshwar Sharon

Nanotechnology Research Centre, Birla College, Kalyan, Maharashtra, India

5.1   Interaction of Photons with Atoms

A laser is a device that creates and amplifies a narrow, intense beam of coherent light. The word *LASER* is an acronym for *light amplification by stimulated emission of radiation*. To understand the process of formation of laser beams, it is necessary to examine the various excitations that can take place when an atom is illuminated with a specific source of light.

How does light interact with individual atoms? Atoms consist of a positively charged core (nucleus) that is surrounded by negatively charged electrons. According to the quantum mechanical description of an atom, the energy of an atomic electron can have only certain values, and these are represented by energy levels. The electrons orbit the nucleus, and those with the largest energy orbit at a greater distance from the nuclear core. An electron within an atom can occupy many energy levels, but only one is considered here—the electrons in the outer orbits of the atom because these can be easily raised to higher unfilled energy states. The three types of interactions an atom and a photon can have are absorption, spontaneous emission, and stimulated emission.

5.1.1   Absorption and Spontaneous Emission

When a photon of light is absorbed by an atom (Fig. 5-1) in which one of the outer electrons is initially in a low energy state denoted by 0, the energy of the atom is raised to the upper energy level, 1, and remains in this excited state for a period of time that is usually less than 10^{-8} second. It then spontaneously returns to the lower state 0 with the emission of a photon of light. This type of absorption in which the energy of the absorbed photon is equal to the difference in energy between the levels 0 and 1 is referred to as a *resonant process*, suggesting that photons of only a particular frequency (or wavelength) will be absorbed. Similarly, the photon emitted will have energy equal to the difference in energy between the two energy levels. Thus, *absorption* occurs when a photon of just the right energy for a particular transition encounters an atom in the ground state and causes the atom (or rather, one of the atom's electrons) to jump into

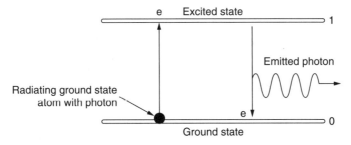

FIGURE 5-1 Schematic diagram of the process of excitation of atom from its ground state. The probability for absorption and spontaneous emission is exactly the same.

an excited state. *Spontaneous emission* occurs when an excited atom spontaneously returns to the ground state, emitting a photon of the almost same energy in the process (see Fig. 5-1).

The lifetime of a typical excited state of electron being about 10^{-8} seconds; it starts to drop down by emitting photons as fast as they are excited to the upper energy level. Hence, no significant collection of electrons can be accumulated by absorption alone. These common processes of absorption and spontaneous emission, thus, cannot give rise to the amplification of light. For every absorbed photon, another photon is emitted, so the number of excited atoms does not exceed the number of ground state atoms (or electrons of the atom).

5.1.2 Stimulated Emission

Stimulated emission is a very uncommon process in nature, but it is central to the operation of lasers. Stimulated emission is the process that can give rise to the amplification of light. In Section 5.1.1, it was stated that an atom in the excited state may return to the lower state spontaneously. However, if a photon of light interacts with the excited atom, it may stimulate this atom's return to the lower state. One photon interacting with an excited atom results in two photons being emitted. Furthermore, the two emitted photons are said to be in phase (i.e., if they are considered as waves, the crest of the wave associated with one photon occurs at the same time as on the wave associated with the other). This feature ensures that there is a fixed phase relationship between light radiated from different atoms in the amplifying medium.

As with absorption, stimulated emission is a resonant process; the energy of the incoming photon of light must match the difference in energy between the two energy levels. Furthermore, if we consider a photon of light interacting with a single atom, stimulated emission is just as likely as absorption; which process will occur depends on whether the atom is initially in the lower or the upper energy level. However, under most conditions, stimulated emission does not occur to a significant extent. The reason is that, under most conditions (i.e., under conditions of thermal equilibrium), there are far more atoms in the lower energy level, 0, than in the upper level, 1, so that absorption is much more common than stimulated emission. If stimulated emission is to predominate, more atoms must be present in the higher energy state than in the lower one. This unusual condition is referred to as *population inversion*.

5.1.3 Formation of Laser Beams

A population inversion cannot be achieved with just two levels (see Fig. 5-1) because the probability for absorption and for spontaneous emission is exactly the same; however, as in three levels (i.e., when a meta-stable state exists; Fig. 5-2), the probability for absorption and emission are different. The wavelength of stimulated radiation is equivalent to the difference in energy of the short-lived state and the ground state; hence, they are monochromatic.

The condition for population inversion is created in a unique manner. When one of the excited atoms spontaneously emits a photon, the photon is likely to hit another excited atom, stimulating the emission of another photon. If these photons are not allowed to escape the media, then they in turn will probably hit other excited atoms and stimulate more emissions and so on. This process produces a sudden burst of coherent radiation as all the atoms discharge in a rapid chain reaction. As a result, the population of excited atoms becomes greater than the population of the ground state atoms, a condition needed for creating stimulated emission. In this sense, the medium acts as a light amplifier; the original photon gets amplified to a large number of photons, all in phase and traveling in the same direction. This is how a laser beam is produced.

But how is a condition set up so that amplification of photons takes place? Let us assume that we carry out the process of stimulation in a cylindrical-shaped tube that has mirrors fixed on either end of the tube (Fig. 5-3). Various excited atoms will emit photons, which will travel through the substance for a while, perhaps stimulating

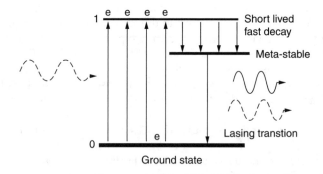

FIGURE 5-2 Schematic diagram of pumping ground state atoms and creating an excited atom, which is then illuminated with photon of energy equivalent to the difference in energy between the ground state and the excited state. These multiplied excited atoms decay fast to a meta-stable state that falls down to the ground state, emitting a monochromatic radiation.

Energy input by pumping

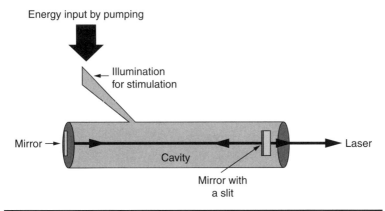

FIGURE 5-3   Schematic diagram showing emission of a laser beam. In a system (known as a cavity) excited atoms are illuminated with suitable light source. The multiplied excited atoms move forward and backward in the tube, and those aligned with the axis of the tube come through a narrow slit as a monochromatic laser beam.

more emissions, and then exit the tube. However, if one of the spontaneous emissions happens to be along the axis of the rod, the photons it stimulates will reflect off the mirror at one end of the tube (known as the *cavity*) and travel back through it, stimulating more emissions and being further amplified. The photons will bounce off the back mirror as well and (as long as they are parallel to the axis) will continue being reflected back and forth and being amplified as they go. Now if one of the mirror is made to partially transmit the light (often as little as 1%), then part of the light will escape from that end of the cavity. This escaped light forms the laser beam.

The light thus produced has three important characteristics:

1. Only photons traveling extremely close to parallel with the axis of the cavity will be amplified enough to escape through the mirror. Hence, the laser light will only exist in a narrow beam.

2. Because all the photons are coming from the same transition, they have the same energy and hence the same color, so the laser light will be monochromatic.

3. Because stimulated emission produces another photon in phase with the incoming one, all the photons in the laser beam will be in phase, so the laser light will be coherent.

These three properties distinguish laser light from light produced by other sources. As a result of these phenomena, the light from a laser comes from one atomic transition with a single precise wavelength. Therefore, the laser light has a single spectral color and is almost the purest monochromatic light available. Because of spectral emission line from which it originates, it has a finite width (Fig. 5-4).

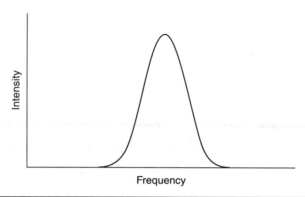

FIGURE 5-4   A typical laser beam spectra showing its band width.

There are some difficulties in making a laser. First, the laser has to be made from material capable of sustaining a population inversion for a long enough time. Second, there has to be some method of pushing a large number of the atoms into the excited state. There are many methods used to bring this about, depending on the laser material.

In 1962, the first laser was built using ruby crystal, and a pulse of light from a flash tube was used to excite the chromium atoms in the ruby. In a gas laser, such as a helium–neon laser, an electric current is used to excite the atoms of a gas, the same way as in a conventional neon light, and the population inversion comes about because of the way the gasses interact.

In a semiconductor diode laser, such as the laser pointer, the light is produced when electrons in one semiconductor fall into a lower energy state in another semiconductor. The process is the same as the process that occurs in a light-emitting diode (LED), and indeed, at low currents, a laser diode produces incoherent light similar to an LED. A large enough current passing through the semiconductor produces a population inversion and gives laser operation. A laser beam can be of a continuous type or of a pulsed type of frequency 5Hz, 10Hz, 20Hz, and so on. The latter type of laser is known as *pulse* (or *pulsed*) *laser*.

5.2   Overview of the Development of Lasers

Charles Townes and Athur Schawlow of Bell Laboratories invented the LASER. Originally, it was called MASER (M stands for microwave). The concept of the LASER was established after Albert Einstein (1916) postulated the possibility of stimulated emission. Einstein realized that when a light is shone on an atom that is in an excited state, it can induce the atom to make a downward transition (emitting a photon) if the incoming light's frequency matches the atomic transition energy. This led to the construction of first optical laser by Theodore H. Maiman in 1960 using a rod of ruby as the lasing medium. In 1962, Breech and Cross used a ruby laser to vaporize and excite atoms from a solid surface. Smith and Turner

(1965) used a ruby laser to deposit thin films. The technological break-through occurred in early 1980 when installation of continuous as well as pulsing laser became a reality.

Laser gives a monochromatic light, and its energy per pulse can be very high and can be used even to cut steel. Unlike thermal vapor-ization, material vaporized by laser has the same stoichiometry as that from which it is vaporized. Therefore, thin film can be prepared by laser with the same composition as the parent material.

The pulsed-laser deposition (PLD) technique is one of the most sophisticated and yet one of the simplest techniques for depositing thin film from a material without altering the stoichiometric compo-sition. As a result of these advantages, after the discovery of high-temperature superconductivity in 1986, the research interest in deposition of thin film by lasers increased dramatically. In the mid-1980s, this new technique was used to explore the possibility of syn-thesizing fullerene and later on to synthesize carbon nanomaterial such as single-walled carbon nanotubes (SWCNTs) and multiwalled carbon nanotubes (MWCNTs) as well as deposition of amorphous carbon and diamondlike carbon (DLC) films. Vaporization of mate-rial by laser is given a special term, *ablation*, and if deposition is done by a pulsing laser, the process is known as PLD.

5.3   Pulsed-Laser Deposition Technique

The PLD technique is a very simple technique for deposition of thin film and synthesis of carbon nanomaterials (CNMs). In this process, a pulsed laser beam (pulse duration, 10 to 50 ns) is used that hits a target mate-rial. Usually, graphite with 99.99% purity has been used as the target material for deposition of a carbon film; there have also been reports of using carbon soot as pellets for deposition of carbon thin films.

The target material is placed in a vacuum chamber, which is evacu-ated up to 10^{-3} to 10^{-6} Pascal (10^{-5} to 10^{-8} torr). The target is kept at an angle of 45 degrees with the laser beam and perpendicular to the substrate. The

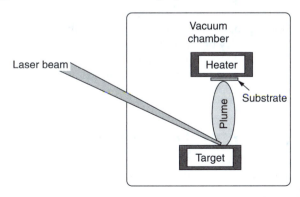

Figure 5-5   Schematic diagram of a pulsed-laser deposition system.

intense pulsed laser hits the surface of the target, producing a flux of evaporated or ablated material that condenses on a substrate to form a film. The strongly forward-directed flux of material is called *plasma plume*. A set of optical lenses is used to focus the laser beam over the target surface. Figure 5-5 shows a schematic diagram of a PLD setup. The deposition can be carried out in the presence of a reactive environment as well as in an inert atmosphere. This can also be operated in conjunction with other types of deposition techniques, such as laser-arc deposition (Scheibe et al., 1996) and magnetron-assisted PLD (Voevodin et al., 1996).

For laser ablation, many types of laser sources can be used. Generally, the Nd:YAG or excimer laser is used for this purpose.

5.3.1   Parameters That Influence Laser Deposition

The laser energy density absorbed by the target material is sufficient to break any chemical bonds of the molecules within the volume. Hence, if it is desired to deposit thin film that has the same composition as the target, the energy density of the laser pulse should be carefully chosen. When the laser beam hits the target material, high-pressure gas is produced in the surface layer. As a result of the pressure gradient, a supersonic jet of particles is ejected normal to the target surface. The particle cloud absorbs a large amount of energy from the laser beam, producing an expansion of hot plasma (called the *plume*). The ablated species confined in the plume condense on the substrate placed opposite to the target, forming a thin film.

The quality of the film deposited on the substrate is influenced by various parameters, such as the energy density (often known as the *fluence*) of the laser pulse, repetition rate of the laser pulse, temperature of the substrate, distance between the target and the substrate, and the environment of the chamber (i.e., the nature of the gas or gases and their pressure or magnitude of the vacuum in the chamber). The magnitude and nature of these parameters depend on the type of target materials used for the ablation. Hence, for each type of material there is a need to find out the best conditions of the parameters.

PLD can take place both in a vacuum and in the presence of a dilute background gas, which is used to influence the composition of the film. Under proper process parameters, the film grows epitaxially, and the stoichiometry of the film is a replica of that of the target. Process parameters also have an effect on the growth rate of the film, but when the fluence and the target–substrate distance have been optimized, the rate of deposition remains nearly constant.

5.4   Types of Lasers Used in Pulsed-Laser Deposition

The different types of pulsed lasers have different specific properties (Table 5-1). The type of material used for lasing generally determines the properties of the lasers. Nd:YAG lasers and excimer lasers are commonly used in the PLD technique. These two types of lasers are discussed in the next sections.

Laser type	Wavelength (nm)
Argon fluoride	193
Xenon chloride	308 and 459
Xenon fluoride	353 and 459
Helium cadmium	325–442
Rhodamine 6G	450–650
Copper vapor	511 and 578
Argon	457–528
Helium neon	543, 594, 612, and 632.8
Krypton	337.5–799.3
Ruby	694.3
Laser diodes	630–950
Ti:Sapphire	690–960
Alexandrite	720–780
Nd.YAG	1064
Hydrogen fluoride	2600–3000
Erbium:Glass	1540
Carbon monoxide	5000–6000
Carbon dioxide	10,600

TABLE 5-1   Types of Lasers and Their Corresponding Wavelengths

5.4.1 Excimer Lasers

The excimer is a gas laser system that emits ultraviolet (UV) radiation. This system can also achieve pulse repetition rates up to several hundred hertz with energies near 500 mJ/pulse. The excimer molecules are formed in a gaseous mixture such as Xe, He, CI, and Ne. Energy is pumped into the gas mixture through an avalanche of electric discharge excitations. The pumping energy creates ionic and electronically excited species that react chemically and produce the excimer molecules. Electron-beam excitation and microwave discharge have also been used to create ionic species from the gas mixture.

The details of the kinetics and chemical reactions leading to the formation of the excimer molecules are quite complex and consist of many steps. Some of the important reactions of krypton fluoride (KrF) laser are listed here, where the* denotes an electronically excited species and X denotes a third body (He, Ne).

$$
\begin{aligned}
Kr + e^- &\longrightarrow Kr^+, Kr^*, Kr^+_2 \\
F_2 + e^- &\longrightarrow F + F^- \\
Kr^+ + F^- + X &\longrightarrow Kr\,F^* + X \\
Kr^+_2 + F &\longrightarrow Kr\,F^* + Kr \\
Kr^* + F_2 &\longrightarrow Kr\,F^+ + F
\end{aligned}
\tag{5-1}
$$

After the excimer is formed, it decays by spontaneous emission and collision deactivation, giving the molecule a lifetime of 2.5 ns. Because the excimer is stabilized by a third body, the fast kinetics involved in producing the excimer require gas pressures in the range of 2 to 4 atm within the discharge volume. Other parameters for producing discharge are electron density ($10^{15}/cm^2$), current density ($10^3 A/cm^2$), and electron temperature (~1200 K). These requirements are met with electric discharge field of strengths of 10 to 15 kV/cm. Consequently, the spacing between the discharge electrodes is limited to 2 to 3 cm. Hence, the discharge voltages will be in the range of 20 to 45 kV.

Gas Mixtures for Excimer Lasers

The gas mixtures for excimer lasers are composed of three components: rare gas (Xe, Kr), halogen (He CI or F_2), and Ne as a buffer gas. Depending on the size of the laser, the desired output energy, and the operating pressure (3000 mbar), the total volume of the laser head could be from 40 to 60 L. The gas fills have a finite lifetime because of slow consumption of the halogen component by impurities arising mainly from degradation of the discharge electrodes and preionization pins.

Gas lifetimes are dependent on the type of halogen being used and the wavelength of laser. Properly engineered lasers have achieved greater than 20×10^6 shots with chlorine operation and greater than 5×10^6 shots with fluorine operation. A decrease in gas lifetime is observed with a decrease in laser wavelength. This occurs because the absorption cross-section of impurities present in the discharge increases with increasing photon energy. Consequently, photochemistry leading to halogen consumption is an important factor for gas lifetime.

5.4.2   Nd:YAG Lasers

The Nd:YAG laser is an optically pumped solid-state laser that can produce very high-power emissions. Nd:YAG lasers are optically pumped using a flash lamp or laser diodes. Population inversion stored in a capacitor is discharged, emitting blue and UV radiation resulting from shining light on this crystal. If the light is intense enough, atoms within the crystal that absorb these lights transition from the ground state to the absorption bands. The optical pumping is done in a quartz tube filled with a noble gas through which high energy flows. (Fig. 5-6).

Nd:YAG is an acronym for *neodymium doped yttrium aluminum garnet* (Nd:$Y_3Al_5O_{12}$), a compound that is used as the lasing medium for certain solid-state lasers. The YAG crystal is doped with triply ionized neodymium, which replaces another element of roughly the same size, typically yttrium. Generally, the crystalline host is doped with around 1% neodymium by weight. The amount of the neodymium dopant in the material varies according to its use. For continual wave output, the doping is significantly lower than for pulsed lasers. The lightly doped yttrium aluminum garnet rods can be optically distinguished by being less colored and almost white; higher doped rods are pink-purplish.

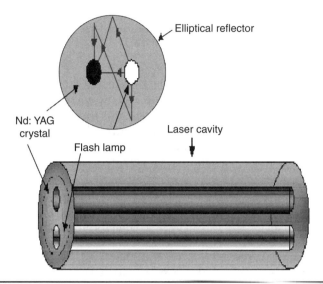

Figure 5-6   Schematic diagram of an Nd:YAG lasing medium.

Nd:YAG lasers typically emit light with a wavelength of 1064 nm (infrared). However, there are also transitions at 940, 1120, 1320, and 1440 nm. Nd:YAG lasers operate in both pulsed and continuous modes. Pulsed Nd:YAG lasers are typically operated in the so-called Q-switching mode. An optical switch is inserted in the laser cavity, waiting for a maximum population inversion in the neodymium ions before it opens. Then the lightwave can run through the cavity, depopulating the excited laser medium at maximum population inversion. In this Q-switched mode, an output of 20 MW and pulse durations of less than 10 ns are achieved.

Krypton flash lamps, with high output at those bands, are therefore more efficient for pumping Nd:YAG lasers than the xenon lamp, which produces white light, wasting a lot more energy.

Other common host materials for neodymium are YLF (yttrium lithium fluoride; 1047 and 1053 nm), YVO_4 (yttrium vandate; 1064 nm), and glass. A particular host material is chosen to obtain a desired combination of optical, mechanical, and thermal properties. Nd:YAG lasers and variants are pumped by flash lamps, continuous gas discharge lamps, or near-infrared laser diodes (DPSS lasers).

Nd:YAG lasers are extensively used for cutting and welding steel and super alloys. For these purposes, the power levels are 1 to 5 kW. Super-alloy drilling for gas turbine parts uses pulsed Nd:YAG lasers (millisecond pulses, not Q-switched). These lasers are also used to make surface markings on transparent materials such as glass and acrylic glass. The most versatile use of the Nd:YAG laser is in the PLD technique, which has been used for deposition of various stoichiometric thin-film superconductors as well as CNMs.

5.5   Influence of Laser Parameters on the Growth of Film

5.5.1   Wavelengths of Lasers

The wavelength of the laser has a significant effect on the yield of the ablated particles. Usually, lasers with short wavelength (UV region) are preferred because at shorter wavelengths the reflectivity of most materials is much lower than the long infrared wavelengths. However, the absorption coefficient of shorter wavelength is larger, which enables more efficient ablation, even with thin film.

5.5.2   Fluence of a Laser Pulse

The fluence of a laser pulse has to be larger than a certain threshold value so that all the species can be stoichiometrically removed from the target. About 2 to $3 J/cm^2$ fluence is used to reduce the number of particulates on the deposited film. On the other hand, larger fluence may result in the ejection of large target fragments, which increases the number of droplets on the film surface.

5.5.3   Ablation Plumes

The material evaporated from the target is hot, so part of the atoms in the vapor is ionized. In addition, this ionized cloud absorbs energy from the laser beam and becomes more ionized. Finally, fully ionized plasma is formed in the vicinity (~50 μm) of the target. The plasma expands away from the target, similar to the rocket exhaust from jet nozzles, with a strongly forward-directed supersonic velocity distribution. The visible part of the particle jet is referred to as an *ablation plume*. The plume consists of several types of particles, including neutral atoms, electrons, and ions. Furthermore, clusters of different compounds of the target elements are observed near the target surface. The visible light of the plume is caused by fluorescence and recombination processes in the plasma. The plume behaves in a different manner in a vacuum and in the presence of an ambient background gas. In a vacuum, the plume does not expand unidirectionally, but backward velocity components appear as well because of the high density of the plasma. The ejected species diffuse in the plume and collide with each other, which leads to a rapid thermalization of the particle's cloud. Moreover, the plume in the vacuum is visible only in the immediate vicinity of the target. Ambient gas molecules scatter and attenuate the plume, changing its spatial distribution, the deposition rate, and the kinetic energy distribution of the different species. In addition, reactive scattering results in the formation of molecules or clusters, which are essential for proper stoichiometry.

The properties of the plume, such as color, shape, and size, depend on several parameters. If the laser spot size on the target is reduced, keeping the fluence constant, less material is removed from the target, and the plume becomes wider and shorter. On the other hand, if the fluence is increased, a longer plume is produced because the initial velocity of the particles is higher. Gas pressure also has an influence on the length of the plume, and it is often the parameter that is simplest to modify slightly during the deposition. Process parameters have to be adjusted in such a way that the plume tip touches the substrate. Too short a plume does not provide enough material to the substrate.

5.5.4 Film Quality

The quality of the deposited film is, above all, determined by the crystallinity of the lattice and the surface smoothness. The generation of particulates during the PLD process is one of the most important factors affecting the smoothness of the resulting film. These particulates can be classified into small droplets (typical size, 0.2 to 3 μm) and large, irregularly shaped outgrowths (diameters >10 μm).

The largest outgrowth of films is believed to originate directly from the target. Breaking of protruding surface features or splashing of a molten surface layer are typical explanations for the occurrence of large target droplets that get deposited on the substrate. Droplets can be totally eliminated from the film surface by properly adjusting different parameters of the PLD conditions.

To obtain epitaxial thin films, the substrate has to be heated to a temperature of approximately 700° to 900°C. The smoothness and crystallinity of the film can be further increased by rotating the target during the deposition so that laser pulses do not strike the same spot on the target surface all of the time. In addition to temperature and pressure, the laser-pulse repetition rate has to be low enough (<10°Hz) so that the ablated species will have time to form a smooth layer between successive pulses. On the other hand, too low a repetition rate has to be avoided because fast chemical reactions may hinder the epitaxial growth of the film. Moreover, fluence and spot size on the target also have to be optimized to minimize the number of particulates.

5.6 Synthesis of Carbon Nanomaterials

5.6.1 Fullerene and Multiwalled Carbon Nanotubes

The PLD technique has been used for synthesis of fullerene (C_{60}). Kroto and coworkers (1985) vaporized graphite in a pulsed jet of helium by focusing a pulsed laser on the surface of a rotating graphite target. MWCNTs have also been grown with the PLD technique by using an Nd:YAG laser of wavelength 266 nm, pulse duration of 7 ns, and repetition rate of 10 Hz. Carbon nanotubes (CNTs) were grown by using graphite mixed with some suitable metal such as Ni and Co as a target. The graphite target material is sintered before attaching it to the target holder. A high-resolution transmission electron micrograph

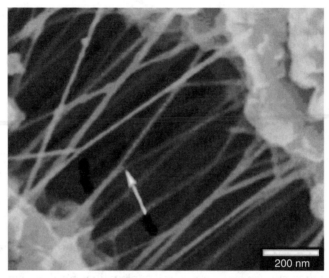

Note: Arrow shows the joining (agglomeration) of two CNT branches forming one branch.

Figure 5-7   High-resolution transmission electron micrograph of carbon nanotubes stretched between two metal agglomerates, synthesized by ablation of graphite target containing Ni and Co.

(HRTEM) of CNT prepared by laser ablation of graphite target that contained Co and Ni is shown in Fig. 5-7. HRTEM shows "closed"-end CNTs arranged parallel to each other. The diameter of these CNT is 5 to 20 nm, and their length is about 1 mm (see Fig. 5-7).

The PLD technique has also been used for the synthesis of SWCNTs (Maiti et al., 1997). To grow SWCNTs, different varieties of substrates have been used. These include freshly cleaved, highly oriented pyrolytic graphite (HOPG), thermally grown SiO_2 onto Si (100), amorphous silicon carbide–coated Si (100), TiN-coated Si, and silicon nitride (SiN).

5.6.2   Single-Walled Carbon Nanotubes

For the growth of SWCNTs, the KrF laser (wavelength, 248 nm; pulse duration, 15 ns) has been used. For synthesis of SWCNTs, Co–Ni nanoparticles have been taken as the catalyst. Co–Ni nanoparticles have been deposited on the substrate by ablating a polycrystalline Co–Ni target (99.95%) using a KrF excimer laser with a laser intensity of 4×10^8 W/cm^2 under a helium background atmosphere at room temperature. The laser beam from the KrF laser was focused at an incident angle of 45 degrees on the surface of the Co–Ni target with the pulse rate of 10 Hz. The characteristics of catalyst, such as particle size and surface density, are known to strongly influence their catalytic behavior for CNT growth (Hovel et al., 2002; El Khakani and Li, 2004; Li et al., 2001).

The best conditions for deposition of Co–Ni nanoparticles as reported are He pressure of 300 m torr, a target–substrate distance of 5 cm, and a number of laser pulses in the 20 to 100 range. These parameters were chosen to yield Co–Ni nanoparticles with diameters of a few nanometers. The second step was achieved by placing the substrates in a quartz tube at the center of a furnace and exposing them to the carbon vapor produced from the KrF laser ablation of a pure graphite target. The carbon ablation was carried out in argon at temperature ranging from 1000° to 1150°C, with a target KrF laser intensity of 3×10^8 W/cm^2. SWCNTs were obtained from the ablation of the graphite target by 5 to 100 laser shots. The scanning electron micrograph (SEM) of the SWCNTs is shown in Fig. 5-8. The SWCNTs grown in the SiO$_2$ substrate show random orientation (see Fig. 5-8). Most of the SWCNT are seen to be relatively straight and have a uniform diameter distribution. In Figs. 5-8B and 5-8C, the top-view SEM images of SWCNTs grown on SiC and TiN substrate are shown, respectively. These show the growth of SWCNT, which are densely entangled bundles. The typical cross-section view of the SWCNT thin films is shown in the Fig. 5-8D (Maiti et al., 1997).

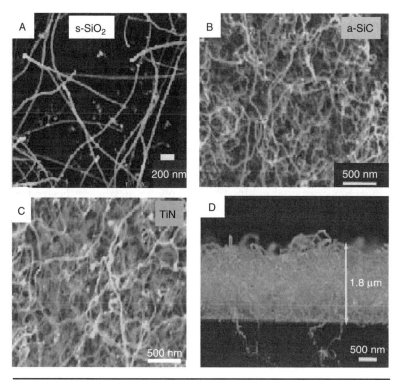

FIGURE 5-8 Scanning electron micrograph of single-walled carbon nanotube (SWCNT) bundles grown on (A) SiO$_2$, (B) a-SiC, (C) TiN substrates and (D) Typical cross-section view of the SWCNT thin films grown by laser.

5.6.3   Diamondlike Carbon Nanotubes, Diamond, and Amorphous Carbon

PLD has become one of the most successful techniques for the deposition of amorphous carbon thin film. In 1985, Marquardt et al. first reported the use of a laser ablation source of carbon ions to produce DLC films at high rates of growth approaching 0.3 μm/h. They determined a critical threshold intensity of 5×10^{10} W/cm^2 on the carbon feedstock, above which the carbon plasma condensed to DLC film; below the critical threshold energy, only soft graphitic layers were deposited that resembled those produced by thermal evaporation of carbon. For this work, near-infrared Nd:YAG (1064 nm) lasers have been used to produce ta-C films with 40 to 70% sp^3 bonding.

A comparative study of ArF and KrF lasers showed that ArF laser irradiation produces the highest quality amorphous DLC films, with higher optical transmission, higher index of refraction, better electrical insulation, and higher sp^3:sp^2 ratios than KrF laser–deposited DLC films. KrF laser generates substantially larger amounts of ultrafine particles and C_2 (and probably larger clusters) compared with those generated with the ArF laser. Apart from these, XeCl pulsed excimer lasers (Stevefelt and Collins, 1991), pulsed ruby lasers, and CO_2 lasers have also been used for the deposition of DLC films.

Efforts have been made to deposit uniform DLC films by combining laser ablation of graphite target with some other auxiliary techniques. Qingrun and Gao (1997) used magnetic field in the PLD process. DLC films deposited by this technique were found to be superior to those produced by the conventional method with regard to their micro hardness and uniformity. It has also been explored that with the suitable condition of PLD semiconducting, amorphous carbon thin film can be deposited. A schematic diagram of PLD setup for deposition of amorphous carbon thin film is shown in the Fig. 5-9. Different precursors enriched in carbon compound may be used for amorphous carbon thin film. Carbon powder may be synthesized by pyrolysis from different natural precursors, which may be used as the target material besides graphite rod.

Boron and nitrogen-doped carbon thin film have been deposited with the PLD technique for the use of electronic devices. Homojunction layer of n-type carbon and p-type carbon has been deposited by PLD and optical lasers, and their electrical properties have been studied. A SEM of undoped carbon thin film is shown in Fig. 5-10A, and boron-doped carbon thin film prepared by PLD is shown in Fig. 5-10B.

5.7   Summary

Lasers are formed by irradiating a material that can generate a large concentration of inversion layer to produce stimulated radiation. This gives a monochromatic radiation. Lasers may be continuous or have pulsing type of frequencies ranging from 5 to 20Hz.

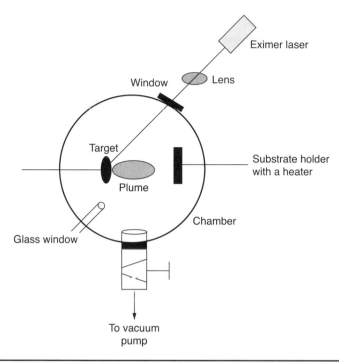

Figure 5.9   Schematic diagram of XeCl PLD unit.

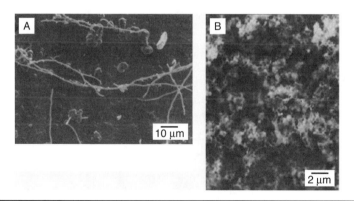

Figure 5-10   (A) Scanning electron micrograph of amorphous carbon thin film prepared by pulsed-laser deposition showing deposition of uniform carbon thin film (Douglas et al., 1995). (B) Boron-doped amorphous carbon thin film showing formation of clusters (Voevodin et al., 1996).

The PLD technique may be used to grow thin-film CNMs, DLC, and amorphous carbon.

Various laser parameters that influence the deposition of thin film are the laser's wavelength, the fluence of the laser, the ablation plume, and the film quality.

CHAPTER 6

Thermal Vapor Deposition for Thin Film of Carbon Nanomaterial

Pravin Jagadale and Maheshwar Sharon

Nanotechnology Research Centre, Birla College, Maharashtra, India

6.1 Introduction

Thin film can be defined as a low-dimensional material (~1000 nm) created by one-by-one condensing of the atomic, molecular, or ionic species of matter onto a substrate. In the early 20th century, thin film technologies were developed to make various gadgets using a smaller quantity of material, thus reducing the cost.

Thin films can be prepared with many different methods, including evaporation of material by either thermal or microwave heating under a very high vacuum. Since the development of lasers, thin film deposition by lasers has also been possible. However, all of these techniques require sophisticated instrumentation and a high-vacuum system. The thermal flash evaporation technique, on the other hand, can be used to deposit thin film at a relatively lower vacuum level (Osofsky et al., 1998; Ramadan, 2000). The deposition unit is also very simple, and thin film can be prepared in few minutes.

6.2   Morphology of Thin Film

The detailed morphology of a film depends on many factors involved in the growth of film, including the chemical and physical properties of the depositing material, substrate temperature, flatness of the substrate, deposition rate, process or residual gas pressure, surface diffusion, film growth mode, residual stress in the film, and match between the film's and substrate's lattice parameters.

6.3   Stoichiometric Composition of Thin Film

The stoichiometry of material in film sets in during deposition of films of chemical compounds such as $CuInSe_2$, Fe_2O_3, and so on. It is possible to deposit thin film of these types of materials by suitably controlling the deposition parameters. Sometimes films with nonstoichimetric composition are formed because of mismatch of the vapor pressure of the constituents. For example, zinc oxide (ZnO) is deposited as a film from bulk of ZnO (using sputtering or thermal evaporation) and loses oxygen, giving a film of nonstoichiometric ZnO_{1-x}. To correct this loss, a reactive deposition technique (e.g., with additional oxygen) is used.

6.4   Uniformity of Thin Film

A film's thickness is one of the most critical parameters needed for obtaining the desired performance of the material. Properties such as antireflection coatings, giant magnetoresistance devices, neutron beam guides, and optical filters depend highly on the thickness of the film. Equally important is the uniformity of thickness across the area of interest. With the exception of atomic layer deposition, all of the other deposition techniques can produce films with some level of uniformity. Approaches to reduce thickness variations in thin film deposition methods include optimizing the bulk material's throw distance to the substrate and speed of the substrate rotation.

The distance between the target material and the substrate has an impact on the quality of uniformity of the deposition. When the target material is evaporated by heating, its vapor is formed in the form of a solid cone (Fig. 6-1). The concentration of material evaporated in vapor changes

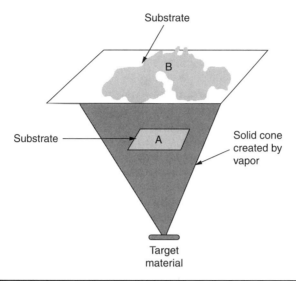

FIGURE 6-1 Schematic diagram showing the effect of distance between the target material and the substrate on the quality of thin film deposited. (*A*) Uniform deposition occurs because the substrate covers the solid cone created by the vapor produced by heating the target material. (*B*) Non-uniform deposition occurs because the substrate is kept very far from the target material, causing some part of the substrate uncovered with vapor forming solid cone

as vapor moves in a direction perpendicular to the target material. Positioning of the substrate in this cone determines the quality of the thin film deposited over the substrate. Positioning of substrate at A can give uniform film, but at B, non-uniform film might get deposited. Rotation of substrate further helps in allowing a uniform deposition of the material on the substrate.

6.5   Preparation of Thin Film of Carbon

Various techniques are being used for making thin film. The selection of techniques depends on the type of materials to be used for making thin film. Broad classification of thin film deposition can be made as the chemical, physical, and electrochemical methods. For all of these processes, the typical steps in making thin films are:

1. Emission of particles from the source (heat, high voltage)
2. Transport of particles to the substrate
3. Condensation of particles on the substrate

Physical methods of deposition cover a number of deposition technologies (evaporation and sputtering) in which material is released from a source and transferred to the substrate.

This method is commonly used for the deposition of metals. It is far more common than chemical vapor deposition (CVD) for metals because it can be performed at lower toxicity level and is more cost effective. The quality of the film is inferior to CVD, especially for metals films, which are of higher resistivity; for insulators, it contains more defects and traps. The choice of deposition method (i.e., evaporation or sputtering) may be arbitrary in many cases and may depend more on what technology is available for the specific material at the time.

In the chemical process, the substance to be deposited is heated with a suitable source to create its vapor. Then the vapor is allowed to deposit on a substrate, giving a thin film.

The electrochemical method follows Faraday's law in which cations are electrochemically reduced at the cathode by applying suitable potential. Thus, a thin film of cation is deposited on the cathode. The thickness of film and its uniformity depend on many factors, but the rotation of the electrode and the magnitude of the current used for the deposition control the uniformity of the deposition.

Because the goal of this chapter is to discuss the thermal flash evaporation, it would be beyond the scope of this chapter to discuss each of the techniques described above. Hence, efforts are made to discuss the thermal evaporation technique in detail even though carbon thin film can be prepared by all of the other methods as well.

6.5.1 Thermal Flash Evaporation Technique

The thermal flash evaporation technique has been of interest particularly in research and development applications in which low installation costs and inexpensive, disposable evaporant "containers" are required. In this technique, a quartz boat containing the substrate is placed inside the tungsten coil (which is used for heating). The source material is heated by passing a current through the tungsten coil to evaporate the substrate. Vacuum is required to allow the molecules to evaporate freely in the chamber and subsequently condense on the substrate kept near the coil. Many materials are restrictive in terms of what evaporation method can be used (e.g., platinum is quite difficult to evaporate using resistive heating). Such materials cannot be used for making thin film with the flash evaporation technique.

A typical design of a thermal flash vaporization unit is shown in Fig. 6-2. Thin film of carbon material has been synthesized using this technique. The deposition is carried out under vacuum in the chamber; the vacuum is created using a rotary pump. A quartz plate or any other ceramic material is taken as the substrate (B), which can withstand high temperature. The substrates are precleaned with distilled water, trichloroethylene, and acetone. The thoroughly cleaned quartz plate or substrate is etched by 50% nitric acid solution. The precursor is thermally cracked by keeping it in a tungsten coil (C) inside the

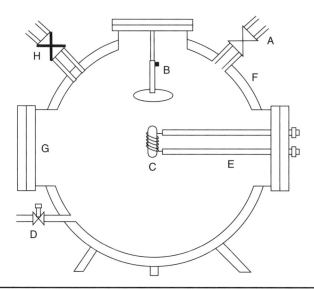

Figure 6-2 Schematic diagram of the thermal flash vapor deposition unit. A and H are two valve systems connected to vacuum pump and pressure gauge. B is a substrate holder attached to a thermocouple to measure the temperature. C is the tungsten coil holding a quartz boat for holding precursor. D is a valve to allow any reactive gas into the chamber. E is electric power supply connections. F is the chamber of flash vaporization unit. G is a glass window used to view the deposition process.

chamber, which is heated by passing a very high current (E). The substrate holder is kept at the minimal distance from the heating coil so that the substrate also gets heated. Because of the high temperature, the oil is thermally cracked and gets deposited on the substrate. The deposition temperature is measured by a thermocouple attached to the chamber. With this technique, carbon thin film has been synthesized from polyvinyl alcohol, camphor, turpentine, clove oil, and different plant-derived precursors. The band gaps of thin films obtained by using this process with different precursors are shown in Table 6-1.

Name of precursor	Energy gap (Eg in eV)
Camphor	1.40
Terpentine	1.34
Kerosene	2.58
Clove oil	1.33
Polyvinyl alcohol	1.65
Polyethylene glycol	4.2

TABLE 6-1 Carbon Thin Film Obtained with the Thermal Flash Evaporation Technique and Their Corresponding Band Gaps

Table 6.1 shows the impact of various precursors on getting thin film of carbon of various band gaps. These films were prepared under similar conditions using different precursors. The variable parameters of this technique are: voltage applied to coil (or current passing through the coil), distance between the coil (i.e. quartz boat) and substrate, vacuum level in the chamber, and type of gas. These parameters are adjusted to get the desired product.

6.6 Why Thin Film of Semiconductor Carbon?

Silicon-based photovoltaic solar cell is based on group IV elements in the periodic table. About 90% of the world's integrated circuit production and solar cell production is based on silicon. Silicon has a number of advantages, including abundance, nontoxicity, performance stability, simplicity (it is a monoelemental semiconductor), relative toughness, and the fact that it forms a stable oxide (SiO_2). However, the purification and production of silicon is an expensive process.

Certain similarities in the properties of carbon and silicon have attracted researchers to using carbon for making semiconducting thin film of carbon, which may reduce the cost of existing solar cells. Amorphous carbon (a-C) thin films have the added advantage that they can be deposited at low substrate temperatures, making it possible to achieve hard films with thicknesses of less than 1 nm over large areas. The cost of carbon thin film can be far less than the cost of crystalline silicon film.

Unlike other elements of group IV, the different hybridizations of C atoms allow for the formation of sp^2 as well as sp^3 bonding. Carbon has long been considered the only element that exists in allotropic crystalline modifications of diamond (sp^3 hybridization with a band gap of 5.5 eV) and graphite (sp^2 hybridization with a band gap of ~0.25 eV). Thus, preparation of thin film of carbon with controlled composition of sp^3 and sp^2 carbon may provide a method of producing semiconducting carbon with a band gap in the range of 5.5 eV to 0.25eV. Such flexibility does not exist with any other inorganic materials. The discovery of carbines (sp hybridization) in 1960 also attracted scientists (plural) to explore its application. Thus, techniques for developing carbon thin film with various compositions of sp^3 and sp^2 are challenging and developing a suitable solar cell from them is another challenging task, making carbon an interesting material to work with.

6.7 Various Applications of Thin Films

Thin films have been used in a variety of applications. A few examples of products that could not have been developed without vacuum thin film deposition processes include:

- **Microelectronics:** Electrical conductors, electrical barriers, diffusion barriers, Micro-Electro Mechanical Systems (MEMS), organic electronics and displays, quantum dots

- **Electronic gadgets:** CDs and DVDs, computer hard drives and Giant Magneto Resistance (GMR) read heads, headlamp reflectors

- **Sensors:** Magnetic materials, gas, magnetic memory, accelerometer sensors

- **Tailored materials:** Optical application, corrosion protection, wear resistance

- **Semiconductors:** Solar cells, transparent conducting layers, semiconductor chips

- **Military equipment:** Aircraft guidance control assemblies, aircraft canopies, night vision goggles, forward-looking infrared (FLIR) vision devices

- **Medical and diagnostic equipment:** Excimer lasers for eye surgery, passivated prosthetic joints, coated stents, superinsulation for magnetic resonance imaging magnets

- **Consumer products:** Aluminized plastic food packaging materials; camera lenses; mirrors; optical coatings for windows, glasses, and sunglasses; architectural glass; tool hard coatings

- **Research:** Diamondlike films, neutron beam guides, laser mirrors, optical filters, super lattices, rugates, high- and low-Tc superconductors Superconducting Quantum Interference Device (SQUID) magnetic detectors

- **Ultrathin carbon** foils are key components of numerous instruments using thin films of various materials currently used in space and proposed for future space missions. These foils are approximately 0.5 to 2 μg/cm, equivalent to approximately 3 nm or approximately 20 atoms thick. As accelerated atoms pass through the thin foils, they emit secondary electrons from the entrance and exit surfaces and change their atomic charge states.

- **Space research** institutes are working on a number of space using thin films of various materials.

6.8   Summary

The preparation of thin film of carbon material is one of the basic requirements for developing various types of gadgets (e.g., carbon solar cells, electron field emission). Thin film preparation of carbon nanomaterial by techniques such as thermal vacuum evaporation,

microwave plasma technique, and laser ablation requires sophisticated instrumentation. Moreover, none of these methods can be used
for making a continuous production of thin film. These techniques
give thin film by a batch process. On the other hand, the thermal flash
evaporation technique can be used to make thin film, either as a batch
process or as a continuous process. In addition, the precursor can be
used directly for depositing thin film of carbon nanomaterial, which
is not available with other techniques. The precursor can be kept in
the boat of the thermal flash evaporation. When electrical current is
given to the heating coil, the precursor decomposes, giving vapor of
the decomposed product, which gets deposited on the substrate, giving a thin film of carbon nanomaterial. By controlling the current of
the coil, the environment of the chamber (e.g., oxidizing or reducing
gas), and distance between the substrate and target, a desired type of
carbon nanomaterial can be obtained directly from the precursor.

CHAPTER **7**

Natural Precursors for Synthesis of Carbon Nanomaterial

Shrikant Kawale

Nanotechnology Research Centre, Birla College, Kalyan, Maharashtra, India

Madhuri Sharon

MONAD Nanotech Pvt Ltd, Mumbai, India

7.1   Introduction

A new form of carbon, Buckminster fullerene (C_{60}), was discovered in 1985 by a team headed by Smalley and coworkers and led to the Nobel Prize in chemistry in 1997. C_{60} is a soccer ball–like molecule made of pure carbon atoms bonded in hexagon and pentagon configurations. Besides diamond, graphite, and C_{60}, quasi one-dimensional nanotube is another form of carbon first reported by Ijima in 1991 when he discovered multiwalled carbon nanotubes C (MWCNTs) in carbon soot made by an arc-discharge method. About two years later, he made the observation of single-walled carbon nanotubes (SWCNTs). Since then, nanotubes have captured the attention of researchers worldwide. A significant amount of work has been done in the past decade to reveal the unique structural, electrical, mechanical, electromechanical, and chemical properties of CNTs, and to explore what might be the key applications of these novel materials.

Various methods of preparing CNTs have been discussed in earlier chapters of this book. In this chapter, emphasis is given on use of plant-derived materials as precursor or raw material for synthesis of carbon nanomaterials (CNMs) with the chemical vapor deposition (CVD) method.

The CVD technique for synthesis of CNMs was developed by Cheng et al. (1998) using hydrocarbons as precursors. Using the same technique in the gas phase, made SWCNTs has been synthesized by using carbon monoxide as a source of carbon. Most scientists have used organic chemicals, such as acetylene, methane, benzene, and cyclohexane, as precursors (Li et al., 1996). In addition to being expensive, all of these precursors are related to nonregenerative fossil fuels. Hence, efforts of Sharon's group in using plant-derived precursors have been developed and used successfully (Afre et al., 2005; Bhardwaj et al., 2008a, b; Jagdale et al., 2007; Kshirsaga et al., 2006; Mukhopadhyay et al., 1994, 1996; Sharon et al., 2007; Sharon and Sharon, 2006).

7.2   Criteria for Selecting Precursors

Apart from being rich in carbon and comparatively low in oxygen and metal content, there have been certain other requirements envisaged for selecting a precursor for synthesis of CNMs, especially when plant-derived materials are to be used as raw material for their synthesis:

1. The precursors should not generate a large quantity of waste material after completion of the process, which would create the problem of purification.

2. Precursors should not be a hazardous to human beings or the environment.

3. Precursors must be available on a continual basis (e.g., agrowaste or agricultural products that can be cultivated as per demand are suitable precursor material).

4. Precursors should give a high yield of desired material (i.e., CNMs) under suitable conditions.

7.3   Why Is Plant Material A Good Precursor?

Most of the CNMs have been synthesized from precursors based on the fossil fuels such as petroleum products and acetylene. These precursors are destined to get depleted one day. Moreover, the cost of these raw materials is expected to increase day by day. Therefore, it is necessary to look for precursors that are plant based. Natural precursors are available in abundance and can be produced in large amounts as and when required. It would be an added advantage if plant wastes are used (e.g., bagasse), or products from plant that have no other use and can be cultivated on wasteland.

While preparing carbon nanobeads (CNBs) or even of CNTs, most of carbon source materials contain almost 100% sp^2 carbon atoms. To form a three-dimensional spherical shape, conversion of some of the sp^2 carbon atoms to sp^3 carbon atoms is required. As a result, most of preparative conditions with conventional sources are stringent. If the starting material contains both sp^2 and sp^3 carbon atoms and is aromatic in nature, then it is easier to form three-dimensional spherical-shaped carbon material. Sharon and colleagues have shown that this condition can be met by using a natural source such as camphor (Afre et al., 2005; Mukhopadhyay et al., 1994; Sharon and Sharon, 2006; Sharon et al., 2007).

Plant-derived materials that have been successfully used so far are:

- Resin-derived material and volatile oil
- Vegetable oils
- Fibrous plant tissues
- Seeds

7.4   Carbon Material Produced by Various Plant Materials Using the Chemical Vapor Deposition Process

With pyrolysis, plant-derived precursors such as fibers, seeds, oils, and camphor, have yielded different forms of carbon. The study of ultrastructures of plant material and carbon produced by them have revealed that plant-derived precursors have some basic carbon compound containing skeletal structures; carbon produced by them shows similarities to those structures, suggesting that they remain as is after pyrolysis. These structures are otherwise very difficult to synthesize and cannot be produced from any artificial or synthesized precursors.

The structures of various plant materials, tissues, and products and CNMs derived from them are discussed in the next sections.

7.4.1 Resin-Derived Materials and Volatile Oils

Resins, camphor, and volatile oils such as turpentine oil and eucalyptus oil are good precursors for obtaining high-purity, aligned CNTs.

Camphor
The first step in getting CNMs for natural precursors was made by Sharon's group (Mukhopadhyay et al., 1994) using camphor as a source of carbon. An aromatic crystalline compound, $C_{10}H_{16}O$, is obtained naturally from the wood or leaves of the camphor tree (*Cinnamomum camphora*). Camphor is a bicyclic, saturated terpene ketone. It exists in the optically active dextro and levo forms and as the racemic mixture of the two forms. All of these forms melt around 178°C (352°F). The principal form is dextrocamphor, which occurs in the wood and leaves of the camphor tree. Camphor is also synthesized commercially on a large scale from pinene, which yields mainly the racemic variety. The structural formula of the molecule is shown in Fig. 7-1.

Camphor has a characteristic odor; it crystallizes in thin plates and sublimes readily at ordinary temperatures. It is a white or transparent crystalline solid with a strong aromatic odor. Both trees are being cultivated in large quantities and form a large part of subtropical forests.

Sharon's group has reported the formation of fullerene (1994) as well as diamondlike carbon (DLC) (1997c), carbon nanofibers (CNFs) (1999), and CNBs (2000) from camphor.

Raw materials used for preparation of CNBs or CNTs mostly contain 100% sp^2 carbon atoms. To form a three-dimensional spherical shape, conversion of some of the sp^2 carbon atoms to sp^3 carbon atoms is required. Because of this, most of preparative conditions with conventional sources are stringent. If the starting material contains both sp^2 and sp^3 carbon atoms and is aromatic in nature, then it is easier to form three dimensional spherical-shaped carbon material. Sharon et al. (1998)

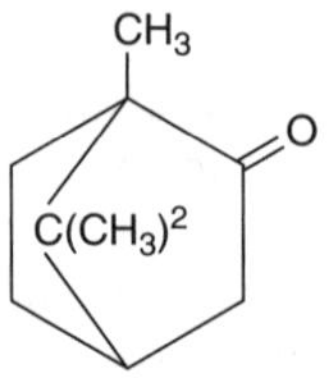

FIGURE 7-1
Structural formula of camphor.

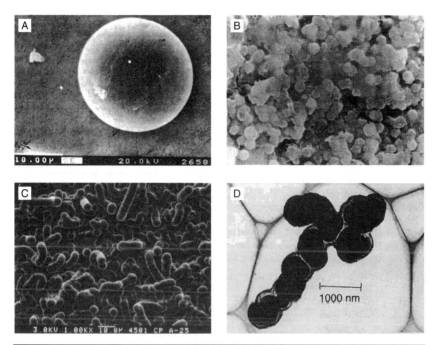

FIGURE 7-2 Carbon nanomaterials synthesized from camphor by Sharon et al. 1998. (*A*) Fullerene. (*B*) Diamondlike carbon. (*C*) Carbon nanofibers. (*D*) Carbon nanobeads.

have shown that this condition can be met by using a natural source such as camphor (Fig. 7-2).

Turpentine

Turpentine is a thin, volatile essential oil, $C_{10}H_{16}$, obtained by steam distillation or other means from the wood or exudates of certain pine trees. These exudates are a sticky mixture of resin and volatile oil from which turpentine oil is distilled.

The preparation of SWCNTs and MWCNTs from turpentine is presented in Chapter 3 (see Figs. 3-11 and 3-12). The purity of CNTs formed is very high. Turpentine has been found to be a good precursor for preparing aligned (Fig. 3-12) CNTs (Kumar and Ando, 2005). The molecular structure of turpentine oil is shown in Fig. 7-3.

Eucalyptus Oil

Eucalyptus oil is a volatile oil (1,3,3-trimethyl-2-oxabicyclo[2.2.2.]-octane) derived from the leaves of eucalyptus. Its molecular formula is $C_{10}H_{18}O$, and its molecular weight is 154.25.

SWCNTs are synthesized from eucalyptus oil (Fig. 7-4). The molecular structure of eucalyptus oil is shown in Fig. 7-5.

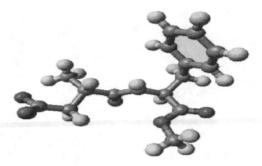

FIGURE 7-3   Molecular structure of turpentine.

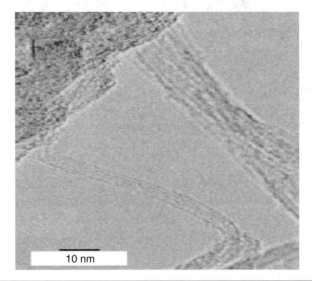

10 nm

FIGURE 7-4   Bundle of single-walled carbon nanotubes synthesized from eucalyptus oil.

FIGURE 7-5   Molecular structure of eucalyptus oil.

7.4.2   Vegetable Oils

Triglycerides are the main constituents of vegetable oils. Triglycerides have lower densities than water (i.e., they float on water) and at normal room temperatures may be solid or liquid. When solid, they are called *fats* or *butters*, and when liquid, they are called *oils*. Triglyceride,

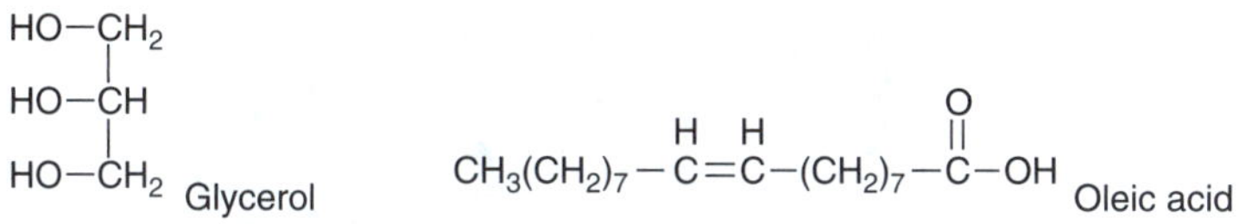

HO—CH$_2$
|
HO—CH   H   H   O
| CH$_3$(CH$_2$)$_7$ — C = C — (CH$_2$)$_7$ — C — OH
HO—CH$_2$ Glycerol   ‖
                                                                 Oleic acid

FIGURE 7-6   Chemical structure of triglycerides.

also called triacylglycerol (TAG), is a chemical compound formed from one molecule of glycerol and three fatty acids (Fig. 7-6).

Glycerol is a trihydric alcohol (containing three -OH hydroxyl groups) that may combine with up to three fatty acids to form monoglycerides, diglycerides, and triglycerides. Fatty acids may combine with any of the three-hydroxyl groups to create a wide diversity of compounds. Monoglycerides, diglycerides, and triglycerides are classified as esters, which are compounds created by the reaction between acids and alcohols that release water as a byproduct. Initial attempts to synthesize CNMs used single-component volatile oils. Recently, Sharon et al. (2006) used different plant-based edible oils for making CNMs.

Mustard Oil

Mustard oil (*Brassica* spp.) is of vegetable origin and is obtained from seeds of black and white mustard plants. It contains 0.30 to 0.35% essential oil (allylisothiocynate), which acts as preservative. It has an omega-3 (monounsaturated fat) and six fatty acid compositions (linoleic and α-linoleic acid, respectively). Mustard oil is composed mostly of the fatty acids oleic acid, linoleic acid, and erucic acid. At 5%, mustard seed oil has the lowest saturated fat content of the edible oils. Erucic acid is a monounsaturated omega-9 fatty acid, denoted 22:1 ω-9. It is prevalent in mustard seed, making up 40 to 50% of their oils.

CNBs produced from mustard oil are in the range of 500 nm to 2.5 μm. CNMs of different morphology have also been synthesized from mustard oil with the CVD method at 800°, 850°, and 900°C using transition metals (Ni, Co, and Ni–Co alloys) as catalysts (Fig. 7-7; Table 7-1).

Karanja Oil

Karanja oil (*Pongamia glabra*) is used in pharmacy, agriculture, and soap manufacturing and for making biodiesel and lubricants. In addition to its fatty acid content, crude karanja oil contains some toxic phenolic compounds, such as karanjin and pongamol; this oil is edible only after the removal of these compounds.

When karanja oil is pyrolyzed at 800° to 900°C, it produces CNF along with some DLC or amorphous carbon (Fig. 7-8). These fibers are coiled and have a diameter of approximately 100 to 150 nm.

Figure 7-7   Carbon nanobeads from synthesized mustard oil.

Oil from	Pyrolysis temperature (°C)	Carrier gas	Carbon nanomaterial type
Mustard	800	H_2	DLC
Mustard	850	H_2	Beads
Mustard	900	H_2	DLC
Karanja	800	H_2	Fiber
Karanja	850	H_2	Fiber
Karanja	900	H_2	Fiber
Neem	800	H_2	Fiber
Neem	850	H_2	Fiber
Neem	900	H_2	Fiber
Castor	800	H_2	DLC
Castor	850	H_2	Beads
Castor	900	H_2	DLC
Til	800	N_2	DLC
Til	850	H_2	DLC
Til	900	H_2	DLC
Linseed	800	H_2	DLC
Linseed	850	H_2	DLC
Linseed	900	H_2	DLC

DLC = diamondlike carbon.

Table 7-1   Carbon Nanomaterial Prepared from Various Plant Oils by Sharon et al. 2007

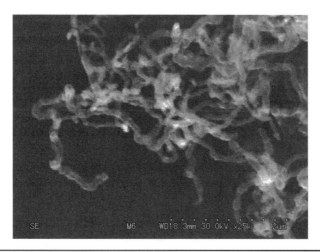

FIGURE 7-8   Carbon nanofibers synthesized by the carbon vapor deposition method from karanja oil.

Neem Oil

Neem oil (*Azadirachta indica*) is light to dark brown, bitter, and has a strong odor. It is made up of mainly triglycerides and large amounts of triterpenoid compounds, which are responsible for the bitter taste. Neem oil also contains steroids (campesterol, β-sitosterol, stigma sterol) and many triterpenoid, of which azadirachtin is the most well known (Fig. 7-9). The average composition of neem oil fatty acids are linoleic acid, oleic acid, palmitic acid, stearic acid, α-linolenic acid, and palmitoleic acid.

Neem oil–derived CNFs with a diameter of 500 nm to 2.5 μm have a rough surface (Fig. 7-10).

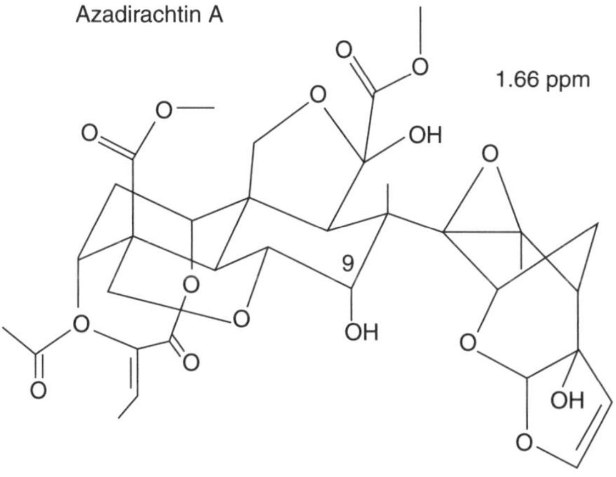

FIGURE 7-9   Chemical structure of azadirachtin A, which is present in neem oil.

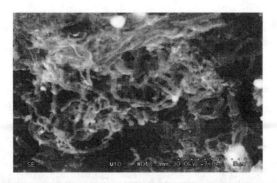

Figure 7-10 A. Coiled carbon nanofibers synthesized from neem oil.

Castor Oil

Castor oil (*Ricinus communis*) is a colorless to very pale yellow liquid with mild or no odor or taste. Its boiling point is 313°C (595°F), and its density is 961 kg·m⁻³. It is a triglyceride in which approximately 90% of fatty acid chains are ricinoleic acid; oleic and linoleic acids are the other significant components. The structure of the major component of castor oil is shown in Fig. 7-11.

Castor oil has been found to be a good raw material for synthesizing CNBs, although along with it, some straight CNF are also produced (Fig. 7-12). CNBs have a diameter ranging from 400 nm to 3 μm. Both free as well as aggregates of CNB are formed. CNBs have rather smooth surface.

Linseed Oil

Linseed oil (*Linum usitatissimum*), also known as flax seed oil, is a clear to yellowish oil derived from the dried ripe seeds of the flax plant. Palmitic acid, stearic acid, arachidic acid, oleic acid, linoleic acid and linolenic are the main fatty acids of this plant. Linseed oil contains no significant amounts of protein, carbohydrates, or fiber (Fig. 7-13).

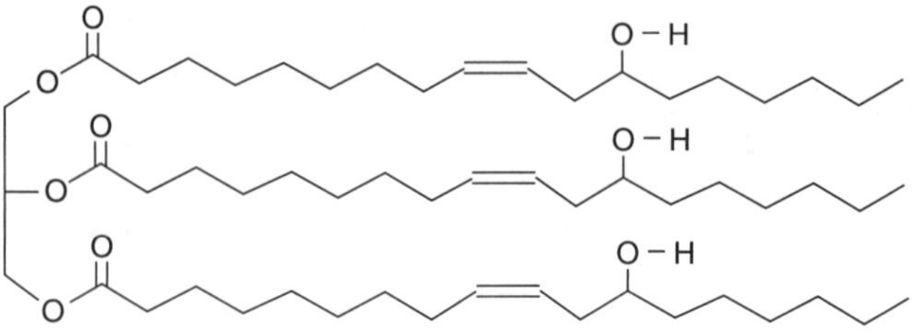

Figure 7-11 Structure of the major component of castor oil.

FIGURE 7-12   Carbon nanobeads synthesized from castor oil.

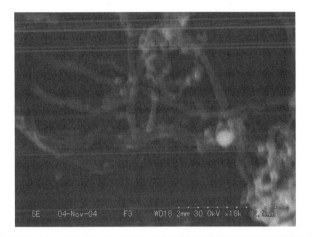

FIGURE 7-13   Coiled carbon nanofibers synthesized from linseed oil.

Pyrolysis of linseed oil in the presence of various metal catalysts provides coiled CNFs with a diameter of approximately 100 to 250 nm, along with some amorphous carbon.

Although edible oils have been found to give good-quality CNTs, these oils are mixtures of different fatty acids, so there is a need to check the composition of the different edible oils for the synthesis of different forms of CNTs.

7.4.3   Fibrous Plant Tissues

Plant fibers mainly come from xylem and phloem fibers. Other parts of the plant composed of fibers include seeds, leaves, and fruits.

These fibrous tissues have elongated cells with thickened walls. Under a microscope, they show a clear channel-like structure.

Sharon et al. (2007) has shown that CNMs prepared from plant fibers show a similar channel-like structure. Plant materials, such as bagasse, cotton grass, rice stem, and jute fibers, have been successfully used to prepare carbon showing various forms and specially nanosized channels and pores, which have found many applications. Zhenhui et al. have also successfully made CNTs from grass.

As can be seen in Table 7-2, both the transverse as well as the vertical section of bagasse fiber show a lot of similarities to the carbon material produced from them. Fiber and carbon synthesized from all the plant fibers show the same results.

7.4.4   Plant Seeds

All of the CNMs derived from plant material show a structure that resembles the tissue structure of that plant material. It appears as if, except for carbon, all of the other ingredients of plant tissues have sublimed, leaving the carbon skeleton intact. Seeds are mostly composed of tissues that stores carbohydrates, oil, and pertinacious material. Similar to fibers, the scanning electron micrographs of the tissue structure of seeds and of CNM show a lot of similarity, as can be seen in Table 7-3.

7.5   Various Applications of Carbon Nanomaterials Synthesized from Natural Sources

Sharon's group (2002, 2003, 2005, 2006, 2008a&b) has reported several applications of CNMs synthesized from the above-mentioned plant sources.

7.5.1   Microwave Absorption

Kshirsagar et al. (2006) synthesized CNMs from mustard oil, karajna oil, and linseed oil. They obtained almost 97% of absorption in the range of 8 to 18 GHz, with CNFs obtained from karanja oil (Fig. 7-14) while CNBs, obtained from mustard oil, or CNTs, obtained from Linseed oil, gave nonuniform absorption in this range.

7.5.2   Carbon Solar Cells

Sharon et al. (1997a,b&d) opened a new avenue of research to develop the carbon solar cell at very low cost compared with silicon, by reporting the formation of n- and p-type semiconducting carbon from the naturally obtained camphoric soot. The same group developed photoelectrochemical carbon solar cells using a camphoric p-carbon semiconducting pellet.

Precursor	SEM after pyrolysis	SEM of plant tissues	Pore size in carbon (nm)
Bagasse			10.1
Coconut fiber			6.4
Bamboo			8.1
Semar cotton			8
Rice straw			6.8

TABLE 7-2 Scanning Electron Micrographs of Various Fibrous Plant Materials and Carbon Nanomaterial Synthesized from Them

Precursor	SEM after pyrolysis	SEM of plant tissues	Pore size in carbon (nm)
Jute straw			1.2
Grass			Data not available

SEM = scanning electron micrograph.

TABLE 7-2   Scanning Electron Micrographs of Various Fibrous Plant Materials and Carbon Nanomaterial Synthesized from Them (*Continued*)

Precursor	SEM of carbon after pyrolysis	SEM of plant tissues before pyrolysis	Pore size (nm)
Soya bean			18.2
Til			13

TABLE 7-3   Scanning Electron Micrographs of Various Seeds and Carbon Nanomaterial Synthesized from Them

Precursor	SEM of carbon after pyrolysis	SEM of plant tissues before pyrolysis	Pore size (nm)
Sunflower			8.5
Castor			4.4
Kardl			1.2
Ambadi			5.8
Corn			6.0

TABLE 7-3   (Continued)

Precursor	SEM of carbon after pyrolysis	SEM of plant tissues before pyrolysis	Pore size (nm)
Rye			4.2
Anacardium			2.7
Almond			7.3
Karanjal seed			8.1
Ritha seed			13.1

TABLE 7-3   Scanning Electron Micrographs of Various Seeds and Carbon Nanomaterial Synthesized from Them (*Continued*)

Precursor	SEM of carbon after pyrolysis	SEM of plant tissues before pyrolysis	Pore size (nm)
Jack fruit seed	100 um		13.0
Supari seed	100 μm		25.1

SEM = scanning electron micrograph.

TABLE 7-3   *(Continued)*

They developed a solar cell of configuration Au–p-C–n-C–Si by depositing camphoric carbon thin films over a single crystal Si (100) surface with electron ion beam depositions. The current–voltage characteristic of their solar cell is shown in Fig. 7-15.

The band gap of these films was found to be 1 eV. Although the efficiency and fill factor obtained from these cells are low, the semi-conducting nature of these carbon films still encouraged high prospects for developing low cost and highly efficient carbon photovoltaic solar cells from environmentally clean and naturally available precursor camphor.

Sharon et al. (1999) has reported that with suitable control of deposition condition of pyrolysis, it is possible to make semiconducting carbon with a band gap in the range of 0.25 eV to 5.5 eV not only from camphor, but also from other plant-derived precursors such as turpentine oil. Kalita et al. (2007) have successfully synthesized thin films of nanocrystalline DLC with the thermal flash vapor deposition technique using turpentine oil with a direct band gap of 1.04 eV. The film had a thickness of 2 μm (Fig. 7-16).

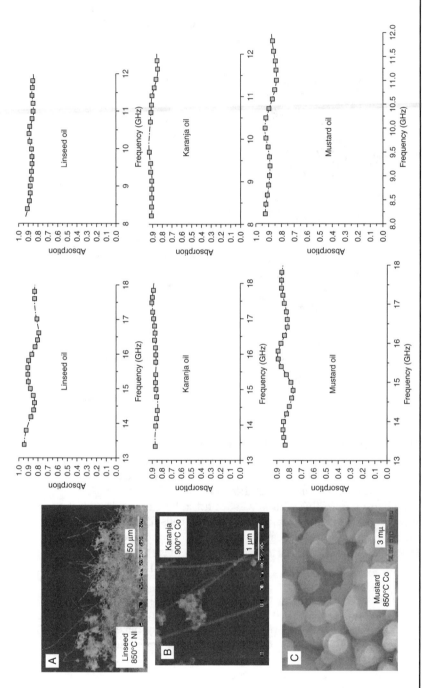

Figure 7-14 Graph of microwave absorption by carbon nanomaterials obtained from linseed oil (*A*) karanjal oil (*B*) and mustard oil (*C*) in the range of 8 to 18 GHz.

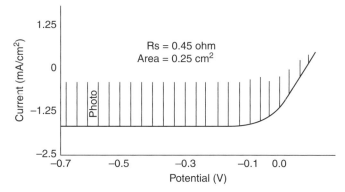

Note: *Vertical lines* indicate the magnitude of photocurrent at the corresponding potential when the cell was exposed to light. Rs = series resistance of the cell.

FIGURE **7-15**   Current–voltage characteristic of carbon solar cells.

FIGURE **7-16**   Scanning electron micrograph of a 2-µm thick (as denoted by cursor) carbon thin film.

7.5.3   Hydrogen Absorption

Hydrogen storage has been an important consideration for various applications when hydrogen is required. CNMs derived from fibrous plant parts and seeds have been found to play very important role in the storage of hydrogen because of their unique, highly porous structures, which provides an increased surface area for enhanced hydrogen adsorption (Sharon et al., 2007). Carbon fibers with a channel-type structure are obtained from bagasse and coconut fibers. It is reported that among the different plant-based precursors studied, carbon from cotton (4.09 wt%) and bagasse (9.0 wt%) have better capacity to store hydrogen at room temperature. (Poirier et al. 2004) have suggested that hydrogen adsorption of 6.5 wt% will be economical to run an electric vehicle powered by hydrogen. CNMs have been envisaged to have a desired diameter, surface area, and the presence of specific combinations of sp^3 and sp^2 carbon into the carbon materials; therefore, they are being tried as a material for hydrogen storage.

7.5.4   Fuel Cells

An approach to using CNM for preparing fuel cells is discussed in Chapter 14. CNM has been found to have good porosity, low resistivity, mechanical stability, and high electroactivity, so it has been proposed to be a suitable material for fuel cell electrodes. CNTs, CNFs, and CNBs have all shown to have potential for use in fuel cells.

The SEMs provided in Fig. 7-17 show that CNBs (prepared from turpentine) are highly dense and interconnected structures with diameters in the range of 500 to 650 nm.

Chronopotentiometric studies, current–voltage characteristics, and measurements of the H_2 dissociation voltage of electrodes with Ni-coated CNBs (synthesized from turpentine) deposited over ceramic oxide have shown a yield of high current densities, suggesting that these CNBs have characteristics similar to those of the platinum electrode. These CNBs may be used in place of the Raney nickel electrode in alkaline fuel cells.

7.5.5   Super Capacitors

Large-scale commercialization of CNT-based super capacitors has been restricted because of the high cost of CNTs. Plant-based natural precursors are the best solutions for this. The original morphology of plant-based materials does not get disturbed or spoiled during pyrolysis,

Figure 7-17   Scanning electron micrographs of carbon nanobeads synthesized from turpentine oil.

thus giving unique structures with a pore size or surface area required for developing supercapacitors.

CNTs synthesized from turpentine oil were found to be highly dense and intrinsically n type with low resistance and high carrier concentration. Such low-resistance CNTs are attractive for application in supercapacitors and other electronic devices. Use of these CNTs in assembling an electrochemical double-layer capacitor has yielded up to 82 F/g specific capacity (based on the total weight of active material).

Sharon et al. 2008a has produced porous carbon materials from different precursors, including seeds of castor (*Ricinus communis*), ritha (*Soapnut sapindus*), anacardium (*Semecapus anacardium*), jack fruit (*Artocarpus heterophyllus*), kardi (*Carthamus tinctorius*), ambadi (*Crotolaria juncea*), bitter almond (*Prunus amygdalus*), til (*Sisamum indicum*), and sunflower (*Helianthus annuus*) and from fibrous materials, including corn stem (*Zea mays*), rice straw (*Oryza sativa*), bamboo (*Bambusoidea bambusa*), and coconut fiber and are studying their properties for developing a low-cost and highly efficient supercapacitor.

7.5.6   Lithium Ion Batteries

The application of lithium batteries as a portable energy source is increasing day by day, especially because it can provide as high as a 3-V cell potential. The main disadvantage of Li–ion batteries is the high cost of electrode materials. Sharon et al. (2001, 2002) Bhardwaj et al. (2008c) have been searching for plant-derived precursors, especially the waste parts or non-edible parts of plants. The intercalation of Li-ion with carbon depends on various factors, such as pore size, density, surface area, and activation of CNM. They have also studied the intercalation of lithium with CNBs synthesized from camphor (Sharon et al., 2002a).

Bhardwaj et al. (2007, 2008a, 2008b) have studied the lithium intercalation by carbon materials obtained from the pyrolysis of some selected oil seeds and stems, including soap-nut seeds, jack fruit seeds, date seeds, neem seeds, tea leaves, bamboo stem, and coconut fiber. SEMs of carbon from tea leaves (Table 7-4) show an appearance similar to beehives filled with small pebbles, which are made of clusters of CNBs of about 500 nm in diameter. Carbon from coconut fiber is heavily porous and looks like a strainer. Carbon from bamboo has a structure composed of bundles of channels lined together of approximately 10 μm in diameter.

Soap-nut seeds yields porous carbon block that looks like cotton balls. Carbon fibers obtained from jack fruit seed also have a cotton ball–like appearance with disorganized pores all over the surface with a pore size similar to that of soap-nut carbon.

Carbon from date seeds looks like fossilized porous rock with large pores (18.25 μm). Carbon from neem seed appears like rectangular

CNM synthesized from	Discharge capacity (mAh/g after 100th cycle)	CNM synthesized from	Discharge capacity (mAh/g after 100th cycle)
Tea Leave	33.76	Jackfruit	15.90
Coconut	37.5		115.07
Bamboo	92.74	Neem Seed	10.31
Soap-nut seed	130.29		154.80

CNM = carbon nanomaterial.

TABLE 7-4   Lithium Intercalation Capacity Obtained with Carbon Nanomaterials Obtained from Some of the Plants' Precursors

block with some cavities on the surface. Details of the lithium intercalation capacity obtained with different CNMs after running the battery for 100 cycles are given in Table 7-4. It should be noted that carbon fibers obtained from bagasse, cotton, and bamboo seem to give a high capacity. If these materials are properly activated, their capacity can be further improved.

7.5.7   Drug Delivery

CNBs are being thought of as a better drug delivery vehicle than CNTs. Details about the possibilities of using various CNMs are discussed in Chapter 22.

7.6   Summary

In this chapter, various plant tissues and products as the possible precursors for synthesis of CNM have been discussed. From the data available, it has been noticed that the inherent structure of plant tissues plays an important role in determining the type of CNMs that are produced by their pyrolysis. Plant products such as oil and resins are also good precursors for synthesis of CNFs, CNBs, and CNTs.

CHAPTER 8

Waste Plastic as Precursors for Synthesizing Carbon Nanomaterials

Neeraj Mishra

Nanotechnology Research Centre, Birla College, Kalyan, Maharashtra, India

Maheshwar Sharon

Nanotechnology Research Centre, Birla College, Kalyan, Maharashtra, India

Madhuri Sharon

MONAD Nanotech Pvt Ltd, Mumbai, India

8.1 Introduction

Plastic has become an indispensable need for human beings in day-to-day life. Berzelliuz discovered the plastic in 1833, and Alexander Parks revealed the first fabricated plastic in 1862 at the Great International Exhibition in London. After this milestone discovery for the developing world, plastic and plastic industries never looked back. The considerable growth in plastic use is because of its beneficial properties, including:

1. Plastic is extremely versatile and has the ability to be tailored to meet very specific technical needs; it can be molded in to any size and shape. Hence, it was taken up by most of the packing and processing industries for various applications. It can be easily colored or even kept as transparent; therefore, the use of plastics can completely alter the packaging option as per desire, which is not as easy with paper or metal.

2. Plastic is lighter in weight than competing materials, so it is easy to transport.

3. Plastic is extremely durable and does not face any type of corrosion or degradation problems.

4. Plastic is resistant to chemicals and does not react with water.

5. Plastic is a very good thermal and electrical insulator.

6. Plastic possesses good tensile strength and rigidity with or without elongation or an impact on strength.

7. The main advantage of plastic is that it can be easily recycled.

Some of the plastic products are disposable. Moreover, plastics are nondegradable, they take a long time to break down, possibly up to hundreds of years, causing damage to the soil and water where they are disposed.

The world's annual consumption of plastic materials has increased from around 5 million tons in the 1950s to nearly 100 million tons today. Various uses of plastics are provided in Fig. 8-1.

8.1.1 Chemistry of Plastic

Plastic is a synthetic polymer. *Plastic* is a Greek word meaning a material that can be mounded or formed into any shape of one's choice. It is a synthetic polymer that is similar in many ways to natural resins found in plants. *Webster's Dictionary* defines plastic as "any of various complex organic compounds produced by polymerization, capable of being molded, extruded, cast into various shapes and films, or drawn into filaments and then used as textile fibers."

The raw material for plastic synthesis is crude oil rich in carbon and hydrogen. Plastics are large carbon-containing polymers composed

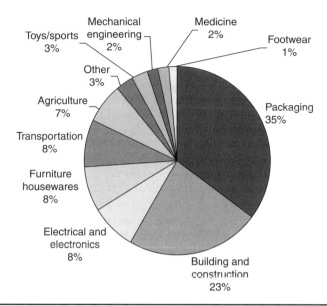

FIGURE **8-1**   Chart showing the percentage of usage of plastics.

of repeating units of shorter carbon-containing monomers. As already mentioned, plastics are resistant to chemicals and water. However, they are sensitive to heat. Based on thermal sensitivity, they are classified as thermoplastic polymers, which soften on heating and stiffen on cooling (e.g., polyethylene [PE], polyvinyl chloride [PVC]), and thermosetting polymers, which become an infusible and insoluble mass after heating (e.g., electric switches).

8.1.2   Types of Plastic

There are about 50 different groups of plastics, with hundreds of different varieties. Almost all types of plastic are recyclable. To make sorting and thus recycling easier, the American Society of Plastics Industry developed a standard marking code to help consumers identify and sort the main types of plastic. These types and their most common uses are shown in Table 8-1.

8.2   Disposal of Plastic

The conventional methods to eliminate plastic wastes are landfill, combustion, and recycling.

8.2.1   Landfill

With more and more plastics products, particularly plastics packaging, being disposed of soon after their purchase, the landfill space required by plastics waste is becoming a growing concern. Landfilling

Type	Uses
Polyethylene terephthalate (PET)	Soda bottles, oven-ready meal trays
High-density polyethylene (HDPE)	Milk bottles, dish detergent bottles
Polyvinyl chloride (PVC)	Food trays, plastic wrap, water bottles, shampoo bottles
Low-density polyethylene (LDPE)	Plastic bags, trash bags
Polypropylene (PP)	Margarine tubs, microwaveable meal trays
Polystyrene (PS)	Yogurt containers, foam meat and fish trays, hamburger boxes, egg cartons, vending cups, plastic cutlery, protective packaging for electronic goods and toys
Other	Any other plastics that do not fall into any of the above categories, such as melamine, which is often used in plastic plates and cups

TABLE 8-1   Uses of Different Types of Plastics

of waste plastic wastes many hectares of lands. Similarly, additives such as colors and stabilizers pollute groundwater. Furthermore, methane produced from buried plastic wastes pollutes the air and causes explosions.

8.2.2   Combustion
Combusting plastic wastes produce dioxin, a toxic material that is carcinogenic. Moreover, incineration is becoming a more expensive and less accepted method for degradation of plastic.

8.2.3   Recycling
Although, recycling plastic wastes is gaining increasing importance, recycling of waste plastic requires further conversion of plastic into plastic; therefore, recycling does not solve the problem of removing or degrading plastics. Nevertheless, according to a 2001 Environmental Agency report, only 7% of plastic waste is recycled, and the remaining waste is dumped in soils or seawater. Today, the main plastics wastes are PE, polypropylene (PP), polystyrene (PS), and PVC.

8.3   Pyrolysis of Waste Plastic to Carbon Nanomaterial

Scientists have recently sought a novel method of converting plastics into valuable carbon nanomaterials (CNMs) by pyrolyzing them. Pyrolysis of plastic produces lower molecular weight fragments of monomers. By adjusting operating conditions, the rate and extent of decomposition are controlled. Researchers at the University of Melbourne, Australia, have designed a waster plastic pyrolysis plant to yield CNMs. However, it must be mentioned here that in all the trials of converting waste plastic to degradable products, CNMs are formed as a byproduct, and the main products are wax or different types of gasoline or fuel.

8.4   Synthesis of Carbon Nanotubes from Plastic

Production of multiwalled carbon nanotubes (MWCNTs) and singled-walled carbon nanotubes (SWCNTs) from waste plastic has been reported by Arena et al. (2005) using a bubbling fluidized bed reactor. The reactor had an inner diameter of 100 mm and was made up of stainless steel 304. The researchers mixed recycled plastic with a catalyst and placed it in the reactor, passed nitrogen gas on the fluidized bed, and heated it to 800° to 900°C. Due to thermal cracking, bond breaking took place, and MWCNTs were formed. When plastic was heated up to 1100°C along with a catalyst, SWCNTs were produced. The structure was determined by scanning electron micrography, transmission electron microscopy, X-ray diffraction, RAMAN, and thermo gravimetric analysis (TG) and differential thermal analysis (DTA).

Sharon's group has also produced CNTs from different plastic using a stainless steel (316) furnace reactor. Plastic was pyrolyzed along with a catalyst while Ar, N_2, or H_2 gas was passed through the reactor throughout the whole reaction. The catalysts used were nano-metals such as Ni, Co, Fe, and zeolite. Catalysts had to be in the nano-form for the synthesis of CNMs. Before placing catalyst inside the furnace, the plastic and catalyst were thoroughly mixed. The researchers observed that in addition to temperature, gas flow rates and stainless steel reactor dimensions also affected the growth of various carbon nanostructures. Moreover, the residence time of plastic in the reaction zone is also an important parameter for the growth of different types of carbon nanostructures materials. A setup of the unit for synthesizing CNT is presented in Fig. 8-2.

8.5   Chemistry of Conversion of Plastic to Carbon Nanotubes

The conversion of plastic to CNTs involves either thermal cracking or pyrolysis of polymer at a high temperature.

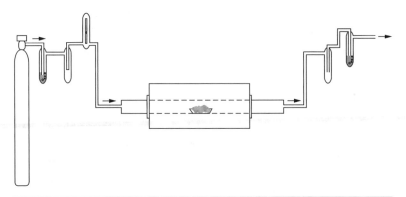

FIGURE 8-2   Schematic diagram of a unit used for synthesizing carbon nanotubes with the carbon vapor deposition method.

8.5.1   Thermal Cracking

In thermal cracking, polymers such as PE, PP, PS, or polycarbonate are placed in a solid-gas fluidized bed reactor operated at specific thermal (500° to 700°C) and hydrodynamic conditions. The bed material and polymer interact, resulting in formation of MWCNTs (Arena and Mastellone, 2005).

This phenomenon promoted thermal cracking of the polymer chain with a concurrent production of light hydrocarbons, aromatics, waxes, and CNTs. When the polymer faces high thermal energy, the polymer surface starts to soften. Several particles stick to the surface of the softened polymer, forming an aggregate that has an external shell made of bed particles and an internal core of not-yet-molten polymer (Arena and Mastellone, 2005). With a further increase in temperature, the surface of pellet melts, and polymer flows throughout the bed particles of the external shell, forming a uniform coating over and between the particles. Heating leads to breaking of the carbon bonds of the polymer chain (i.e., when the polymer has already covered the bed particles, the process of pyrolysis starts). Therefore, only the layer of polymer that coats and adheres to the surface of single-bed particles starts getting pyrolyzed. It is worth noting that this mechanism is only a hypothesis and needs evidence to confirm.

8.5.2   Pyrolysis

In pyrolysis, the polymeric materials are heated to high temperatures in the presence of a catalyst so that macromolecular structures are broken down into smaller molecules and a wide range of hydrocarbons such as paraffin, olefins, naphthenes, and MWCNTs (Demibras, 2004). Four types of mechanism of plastic pyrolysis have been proposed by Cullis and Hirschler (1981) that are primarily, to some extent, related to bond dissociation energies, the chain defects of polymer aromaticity degrees, and the presence of halogen and other hetero atoms in the polymer chain (Bucken and Huang, 1998). These mechanisms are:

- End-chain scission or depolymerization
- Random-chain scission
- Chain stripping
- Cross-linking

Pyrolysis of PS occurs by both end-chain and random-chain scission. Here the monomer recovery is only 45% (Mertens et al., 1982). PS thermal degradation is a radical chain process including initiation, transfer, and termination steps (Kiran et al., 2000). The product

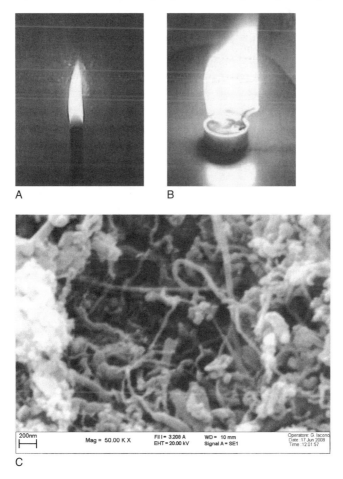

FIGURE 8-3   Pyrolysis of plastic waste giving (A) burnable gas (B) Burnable wax and (C) Carbon nanotubs.

formed depends on the molecular weight of the PS sample (Faravelli et al., 2000).

Pyrolysis of PE and PP occurs through a random-chain scission, forming a whole spectrum of hydrocarbons. The thermal degradation of PE consists of free radical formation and the hydrocarbon abstraction step (Kiran et al., 2000; Schoeter and Buekens, 1979).

It is worth noting that though plastic disposal has created a environmental problem, if processes such as pyrolysis of plastic becomes economically viable, we would be creating four valuable degradable products (i.e. CNMs, wax, low density oil and burnable gas) from one raw material, which is plastic (Fig. 8.3). Sharon and research group are working on this aspect to make this process commercially viable.

8.6   Summary

Plastics are polymers that are a rich source of hydrocarbons. When they are pyrolyzed or thermally cracked in the presence of a catalyst at a high temperature, they can produce MWCNTs.

PART III

Characterization of Carbon Nanomaterials

CHAPTER 9

Characterization of Carbon Nanomaterials by Scanning Electron Microscopy

Dillip Kumar

Department of Physics, IIT Bombay, Mumbai, India

Madhuri Sharon

MONAD Nanotech Pvt. Ltd., Mumbai, India

9.1   Introduction

Scanning electron microscopy (SEM) is widely used for initial characterization of nanostructured materials. It has contributed immensely to characterizing various forms of carbon nanomaterials (CNMs), including carbon nanofibers (CNFs), carbon nanobeads (CNBs), carbon nanotubes (CNTs), single-walled CNTs, (SWCNTs,) and multi-walled CNTs (MWCNTs). However, it has limited application in characterizing SWCNTs. The main problem is in differentiating between SWCNTs and MWCNTs. This is mostly because of the tendency of SWCNTs to strongly adhere to each other, forming bundles or ropes 5 to 20 nm in diameter. In contrast to transmission electron microscopy (TEM), SEM cannot resolve the internal structure of these SWCNT bundles. Nevertheless, SEM can yield valuable information regarding the purity of a sample, as well as insight on the degree of aggregation of raw and purified SWCNT materials and the presence of impurities such as catalysts and amorphous carbon.

SEM is used to observe the morphology of a sample at higher magnification, higher resolution, and higher depth of focus than with an optical microscope. It uses electrons instead of light to form images. The historical development of microscopy technology can be enumerated as follows:

- 1590: Dutch spectacle makers Hans Janssen and his son Zacharias Janssen are claimed by later writers (Pierre Borel, 1620–1671 or 1628–1689, and Willem Boreel, 1591–1668) to have invented a compound microscope, but this is disputed.

- 1609: Galileo Galilei develops an occhiolino, or a compound microscope with a convex and a concave lens.

- 1612: Galileo presents occhiolino to Polish King Sigismund III.

- 1619: Cornelius Drebbel (1572–1633) presents in London, a compound microscope with two convex lenses.

- 1622: Drebbel presents his invention in Rome.

- 1624: Galileo presents his occhiolino to Prince Federico Cesi, founder of the Accademia dei Lincei (in English, The Linceans).

- 1625: Giovanni Faber of Bamberg (1574–1629) of the Linceans coins the word *microscope* in analogy with *telescope*.

- 1665: Robert Hooke publishes *Micrographia*, a collection of biological micrographs. He coins the word *cell* for the structures he discovered in cork bark.

- 1674: Anton van Leeuwenhoek invents the simple microscope.

- 1931: Ernst Ruska builds the first electron microscope.

- 1942: Zworykin develops and describes the first true scanning electron microscope.

- 1965: R.F.W. Pease and W.C. Nixon create the SEM V, which becomes the basis for the first commercial scanning electron microscope.

- 1981: Gerd Binnig and Heinrich Rohrer develop the scanning tunneling microscope.

9.2   Principles of Scanning Electron Microscopy

A schematic diagram of SEM is given in Fig. 9-1. Basically, an electron gun is used to create an electron beam, helping to magnify the electrons. The magnified electrons are allowed to pass through a magnetic lens to make a parallel beam. This is then directed to a scanning coil that concentrates the beam and focuses on the sample. Its focusing is controlled by software. The focused beam falls on the sample, producing backscattered electrons that are detected by an electron detector. These electrons are sent to the monitor screen. At the same time, there is an emission of secondary electrons from the samples, which are also sent to a TV screen. This is the basic functioning operation of a SEM system.

When the focused monoenergetic (~30 KeV) electron beam (electron gun) passes through various stages of magnifications and is incident on a solid surface, various signals are given, such as secondary electrons, primary backscattered electrons, Auger electrons, X-rays, and photons. Backscattered electrons and secondary electrons are of prime interest in the case of SEM. The signal collected is amplified and used to see the sample through a cathode-ray tube (CRT). The electron beam scans the surface with the help of coils and the signals transferred from the specimen on a point-to-point basis, giving the whole surface morphology of scanned area.

The scan coils are energized (by varying the voltage produced by the scan generator) and create a magnetic field, which deflects the beam back and forth in a controlled pattern. The varying voltage is also applied to the coils around the neck of the CRT, which produces a pattern of light deflected back and forth on the surface of the CRT. The pattern of deflection of the electron beam is the same

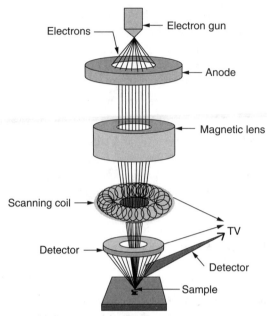

Note: TV = Television screen.

FIGURE 9-1   A schematic diagram of scanning electron micrography showing the passage of electrons generated by the cathode to various stages of amplification. These electrons finally converge to the sample, showing the morphology of the sample.

as the pattern of deflection of the spot of light on the CRT. The electron beam hits the sample, producing secondary electrons from the sample. These electrons are collected by a secondary detector or a backscatter detector, which are converted to a voltage and amplified. The amplified voltage is applied to the grid of the CRT and causes the intensity of the spot of light to change. The image consists of thousands of spots of varying intensity on the face of a CRT that correspond to the topography of the sample.

9.3   Components of Scanning Electron Microscope

The basic components of a scanning electron microscope include:

- Electron gun
- Magnetic lenses (condenser and objective) and scanning coil
- Vacuum system (sample chamber)
- Detectors (scintillation or solid state) and image recorder (CRT and camera)

9.3.1 Electron Gun

The electron gun (see Fig. 9-1) is made up of a filament (cathode), a Wehnelt cylinder, and an anode. Together, these three elements form a triode gun, which is a very stable source of electrons.

The electron beam comes from a filament made of various types of materials. Filament (cathode) is commonly made of tungsten wire. But other materials, such as lanthanum hexaboride (LaB_6), field emission guns, and Schottky (thermal field), are also used as a cathode material.

The most common electron gun is the tungsten hairpin gun. The filament is a loop of tungsten that functions as the cathode. A voltage is applied to the loop, causing it to heat up (mostly in the 5- to 30-kV range for SEM). The anode, which is positive with respect to the filament (cathode), forms a powerful attractive force for electrons. This causes electrons originating from the cathode to accelerate toward the anode. Some are accelerated by the anode, and some are accelerated while passing down the column to the sample.

The most important parameter of the electron gun is its brightness (β):

$$\beta = \text{current/area x solid angle} = I/(\pi d_o^2/4)\alpha \, \pi o^2 \qquad (9.1)$$

where β is the measure of the current I focused on the area, $(\pi d_o^2)/4$ entering and exiting this area through solid angle $\alpha\pi o^2$.

Increasing β improves the performance of the scanning electron microscope. The value of β is a function of the filament material, its operating temperature, and its voltage.

For comparison, a few cathodes are used for SEM are given in the Table 9-1. Schottky and field emission cathodes have the lowest source sizes and are frequently used for high-resolution imaging for characterizing nanostructures or materials.

	Tungsten filament	LaB6	Schottky (TF)	Field emission
Apparent source size	100 μm	5 μm	<100 nm	<100 nm
Brightness (A/cm² sr)	1	20–50	100–500	100–1000
Vacuum required (torr)	10^{-5}	10^{-6}	10^{-8}	10^{-9}

TABLE 9-1 Comparison of Different Types of Scanning Electron Microscopy Cathodes

9.3.2 Magnetic Lenses (Condenser and Objective) and Scanning

The basic function of the condenser lens (see Fig. 9-1) is to focus the electron beam emerging from the gun on the specimen and to provide optimal illuminating conditions for visualizing and recording images. Electron microscopes have magnetic lenses that are similar to simple solenoid coils. A coil of copper wire produces a magnetic field that is shaped by the surrounding iron fixture into an optimum geometry to produce the lensing action. The lenses in a scanning electron microscope reduce the diameter of the electron beam to a very small size on the sample surface. In the electromagnetic lens, the focal length depends on two factors, the gun voltage and the amount of current passing through the coil:

$$fc = kV/Lc^2 \qquad (9.2)$$

where fc = focal length of the condenser
k = a constant that depends on the geometry of the pole pieces and number of turns in the coil
V = accelerating voltage
Lc = lens current.

The two sets of scan coils located in the bore of the objective lens cage perform the scanning function. These coils cause the beam to scan over a square area of size marked by the cursor as X on the sample surface. The scanned area is generally termed the *raster*.

The condenser lens is associated with three interrelated properties: the aperture angle of the illuminating pencils, the intensity of the illumination, and the depth of field.

9.3.3 Vacuum System (Sample Chamber)

When SEM is used, the column hosting gun, anode, coils, others are kept under a vacuum. In general, a sufficiently low vacuum for a SEM is produced by either an oil diffusion pump or a turbo molecular pump. In each case, the system is backed by a rotary prevacuum pump. These combinations also provide reasonable exchange times for the specimen, filament, and aperture without the need to use vacuum airlocks. The column and specimen chamber also, must operate under vacuum conditions because the electron beam would otherwise get scattered by gas atoms present in the column. Therefore, the sample and mounting compound used to hold it must not have high vapor pressure; otherwise, they will get evaporated and spoil the vacuum system. A typical vacuum level for SEM is 10^{-6} torr.

9.3.4 Detectors (Scintillation or Solid-State Type), Cathode-Ray Tube, and Camera

Electrons that pass through the screen are accelerated by high voltage into a quartz light pipe coated with a scintillator material. Detectors used for backscattered electrons and secondary electrons are usually either a scintillation detector or a solid-state detector. In the former case, electrons strike a fluorescent screen, which emits light photons. These photons are allowed to fall on a photocathode, which converts the photon into electrons. These electrons are amplified by allowing them to pass through a series of anodes in increasing potential. In this way, one photon gets amplified to nearly as many as 10^6 electrons. These electrons are finally collected at the anode to give an electrical signal. This entire process is done by a *photomultiplier tube*. In a solid-state detector, a similar operation is done but by a system called a *photodiode* made by forming a p–n junction. When light photons falls at the p–n junction, corresponding electron are produced, forming an electrical signal.

9.4 Specimen Preparation

Preparation of specimen of sample is very important in getting a good picture with SEM. The two basic types of SEM are a regular SEM and environmental SEM (ESEM). SEM requires a conductive sample, but ESEM may not necessarily need the conducting sample. The requirements for preparing samples for a normal SEM are discussed here.

SEM can be done for both solid and liquid samples having low vapor pressures. Water, solvents, and other materials that could vaporize while in the vacuum should be removed from the sample. The sample to be examined by SEM is allowed to adhere over the sample holder. Double-sided graphite tape that has adhesive on both sides of the graphite is normally used for holding the sample over the sample holder. The double-sided graphite tape (Fig. 9-2) is first put over the sample holder by removing the transparent tape from one side of the tape. Then transparent tape from the top side of the graphite tape is removed. Over this tape, the sample to be examined is placed, and it is adhered over the graphite tape. This method can be used for either powder or thin film of metal strip. Graphite helps to give a good conductivity of the sample with the base to sample holder.

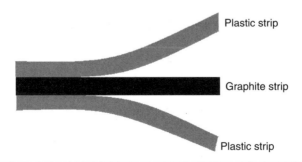

Plastic strip

Graphite strip

Plastic strip

FIGURE 9-2   A schematic picture of a double-sided graphite tape.

If the sample is adhered by a nonconducting tape, then the sample gets damaged when the electron beam in the microscope falls over the sample. Nonmetallic samples, such as bugs, plants, fingernails, and ceramics, are generally coated with a thin layer of carbon, gold, or gold alloy to make them electrically conductive. Metallic samples can be placed directly to be used for SEM with the help of double-sided graphite tape. Depending on the facilities provided by the sample holder of the SEM, samples as large as 1.5 to 2.0 cm can be placed in the microscope.

9.5   Resolution of Scanning Electron Microscopy

To resolve a feature of size R, the beam diameter should be less than the size of the object in the sample to be measured.

$$\text{i.e., } \lambda = 12.3\text{Å}/\sqrt{v} < R \tag{9.3}$$

According to this formula, the wavelength λ of a beam of electrons is dependent on its accelerating potential v. Higher voltage v reduces the wavelength λ, thus giving a good resolution. The depth of resolution d can be defined as:

$$d = 0.612\lambda/n \sin \alpha \tag{9.4}$$

For a given wavelength, the depth of field is given as:

$$\text{Depth of field} = \text{constant}/\alpha \times M \tag{9.5}$$

where M = magnification
       α = divergence of the beam striking the sample.

The value of α is determined by the working distance (WD) and the diameter of the final aperture, da:

$$\alpha \approx (\text{da}/2)/\text{WD} \tag{9.6}$$

9.6   Limitations of Scanning Electron Microscopy

Electron microscopes possess some inherent defects, including spherical aberration and chromatic aberration, which limit the resolution.

9.6.1   Spherical Aberration

This is the most common form of defect that belongs to the class called *geometric aberrations* because it is produced by an error in the geometry of the lens. It cannot be corrected, so this is one of the principal factors limiting resolution.

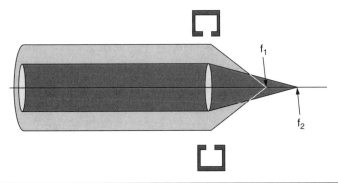

FIGURE 9-3   A schematic diagram showing two focal points (f_1 and f_2) caused by spherical aberration.

Spherical aberration is created because of different zones of lens having different refractive powers. As a result, different focus points (f_1 and f_2) are observed (Fig. 9-3). The limit of resolution due to the presence of the spherical aberration is given by:

$$ds = ks f \alpha_0^3 \tag{9.7}$$

where   ds = the separation of two object points, which has to be resolved
        ks = a dimensionless proportionality constant
         f = focal length
         α_0 = objective aperture angle.

This equation indicates that the aperture angle and focal length are the critical factors that determine the magnitude of spherical aberration.

9.6.2   Chromatic Aberration

When electrons of different velocities pass through the lens surrounded by the electromagnetic field, the electrons with greater velocity converts at a shorter distance than the electrons with lesser velocity. The difference in velocity and therefore in wavelength of electrons is the source of chromatic aberration. This effect also causes two focal points, as discussed earlier concerning spherical aberration. The equations for the limit of resolution in the presence of chromatic aberration are:

$$dc\,v = kc f \alpha_0\, \Delta v/v, \; dc\,I = 2kc f \alpha_0\, \Delta I/I \tag{9.8}$$

where   dc v and dc I = the separation of two object points, which is
                resolved, considering voltage and current,
                respectively
         Δv = maximum departure from V
         ΔI = maximum departure from I.

9.7 Applications of Scanning Electron Microscopy

Improvisation of SEM has resulted in various improvements and additions to data analysis. The most prevalent applications are:

- SEM of some CNMs
- Energy-dispersive analysis of X-ray (EDAX)
- Energy-dispersive analysis of total spectrum of X-ray (EDS)
- Environmental scanning electron microscope (ESEM)

9.7.1 Scanning Electron Microscopy for Morphologic Study of Carbon Nanomaterial

It is possible to differentiate the morphology of various types of CNMs by taking their SEM micrographs (Fig. 9-4). With the pyrolysis of sugar cane, bagasse, or rice straw, special types of channel structures of carbon have been obtained (Sharon et al. 2007).

Similarly, under some special pyrolysis conditions, it is possible to get carbon CNBs, CNTs, CNFs, or bamboo-type fibers from camphor (see Fig. 9-4). Thus, SEM can help to distinguish such type of CNMs. Light microscopy cannot provide such differentiation (Fig. 9-5).

9.7.2 Energy-Dispersive Analysis of X-Ray

When an electron beam strikes a solid surface, electrons and X-rays are emitted from the surface. In addition to the secondary electron detector, most scanning electron microscopes are equipped with an X-ray detector, and specialized backscattered detectors are available.

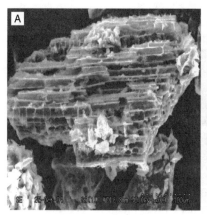

Figure 9-4 Scanning electron micrographs of carbon nanomaterials obtained by the pyrolysis of bagasse (*A*) and rice straw (*B*), both showing a channel-type structure.

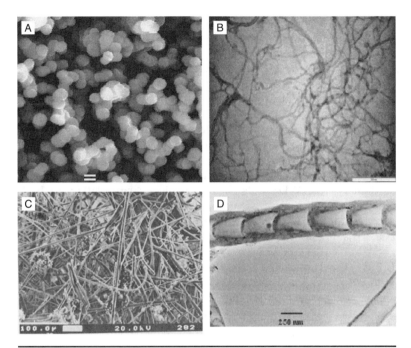

FIGURE 9-5 Scanning electron micrographs of carbon nanobeads (*A*), carbon nanotubes (*B*), Carbon nanofibers (*C*), and bamboo like fibers (*D*) obtained from the pyrolysis of camphor.

The addition of an X-ray detector allows determination of the energy of the emitted characteristic X-ray. Because each element in the periodic table has a different characteristic energy, the X-ray analyzer enables determination of the chemical analysis from point to point on the sample surface. Addition of an X-ray detector to a scanning electron microscope converts the microscope into an electron probe microanalyzer.

Most scanning electron microscopes are currently being equipped with EDS. The EDS is limited to analysis of elements with atomic number Z, above sodium. Thin window EDS are available that allow detection of elements with Z down to carbon, although their sensitivity below Z 8 (oxygen) is poor.

9.7.3   Energy-Dispersive Analysis of Total Spectrum of X-Ray

The X-ray signal from the sample passes through a thin beryllium window into a cooled, reverse-bias p-i-n lithium drifted silicon detector. Absorption of each individual X-ray photon leads to the ejection of a photoelectron, which gives up most of its energy to the formation of electron–hole pairs. They in turn are swept away by the applied bias to form a charge pulse, which is then converted to a voltage pulse by a charge-sensitive preamplifier. The signal is further amplified and shaped by a main amplifier and finally passed to a multichannel analyzer, where the pulses are sorted by voltage. The voltage distribution can be displayed on a CRT or an X-Y recorder.

The second kind of X-ray detector is a wavelength-dispersive (WDS) one, which can detect all elements from Z 5 (boron) and above, improves sensitivity, and enables analysis of samples in which characteristic X-ray peaks from different elements overlap. However, WDS are large and work slowly.

EDS provides observation of the entire X-ray spectrum of interest. This allows for rapid qualitative analysis of major and minor constituents, while WDS must be mechanically scanned through its wavelength range, with several crystal changes to cover the same energy range as EDS. Superior resolution of WDS easily separates peaks that are poorly resolved in EDS.

9.7.4   Environmental Scanning Electron Microscopy

ESEM is a revolution in microscopy. It allows the examination of specimens surrounded by a gaseous environment, meaning that a specimen viewed in the microscope does not need to be coated with a conductive material. Even liquids can be viewed in the microscope. The primary electron beam hits the specimen, which causes the specimen to emit secondary electrons. The electrons are attracted to the positively charged detector electrode. As they travel through the gaseous environment, collisions occur between an electron and a gas particle, resulting in emission of more electrons and ionization of the gas molecules. The increase in the amount of electrons amplifies the original secondary electron signal. The positively charged gas ions are attracted to the negatively charged specimen and offset the charging effect (Fig. 9-6).

Whereas conventional SEM requires a relatively high vacuum in the specimen chamber to prevent atmospheric interference with primary or secondary electrons, an ESEM may be operated with a poor vacuum (≤ 10 torr of vapor pressure, or $1/76$ atm) in the specimen chamber. In such an imaging mode, the specimen chamber is isolated (by valves, pressure-limiting apertures, and a large-diameter bypass tubes) from the rest of the vacuum system.

Water is the most common imaging gas, and a separate vacuum pump permits fine control of its vapor pressure in the specimen chamber. When the beam (primary electrons) ejects secondary electrons from the surface of the sample, the secondary electrons collide with water molecules, which in turn function as a cascade amplifier, delivering the secondary electron signal to the positively biased gaseous secondary electron detector (GSED). Because they have lost electrons in this exchange, the water molecules are positively ionized, so they are forced or attracted toward the specimen (which may be

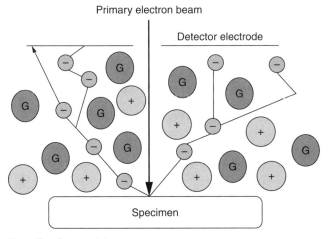

Note: G = Gas particles.

FIGURE 9-6   Schematic diagram of environmental scanning electron micrography function.

nonconductive and uncoated), serving to neutralize the negative charge produced by the primary electron beam.

Field-emission environmental scanning electron microscopy (ESEM-FEG) represents several important advances in SEM. The field-emission gun produces a brighter filament image (primary electron beam) than both tungsten or lanthanum hexaboride (LaB6) sources, and its accelerating voltage may be lowered significantly, permitting nondestructive imaging of fragile specimens. ESEM-FEG also retains the capabilities for conventional secondary and backscattered electron detection, and the field-emission source permits very high-resolution imaging of coated or naturally conductive samples under normal high-vacuum and high-voltage conditions.

9.8   Summary

The history of the development of SEM, its operating principles, and the limitations and methods for sample preparation has been discussed.

Some of the special features of SEM, such as identification of an element and the morphology of CNMs (e.g., CNBs, CNTs, CNFs), have been explained.

The method for taking SEM micrographs of samples containing some water has been explained, and various applications using SEM (EDAX, EDS, ESEM) were discussed.

CHAPTER 10

Characterization of Nanomaterials by X-Ray Diffraction

Dillip Kumar

Department of Physics, IIT Bombay, Mumbai, Maharashtra, India

Maheshwar Sharon

Nanotechnology Research Center, Birla College, Kalyan, Maharashtra, India

10.1 Introduction

Wilhelm Conrad Röntgen surprised the world in 1895 with the first X-ray picture of his wife's hand (Fig. 10-1). The first kind of scatter process to be recognized was discovered by Max von Laue, who was

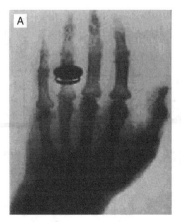
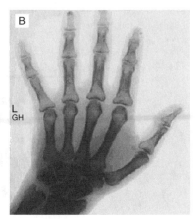

Bertha Rongen's hand (1895)   Modern radiograph of hand

FIGURE 10-1   First picture of Bertha Röntgen's hand taken by X-ray (*A*) and a modern radiograph of a hand (*B*).

awarded the Nobel Prize for physics in 1914 for his discovery of the diffraction of X-rays by crystals. The next year, the father and son team of Sir William Henry and William Lawrence Bragg (Fig. 10-2) were awarded the Nobel Prize for physics for their analysis of crystal structure by means of X-rays. These men were responsible for the famous Bragg's law, which describes the mechanism by which X-ray diffraction (XRD) occurs (Fig. 10-3).

Bragg's law was an extremely important discovery that formed the basis for the whole of what is now known as crystallography. This technique is one of the most widely used structural analysis techniques and plays a major role in fields as diverse as structural biology and materials science.

Max von Laue   Sir William Henry   William Lawrence
(1897–1960)   Bragg (1862–1942)   Bragg (1890–1971)

FIGURE 10-2   Photographs of Max von Laue (*A*), Sir William Henry Bragg (*B*), and William Lawrence Bragg (*C*).

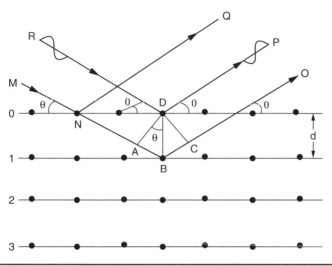

Figure 10-3 Schematic diagram showing diffracted X ray by the ions or atoms lying in the parallel planes separated by a distance d.

10.2   Bragg's Law of Diffraction

When a wave front of X-rays impinges on a crystal, each atom scatters the X-rays in all directions. Let us consider X-rays MN and RD (see Fig. 10-3) incident on a surface of a crystal for explaining the Bragg's law of diffraction. The incoming X-rays MN and RD are incident at an angle of θ to the plane of the surface of the sample. X-ray RD is deflected at an angle θ, forming X-ray DP. These two X-rays are deflected such that X-rays NQ and DP are parallel to each other (i.e., both are forming an angle θ with the surface of the sample). On the other hand, X-ray MN can either get reflected from the surface forming X-ray NQ at an angle θ or it can enter the lattice of the crystal and get reflected by the lattice point B, forming X-ray BO. We need to examine the conditions under which the scattered ray BO comes out of crystal's surface at an angle θ.

Because the total path lengths of the rays MNQ and RDP are the same, these rays are said to scatter in phase with each other; that is, the waves of the individual rays arriving at Q and P form a common wave front. This is the condition for scattering in phase by one plane in a crystal. If they were not in phase, destruction in intensity would take place, causing intensities of reflected or diffracted rays of almost 0 intensities or very low intensities.

To make RDP, MNQ, and MBO in phase, the distance traveled by the rays MBO should be such that all emerge out of the crystal in same angle θ. Because MBO has traveled different distance than X-ray RDP, we need to find out the condition that makes both rays emerge at the same angle θ.

Let us assume that d (the spacing between two consecutive planes) is the hypotenuse of the right triangle ABD. With the help of trigonometry, d and θ can be related to the extra distance (i.e., AB + BC) that MBO has traveled. Because the distance AB = BC, we have AB = BC = BD sin θ = d sin θ.

Because BD distance is the spacing between the two consecutive planes (i.e., d), the path difference is $2d$ sinθ.

If MBO is to arrive at Q, P, and O in phase with rays MNQ and RDP, that is, if the two planes are to scatter in phase, then the path difference (i.e., 2BC) must be an equal integral number of wavelengths, $n\lambda$, where $n = 0, 1, 2, 3, 4 \dots n$. Thus, the condition for in-phase diffraction by a set of parallel planes of a crystal is:

$$n\lambda = 2d \sin \theta \dots \tag{10.1}$$

This is the Bragg's law of diffraction.

Thus, if θ and λ are known, the spacing between the two planes can be found out. Equation 10.1 suggests that the diffraction intensities can build up only at certain values of θ, corresponding to a specific value of λ and d. This is so because the wavelets scattered from various points in the crystal have a common wave front only at these angles. Consequently, the amplitudes of all the individual wavelets add up to give a resultant wave having a maximum amplitude possible, as shown in Fig. 10-3. The diffraction can take place by atoms or ions present in various planes. But the most intense wave would appear when it is diffracted from plane 1 (i.e., when $n = 1$).

Therefore, for general purposes, n is taken as equal to 1. Moreover, the intensities of the wave would be proportional to the number of atoms or ions present in the plane of interest. Thus, depending on the arrangement of atoms in the unit cell of the crystal, the magnitude of the intensities of the X-rays can be observed.

A typical spectrum of intensities of X-ray versus θ angle is shown in Fig. 10-4. This figure suggests that the plane corresponding to 2θ at 10, 26, and 32 contains the largest number of atoms of all the planes.

10.3   X-Ray Equipment

In Fig. 10-5, a photograph of an XRD machine is shown. It has the facilities for either a rotating detector or an X-ray tube to keep the sample holder fixed.

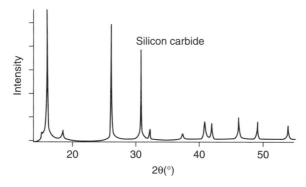

FIGURE **10-4**   X-ray diffraction of cubic SiC, which is plotted between intensities of X-ray waves obtained at different 2θ angles.

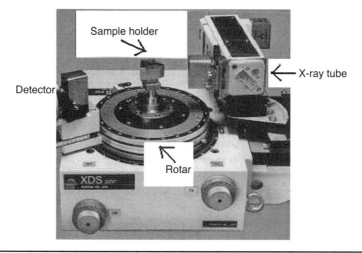

FIGURE **10-5**   A photograph of an X-ray diffraction machine. This diffractometer includes a two-circle (2θ, ω) rotor (goniometer), Cu target X-ray tube, sample holder that can be rotated 360 degrees, and detector.

10.4   Choice of Wavelength

The choice of wavelength depends on the crystal whose XRD is to be taken. Wavelengths of X-ray should be greater than the d spacing; otherwise, there would be no reflection of X-rays. Laboratory XRD equipment relies on the use of an X-ray tube, which is used to produce the X-rays. In general, X-ray tubes use a copper anode, but cobalt and molybdenum are also popular. In some cases, Mo (Kα) is used because its wavelength is 0.071 nm, and it can be used for materials whose d spacing is smaller than 0.15412 nm. The wavelengths in nanometers for various sources are given in Table 10-1.

Element	$K_{\alpha 1}$ (very strong)	$K_{\alpha 2}$ (strong)
Cr	0.228970	0.229361
Fe	0.193604	0.193998
Co	0.178897	0.179285
Cu	0.154056	0.154439
Mo	0.070930	0.071359

TABLE 10-1   X-Ray Wavelengths from Different Target Materials

10.5   Preparation of a Powdered Sample for X-Ray Diffraction

There are many ways of making specimens to suit particular requirements. For most of the work, fine powder may be thinly and evenly spread over double-sided adhesive tape (Fig. 10-6). The other side of the tape is attached to a glass plate of small size depending on the dimensions required by the XRD machine.

After spreading the powder over the tape, the glass plate is tapped to remove any excess powder which is adhered to the tape. It is advisable to make fine powder of the sample by rubbing it in a mortar and pestle for few minutes. If a substance is deliquescent, efflorescent, or otherwise unstable under normal atmospheric conditions, it must be enclosed in a narrow long tube (preferred diameter, 0.3 to 0.5 mm) that is transparent to X-rays (e.g., borosilicate glass). For high-temperature work, silica glass is used. Metal and alloys can be directly measured by using their fine wire.

If some comparative work is to be carried out with several samples, care must be taken to insert the specimen to the XRD machine at exactly same position, and its inclination with respect to the X-rays should also be the same; otherwise, there is a possibility of observing a slight shift in the 2θ angle.

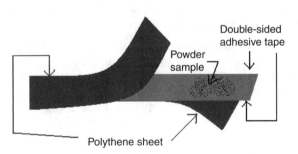

FIGURE 10-6   Double-sided adhesive tape used for powdered specimen preparation for X-ray diffraction.

10.6   X-Ray Diffraction of a Powdered Sample

Many nanomaterials are obtained only in powder form where crystals are randomly oriented. Diffraction patterns from such samples consist of lines and are called *powder patterns*. For a given wavelength, incident rays reflected by a particular set of planes are deviated by 2θ, where θ is the Bragg angle. In Fig. 10-7, it is shown that for the reflected X-ray to make an angle similar to the incidence X-ray (i.e., θ), the reflected X-rays will make a 2θ angle with respect to the incidence X-rays. Therefore, a system of randomly oriented crystals reflected from corresponding sets of planes will all be deviated by 2θ from the direction of the primary beam.

The intensities of the reflected rays could be measured either by a photomultiplier tube and plotting the number of photons per second versus 2θ, as shown in Fig. 10-4, or by exposing an X-ray film (Fig. 10-8).

It is worth noting that in taking XRD of powdered specimens, it is assumed that there are sufficient crystals in all orientations to give rise to smooth lines. For this, it is expected that the size of the crystal would be smaller than 1 μm. When the grains are too large, there is not enough room within the irradiated volume for a sufficient number of crystals to lie in all possible orientations. The reflection will still lie on a curve of constant θ, but the resulting powder reflection will no longer be a smooth line, and with a very coarse-grained specimen, the line may even be a series of spots. By rotating the powder during the exposure to X-ray, the number of crystals contributing to the formation of each powder lines is considerably increased, and the lines become smoother.

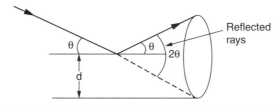

FIGURE 10-7   A schematic diagram showing formation of cone by the reflected rays.

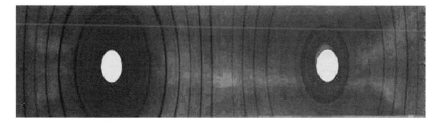

FIGURE 10-8   X-ray film obtained from an X-ray camera.

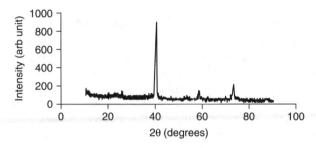

FIGURE 10-9 A typical X-ray diffraction spectrum of diamond thin film showing one strong peak at 2θ angle of 40 degrees and other two small peaks at the other two angles.

10.6.1 Measurement of Bragg's Angle

XRD instruments are provided with photomultiplier tubes to record the intensities of the diffracted X-rays for each 2θ angle. In Fig. 10-9, a typical XRD of diamond film is shown.

10.6.2 Crystallite Size

Line profile analysis is a diffraction technique used to obtain microstructural information of the sample averaged over the diffraction volume. A polycrystalline material that does not contain lattice strain and consists of particle sizes larger than 500 nm shows sharp lines in a powder diffractogram. Imperfections in the structure of the crystallites constituting a sample cause broadening of the diffraction line. Large crystallites give rise to sharp peaks; as the crystallite size reduces, the peak width increases, and the intensity decreases (Fig. 10-10). Peak broadening may also originate from variations in lattice spacing caused by lattice strain. Several analysis methods exist to calculate crystallite sizes and lattice strain separately from diffraction line broadening.

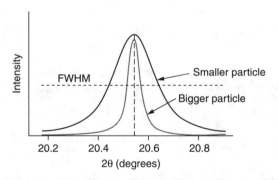

FIGURE 10-10 A graph showing the variation in size of crystal and its effect on the width of 2θ peak. FWHM = full width at half maximum.

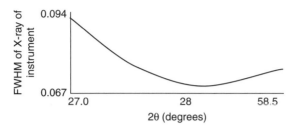

FIGURE 10-11   Full width at half maximum (FWHM) for Cu (K_α).

From the broadening of the peak, it is possible to determine an average crystallite size with the Debye-Scherrer formula:

$$D_{hkl} = \frac{k\lambda}{\beta_\theta \cos\theta} A^0 \tag{10.2}$$

where k varies in the range of 0.8 to 1.39 (usually close to unity), λ-wavelength of the radiation, β_θ is FWHM (full width at half maximum, or half-width) in radiance of XRD peak obtained at 2θ, and D_{hkl} is the crystallite size in Å.

To be more accurate, it has been suggested that the FWHM for the X-ray of the instrument should also be determined and that the value for the corresponding 2θ value should be deducted from the actual FWHM of the spectrum. In other words, β should be calculated:

$$\beta = (FWHM^2_{measured} - FWHM^2_{instrument})^{1/2} \tag{10.3}$$

From Fig. 10-11, the FWHM for the X-ray can be determined for the desired 2θ and inserted into Eq. (10.4) to obtain the actual β value. An error for the crystallite size by this formula can be up to 50%.

10.6.3   Indexing from an X-Ray Diffraction Pattern

The indexing method depends whether one prefers to index the XRD patterns as shown in Figs. 10-4 and 10-9. Indexing of lattice parameters from XRD spectra (see Fig. 10-4) is complicated, but computer programs available for this purpose make the calculation simple. From the XRD pattern (see Figs. 10-4 and 10-9), d values are calculated for each 2θ value. For example, in Fig. 10-4, three lines have high intensities, and other eight have very low intensities. Normally, for making the calculation easy, the first three lines that have good intensities are selected. However, the XRD spectra of diamond film shown in Fig. 10-9 have only one line with high intensity, and the other two lines are of low intensity. Nevertheless, in both cases, one would calculate the d spacing for three lines using Bragg's law. It is worth noting that whereas the XRD spectra show the scale in terms of 2θ, in Bragg's law, only the θ value is needed for calculation of the *d* spacing. Thus, one can select the first three lines with the highest intensities and tabulate the *d* and θ values.

If the values of *a*, *b*, and *c* and planes are not known, then the procedure of reiteration is done using the computer program. With the help of computer calculation, the experimentally observed *d* spacing is assigned for the line to which it belongs and the value of the corresponding lattice parameter. If the structure is simple cubic, then the value of a is assigned; otherwise, one would obtain the value of *a*, *b*, and *c* for other types of structures (Table 10-2). For this type of calculation, a suitable imaginary value of a for an imaginary plane such as (100) (i.e., $h = 1$, $k = 0$, and $l = 0$) is fed into the computer program. This calculates the corresponding *d* value. The computer program tries to match the theoretically calculated values with the experimentally observed d values for the particular (*hkl*) plane with the observed XRD spectra and finally gives all the parameters and the type of structure the compound being examined (see Table 10-2).

In this fashion, we can assign each peak of XRD with a particular a value and a particular (*hkl*) plane. The process becomes complicated if the data do not fit into cubic structure because other types of structures are complicated and require a longer calculation (see Table 10-2). A readymade program is available in which one can insert the experimental *d* values and the corresponding θ. The computer tries to calculate all possible combinations and gives the final results, showing the type of structure, its lattice parameters, and the indexing of each *d* value for the plane. Manually, this calculation would take several days to weeks; the computer can do it in few hours. The relationships between $\sin^2 \theta$ and *d* for common structures are shown in Table 10-2.

Lattice	$\sin^2 \theta$
cubic	$[\lambda^2/4a^2](h^2+k^2+l^2)$
Tetragonal	$\dfrac{\lambda^2}{4a^2}(h^2+k^2)+\dfrac{\lambda^2}{4c^2}l^2$
Hexagonal	$\dfrac{\lambda^2}{3a^2}(h^2+hk+k^2)+\dfrac{\lambda^2}{4c^2}l^2$
Orthorhombic	$\dfrac{\lambda^2}{4a^2}h^2+\dfrac{\lambda^2}{4b^2}k^2+\dfrac{\lambda^2}{4c^2}l^2$
Monoclinic	$\dfrac{\lambda^2}{4a^2\,\sin^2\beta}h^2-\dfrac{\lambda^2\cos\beta\ hl}{2ac\sin^2\beta}+\dfrac{\lambda^2}{4c^2\sin^2\beta}l^2+\dfrac{\lambda^2}{4b^2k^2}$

TABLE 10-2   Relationship between $\sin^2 \theta$ and d^2 for a Few Common Structures

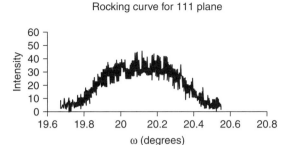

Rocking curve for 111 plane

FIGURE 10-12   Rocking-curve analysis of diamond film obtained for its plane (111).

10.6.4   Quantification Based on Intensity

In this method, XRD of the sample is taken and is compared with the standard XRD of the sample to confirm the material. This method is useful if a standard XRD is available. Most manufacturers of XRD also supply the standard XRD of material available in the literature.

10.7   Rocking-Curve Analysis

Rocking-curve analysis has proven to be particularly valuable because it is relatively fast, requires no sample preparation, and is nondestructive. Rocking-curve analysis is usually used to indicate the quality of the thin film. For this analysis, the detector is fixed at 2θ position. The sample is scanned around θ. The defects in the sample cause the width of the peak to broaden. A rocking curve obtained with diamond film of plane (111) is shown in Fig. 10-12. This curve suggests that the plane (111) is developed, but not very nicely.

10.8   X-Ray Diffraction of Carbon Nanomaterials

One advantage of XRD is getting the structural information about the nature of carbon nanotubes (CNTs). Crystalline CNTs can give a very sharp peak for plane (002). For example, XRD of CNTs synthesized by the pyrolysis of methane on a Ni–Cu–Al catalyst is shown in Fig. 10-13.

Sometimes carbon formed by the pyrolysis of plant-derived precursors does not give a well-defined peak corresponding to any specific plane (Fig. 10-14A). However, after sintering, some materials show the corresponding peak for diamond (Fig. 10-14B and C), whose rocking curve is shown in Fig. 10-12. This is how XRD can help to discover the structural properties of CNMs prepared under different conditions.

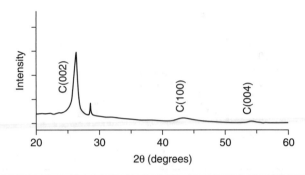

Figure 10-13   X-ray diffraction of CNT showing a characteristic peak of (002) plane.

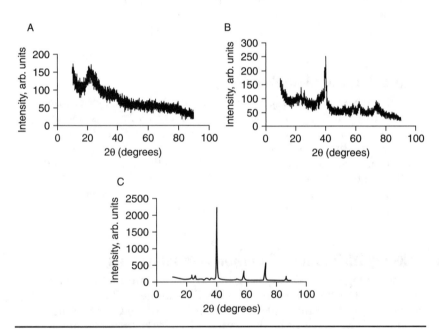

Figure 10-14   X-ray diffraction of carbon film prepared by applying different potentials to tungsten coil of a flash evaporation unit. (*A*) 190 V, substrate temperature of 650°C, showing amorphous carbon. (*B*) 200 V, substrate temperature of 685°C, showing some crystalline structure of the thin film with a low-intensity (111) plane. (*C*) 220 V, substrate temperature of 750°C, showing a well-defined crystalline structure of the thin film with a (111) plane of diamondlike carbon.

10.9   Summary

Synthesis of specific types of CNM requires very accurate setting of the parameters of preparation. Hence, it is necessary to know the structural properties of material obtained at a given condition. XRD is a very simple, quick, and handy technique for such purposes. XRD not only helps to know about the structural property of CNMs, but can also give information about their crystallinity, particle size, and the type of crystallographic planes present in the material. In this chapter, XRD techniques have been explained in detail to familiarize readers with various relevant aspects of this technique.

CHAPTER **11**

Characterization of Carbon Nanomaterials by Raman Spectroscopy

Dillip Kumar

Department of Physics, IIT Bombay, Mumbai, India

Maheshwar Sharon

Nano-Technology Research Centre, Birla College, Kalyan, Maharashtra, India

11.1 History of Raman Spectroscopy

In 1928, the phenomenon of inelastic light scattering was noticed and documented by Indian physicist Sir C. V. Raman, who observed that radiation that had scattered from a sample contained photons identical

to incident photons but also contained photons that had been shifted to a different energy level. The energy-shifted photons were very sparse compared with the number of photons with the same energy as the initial light, but the possibility of this as a quantitative measuring tool was realized soon. Raman spectroscopy became the principal method of nondestructive chemical analysis in the 1930s.

As the 1930s came to a close, scientists had begun to use infrared spectroscopy, a type of absorption spectroscopy, to analyze samples. Most researchers set Raman spectroscopy aside for two reasons. First, the system had to be very carefully aligned and frequently calibrated because the incident wavelength had to be close to perfect to obtain a consistent Raman shift. Wavelength drifting due to excitation fluctuations and alignment issues plagued Raman scientists. Second, Raman's shifted photons had to be observed, almost always in the presence of a large fluorescence background from the sample. Fluorescence is the luminescence that occurs when a sample is illuminated by electromagnetic radiation. The sample absorbs energy, thus exciting the electrons to higher energy orbitals. As the electrons fall back to their ground state, the energy released is in the form of light. This light is emitted over a broadband of frequencies, including the frequencies related to Raman's scattered radiation. This makes it difficult to distinguish between the two types of emitted radiations.

In 1986, the use of a near-infrared (NIR) excitation source sparked a new interest in the process of Raman spectroscopy. With the NIR source, samples are much less likely to fluoresce because there is less energy to excite electrons in the sample. In addition to a lower fluorescence background, the samples can be illuminated longer. With less excitation energy, samples are less likely to photodecompose and heat up over time. A longer illumination corresponds to more Raman-scattered photons and a stronger signal. Because of these factors, Raman is now used in a variety of ways as an analytical tool (Puppels and Jovin, 1990).

11.2   What Is Raman Spectroscopy?

Raman spectroscopy is a spectroscopic technique used in condensed matter physics and chemistry to study vibrational, rotational, and other low-frequency modes in a system. It relies on inelastic scattering, or Raman scattering of monochromatic light, usually from a laser in the visible, NIR, or near-ultraviolet range. Photons or other excitations in the system are absorbed or emitted by the laser light, resulting in the laser photons' energy being shifted up or down. The shift in energy gives information about the phonon modes in the system. Infrared spectroscopy yields similar but complementary information.

Both infrared and Raman spectroscopy measure the vibrational energies of molecules, but these methods rely on different selection rules. In infrared spectroscopy, a continuous beam of infrared radiation is sent through the sample, and the vibrational states are observed from direct absorption of radiation from a lower to an upper vibrational state.

For a vibrational motion to be infrared active, the dipole moment of the molecule must change. Therefore, the asymmetric stretch (rather than the symmetric stretch) is infrared active due to a change in the dipole moment. For example, the symmetric stretch in carbon dioxide is not infrared active because there is no change in the dipole moment (Fig. 11-1).

For a transition to be Raman active, there must be a change in the polarizability of the molecule. On the other hand, Raman spectroscopy deals with the vibrational energies states of materials in terms of energy differences between the incident and scattered radiation caused by inelastic scattering of photons. When light is scattered from a molecule, most photons are elastically scattered (Fig. 11-2A). The scattered photons have the same energy (frequency) and, therefore, the same wavelength as the incident photons (Fig. 11-2B). However, a small fraction of light (~1 in 10^7 photons) is scattered at optical frequencies different from (and usually lower than) the frequencies of the incident photons (Fig. 11-2C). The process leading to this inelastic

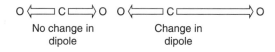

No change in   Change in
dipole   dipole

FIGURE 11-1 Effect of symmetry on the dipole moment. When the bond length of oxygen atoms with carbon are same and in same plane, it gives a symmetrical structure with no dipole, whereas when bond lengths are different, unsymmetrical structure results into giving a dipole.

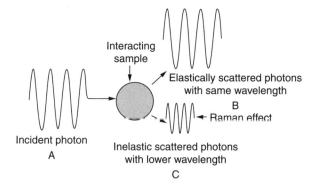

FIGURE 11-2 Schematic diagram showing the effect of interacting light resulting in the same frequency (A) known as Rayleigh scattering (B) and Raman's scattered light with lower frequency due to inelastic scattering (C).

scattering is the termed the *Raman effect*. Because Raman scattering is a two-photon, second-order process and infrared spectroscopy is a one-photon, first-order process, the later is usually more sensitive than the former. Infrared spectroscopy requires more elaborate sample preparation, and the Raman spectrometers can record Raman spectra from almost all kinds of sample. Micro Raman spectroscopy is one of the recent additions in this field.

Vibrational spectroscopy of molecules can be relatively complicated. Quantum mechanics requires that only certain well-defined frequencies and atomic displacements are allowed. These are known as the *normal modes of vibration* of the molecule. A linear molecule with N atoms has $3N - 5$ normal modes, and a nonlinear molecule has $3N - 6$ normal modes of vibration.

Raman spectroscopy is based on the inelastic scattering of photons by molecules. The Raman effect arises when a photon is incident on a molecule and interacts with the electric dipole of the molecule. In classical terms, the interaction can be viewed as a perturbation of the molecule's electric field. In quantum mechanics, the scattering is described as an excitation to a virtual state lower in energy than a real electronic transition with nearly coincident de-excitation and a change in vibrational energy. The scattering event occurs in 10^{-14} seconds or less. The virtual state description of scattering is shown in Fig. 11-2.

The energy difference between the incident and scattered photons is represented by the arrows of different lengths in Fig. 11-3A Numerically, the energy difference between the initial and final vibrational levels, μ, or Raman shift in wave numbers (in cm^{-1}), is calculated by:

$$\mu = 1/\lambda_{incident} - 1/\lambda_{scattered} \qquad (11.1)$$

where $\lambda_{incident}$ and $\lambda_{scattered}$ are the wavelengths (in centimeters) of the incident and Raman scattered photons, respectively. The vibrational

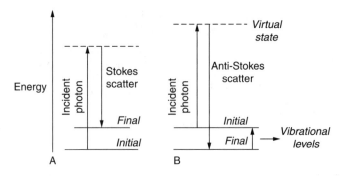

Figure 11-3   Energy level diagram for Raman scattering. (*A*) Stokes Raman scattering. (*B*) Anti-Stokes Raman scattering.

energy is ultimately dissipated as heat. Because of the low intensity of Raman scattering, the heat dissipation does not cause a measurable temperature increase in a material.

In Rayleigh scattering, the emitted photon has the same wavelength as the absorbing photon (see Fig. 11-2B), and in Raman scattering, the energies of the incident and scattered photons are different (see Fig. 11-2C). The energy of the scattered radiation is less than the incident radiation for the Stokes line (see Fig. 11-3A), and the energy of the scattered radiation is more than the incident radiation for the anti-Stokes line (see Fig. 11-3B). The energy increase or decrease from the excitation is related to the vibrational energy spacing in the ground electronic state of the molecule, and therefore the wave number of the Stokes and anti-Stokes lines are a direct measure of the vibrational energies of the molecule. A plot of intensity of scattered light versus energy difference is a Raman spectrum. A simplified energy diagram that illustrates these concepts is given in Fig. 11-4.

In the spectrum (see Fig. 11-4), it can be noticed that the Stokes and anti-Stokes lines are equally displaced from the Rayleigh line. This occurs because in either case, one vibrational quantum of energy is gained or lost. Also, the anti-Stokes line is much less intense than the Stokes line. This occurs because only molecules that are vibrationally excited before irradiation can give rise to the anti-Stokes line. Hence, in Raman spectroscopy, only the more intense Stokes line is normally measured.

The ratio of anti-Stokes to Stokes intensity at any vibrational frequency is a measure of temperature. Anti-Stokes Raman scattering is used for contact less thermometry. The anti-Stokes spectrum is also used when the Stokes spectrum is not directly observable, such as because of a poor detector response or spectrograph efficiency. The typical energy of vibrational levels ranges from 10 to 500 mV depending on the mass of the vibrating atoms and the force constant for different types of bonds between atoms. In gases and liquids, the characteristic levels belong to the atomic vibration in individual molecules; in solids, they may represent vibration of individual atoms as well as vibration of any rigid or semi-rigid molecular unit present in the unit cell. These vibrational states are characteristics features of

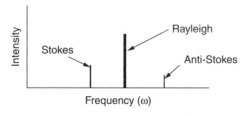

FIGURE 11-4   Schematic diagram showing variation in intensity of Stokes, anti-Stokes, and the Rayleigh lines and their respective frequencies.

these materials; hence, vibrational spectroscopy can be used to characterize or identify materials.

The energy of a vibrational mode depends on the molecular structure and environment. Atomic mass, bond order, molecular substituents, molecular geometry, and hydrogen bonding all affect the vibrational force constant, which in turn dictates the vibrational energy. For example, the stretching frequency of a phosphorus–phosphorus bond ranges from 460 to 610 to 775 cm^{-1} for the single-, double-, and triple-bonded moieties, respectively. Crystal lattice vibrations and other motions of extended solids are Raman active. Their spectra are important in characterizing polymers and semiconducting materials.

Experimentally, we only observe the Stokes shift in a Raman spectrum. Because the Stokes lines will be at smaller wave numbers (or higher wavelengths) than the exciting light, the Raman scattering needs a high-power excitation source such as a laser. Because the energy (wave number) difference between the excitation and the Stokes lines is measured, the excitation source should be monochromatic.

11.3   Instrumentation

A typical Raman scattering setup consists of an excitation source, a spectrometer, a detector, and detection electronics with output devices (Fig. 11-5).

11.3.1   Excitation Sources

The excitation sources are usually continuous wave sources in the visible light region, and the principal requirements are that they should be highly monochromatic (width <0.1 cm^{-1}), directional (beam divergence <1 mrad), and power density (>a few kW/cm^2). The continuous wave (CW) Ar$^+$ or Kr$^+$ lasers with many discretely tunable lines in the full visible range are the main sources. The most common excitation sources are 488.0, 514.5, 532.0, and 789.0 nm laser lines in the visible region.

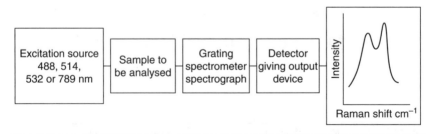

FIGURE 11-5   Schematic representation of various units of Raman spectrometers.

11.3.2   The Spectrometer

The Raman-scattered light is analyzed using a grating spectrometer or spectrograph. The scattered radiation is collected over a large solid angle by a collection lens and is focused on the entrance slit of the spectrometer. This is collimated by a concave mirror to fall on the grating, which spatially separates the Raman lines. These are then focused by another concave mirror on the exit slit, and different Raman lines are recorded by rotating the grating so that they are transmitted by the exit slit in a sequential manner. In a spectrograph, the exit slit is replaced by a broad window with a multichannel detector, which can record the full spectrum over a finite range simultaneously.

The principal requirement of a Raman spectrometer is its stray light rejection capacity. The origin of stray light is from a Rayleigh-scattered radiation, which always accompanies the Raman-scattered light and is approximately 10^3 to 10^5 times stronger and enters the spectrometer and gets scattered in random direction from the mirror and grating and dust particles on various optical components. A small fraction of Rayleigh light may appear at the exit slit position corresponding to low-frequency Raman shift and may completely mask a weak Raman line. However, with the advent of holographic notch filters and sharp, cut-off edge filters, it has become possible to use single-stage spectrometers for Raman spectroscopy. A notch filter has high and sharp absorption about laser light that when placed in front of the entrance slit almost completely blocks the Rayleigh-scattered light, transmitting the Raman-scattered radiation. These filters have finite cut-off range, and Raman shifts below 150 cm^{-1} cannot be observed. Whereas the notch filters allow observation of both stokes and anti-Stokes radiation, edge filters normally allow observation of the Stokes component only.

Another requirement of a Raman spectrometer is the resolving power of the spectrometer, which depends on the number of lines on the grating, the focal length of the monochromator, the number of stages of the monochromator, and the wavelength region. For a typical double monochromator, the wavelength region is 20 cm^{-1}/mm.

11.3.3   Detectors

Until the late 1970s, the most common detector in a scanning spectrometer was a photomultiplier tube (PMT), which had a wide spectral range of 300 to 900 nm and a uniform spectral response. Because the number of Raman-scattered photons is very few, the PMT is used in the photon-counting mode, and the Raman spectrum is represented (see Fig. 11-5) by a plot of intensity as a function of Raman shifts. To get a good signal:noise ratio, the dark count is reduced to a minimum (5 to 10 cps) by cooling the photocathode to about $-30°C$.

The detectors used in present-day (multichannel spectrograph) are charged coupled devices (CCDs) that detect the incident photon by generation of electron–hole pairs in silicon by a photon of wavelength less than 1100 nm. A CCD is a two-dimensional array of potential wells that stores the generated photoelectrons, which are read out in a very complicated fashion. A typical spectroscopic CCD is made up of 1024 (H) × 256 (V) pixels measuring about 25 × 6 mm. The maximum quantum efficiency for a front-illuminated CCD is about 50%; it can be as high as about 95% for a back-thinned device. The multichannel detector systems caused a breakthrough in detecting spectral distributions with a very low intensity level and are a prerequisite for the Raman analysis of interfaces and nanostructures.

11.4   Raman Selection Rules and Intensities

A simple classical electromagnetic field description of Raman spectroscopy can be used to explain many of the important features of Raman band intensities. The dipole moment, P, induced in a molecule by an external electric field, E, is proportional to the field:

$$P = \alpha\, E \qquad\qquad (11.2)$$

The proportionality constant α is the polarizability of the molecule. The polarizability measures the ease with which the electron cloud around a molecule can be distorted. The induced dipole emits or scatters light at the optical frequency of the incident light wave. Raman scattering occurs because a molecular vibration can change the polarizability. The change is described by the polarizability derivative, $\delta\alpha/\delta Q$, where Q is the normal coordinate of the vibration. The selection rule for a Raman-active vibration is that there is a change in polarizability during the vibration:

$$\delta\alpha/\delta Q \neq 0 \qquad\qquad (11.3)$$

The Raman-selection rule is analogous to the more familiar selection rule for an infrared-active vibration, which states that there must be a net change in permanent dipole moment during the vibration. From group theory, it is straightforward to show that if a molecule has a center of symmetry, vibrations that are Raman active will be silent in the infrared rays and vice versa. The scattering intensity is proportional to the square of the induced dipole moment, that is, to the square of the polarizability derivative, $(\delta\alpha/\delta Q)^2$.

If a vibration does not greatly change the polarizability, then the polarizability derivative will be near 0, and the intensity of the Raman band will be low. The vibrations of a highly polar moiety, such as the O–H bond, are usually weak. An external electric field cannot induce a large change in the dipole moment, and stretching or bending the bond does not change this.

Typical strong Raman scatterers are moieties with distributed electron clouds, such as C–C double bonds. The π-electron cloud of the double bond is easily distorted in an external electric field. Bending or stretching the bond changes the distribution of electron density substantially and causes a large change in the induced dipole moment.

Chemists generally prefer a quantum mechanical approach to Raman scattering theory, which relates scattering frequencies and intensities to the vibrational and electronic energy states of the molecule. The standard perturbation theory treatment assumes that the frequency of the incident light is low compared with the frequency of the first electronic excited state. The small changes in the ground state wave function are described in terms of the sum of all possible excited vibronic states of the molecule.

11.5   Polarization Effects

Raman scatter is partially polarized, even for molecules in a gas or liquid, in which the individual molecules are randomly oriented. The effect is most easily seen with an exciting source that is plane polarized. In isotropic media, polarization arises because the induced electric dipole has components that vary spatially with respect to the coordinates of the molecule. Raman scatter from totally symmetric vibrations is strongly polarized parallel to the plane of polarization of the incident light. The scattered intensity from nontotally symmetric vibrations is 75% as strong in the plane perpendicular to the plane of polarization of the incident light as in the plane parallel to it.

The situation is more complicated in a crystalline material. In this case, the orientation of the crystal is fixed in the optical system. The polarization components depend on the orientation of the crystal axes with respect to the plane of polarization of the input light, as well as on the relative polarization of the input and the observing polarizer.

11.6   Application of Raman Spectroscopy in Carbon Nanotechnology

Characterization of carbon-based materials using Raman spectroscopy has created new opportunities for materials research. Carbon nanomaterials (CNMs), such as single-walled carbon nanotubes (SWCNTs), multiwalled carbon nanotubes (MWCNTs), nanocrystals, and nanodiamond films, are difficult to characterize by any other method. Raman spectroscopy has opened an efficient technique to characterize them (Ferrari and Robertson, 2000; Sun, 2000). The biggest advantage of this technique is that it requires practically no sample preparation. With the advent of micro Raman techniques, very little amount of sample is needed and can be used for checking the spatial homogeneity of the samples with submicron resolution. Raman spectroscopy is the simplest and least destructive characterization tool for carbon nanotubes (CNTs) compared with techniques such as high-resolution transmission electron micrography and scanning tunneling microscope (STM), which require elaborate sample preparation.

Based on the symmetry of the CNMs, the A_{2u} and E_{1u} modes are infrared active, but the A_{1g}, E_{1g} and E_{2g} modes are Raman active. It is important to realize that number of infrared- and Raman-active modes is independent of the diameter of the nanotubes. However, the frequencies of these modes do vary with the CNT's diameter. The strong lines between 1550 and 1600 cm^{-1} may be assigned to three E_{1g}, E_{2g} and A_{1g} modes in CNTs with different diameters and in special strong modes of 1567 and 1593 cm^{-1} have been assigned to the E_{2g} and E_{1g} or A_{1g}, respectively. It is important to note that the Raman intensity for graphite in the 1300 cm to 1600 cm^{-1} region is sensitive to the sample's quality. Generally, the intensity of the E_{2g} modes of graphitic materials is sharp and strong when the sample is highly crystalline and free of defects. Disordered graphite and carbons show a broad features around 1350 cm^{-1}. The 1347 cm^{-1} signal may come from a symmetry-lowering effect, due to defects or nanotube caps, bending of the nanotubes, or the presence of carbon nanoparticles and amorphous carbon.

If we examine the Raman spectra of pure diamond and pure graphite (HOPG), it will be noticed that diamond shows one sharp peak around 1350 cm^{-1}, but graphite shows only one peak around 1550 cm^{-1}. Diamond is expected to contain 100% pure sp^3-configured carbon, whereas graphite contains 100% pure sp^2 carbon. Hence, it can be concluded that Raman peak around 1350 cm^{-1} (known as the D band) is an indication of sp^3 carbon and peak around 1550 cm^{-1} (known as the G band) corresponds to sp^2 carbon. A Raman peak around 1350 cm^{-1} in graphite is normally considered a disorder created in graphite. But the nature of the disorder in graphite can be due to the presence of sp^3 carbon. In other words, the intensity of peak (or peak areas) of around 1350 cm^{-1} obtained with graphite may be considered as equivalent to the amount of sp^3 carbon present in the graphite. Similarly, the presence of peak around 1550 cm^{-1} in diamond materials should be considered as the amount of sp^2 disordered present in the diamond. Diamondlike carbon, which is assumed to contain both sp^2 and sp^3 carbon and Raman spectra, does show the presence of both bands (i.e., the presence of peak at 1350 and 1550 cm^{-1}). The E_{2g} mode of graphite at 1582 cm^{-1} thus corresponds to C = C bond stretching motions for one of the three nearest neighbor bonds in the unit cell of CNTs.

At 186 cm^{-1}, a strong A$_{1g}$ breathing mode is found. This peak has strong dependence (i.e., the frequency of the A$_{1g}$ breathing mode) on the CNT diameter. Hence, the frequency of the A$_{1g}$ breathing mode can be used as a marker for assigning the approximate diameter of the CNTs. In a very low-frequency region below 30 cm^{-1}, a strong low-frequency Raman-active E$_{2g}$ mode is expected. However, it is difficult to observe Raman lines in the very low-frequency region, where the background Rayleigh scattering is very strong.

Among the several techniques used to characterize SWCNTs, Raman spectroscopy is perhaps the most powerful tool to get information on their vibrational and electronic structure. The most important observed Raman features of SWCNTs are (1) the radial breathing mode (RBM; Fig. 11-6A), whose frequency varies according to the diameter of the tube; (2) the tangential G band (Fig. 11-6B) in the

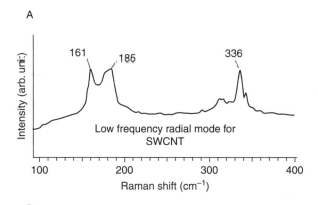

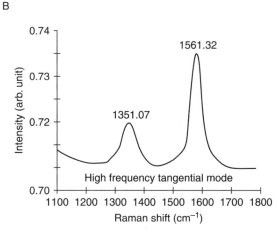

Figure 11-6   Raman spectra of single-walled carbon nanotubes (radial breathing mode; (A) and multiwalled carbon nanotubes (B) at higher frequency showing the G band at 1561.32 cm^{-1} and the D band at 1351.07 cm^{-1}.

range of 1550 to 1605 cm^{-1}; and the disordered-induced D band (Fig. 11-6B) at about 1350 cm^{-1} and its second-order harmonic (G band) at about 2700 cm^{-1}. The RBM has been found to be inversely proportional to the diameter of the tube ($\upsilon = 248/d_t$) and is very sensitive to the charge transfer and tube–tube interaction in a bundle. This makes the RBM a valuable probe for the structure and properties of SWCNTs and SWCNT-based materials. The RBM frequencies of all types of SWCNTs fall on a common line according to the expression:

$$\upsilon = 223.75/d \qquad (11.4)$$

where υ = frequency shift in units of cm to the power –1
d = CNT diameter in nanometers.

However, a slight correction is being made in this formula, taking the intertube coupling when the CNTs are present in a bundle:

$$\upsilon = 238/d^{0.93} \qquad (11.5)$$

where υ is in units of cm^{-1} and d is the CNT diameter in nanometers.

Despite all the interesting information that can be obtained from Raman spectroscopy, this technique has an intrinsic limitation related to its resonant character, which gives information only of the CNTs that are in resonance with the incident or scattered light. Therefore, a full characterization of all the SWCNTs and MWCNTs in the sample with different diameters and chiralities cannot be achieved.

11.7   Summary

Analysis of material by Raman spectra, developed by C. V. Raman, is a nondestructive technique that uses very little amount of material with minimal sample preparation. The equipment is composed of an excitation source, a spectrometer, and a detector. There are certain fixed rules to select and analyze the Raman spectra. Raman scatter is partially polarized. The polarization components depend on the orientation of the crystal axes with respect to the plane of polarization of the input light.

This technique gives a clear picture about the CNTs, revealing whether the material is pure multiwalled or single-walled or mixed with amorphous carbon. The analysis of Raman spectra taken between 1100 to 2000 cm^{-1} gives information about the nature of carbon structure (i.e., whether it is pure diamond or is mixed with graphitic nature). The ratio of peak obtained at the D band and G band and the shift in their peak position with respect to spectra obtained with pure diamond and pure graphite can also reveal structural information of the materials. Raman spectra taken in the range of 100 to 400 cm^{-1} gives information whether the materials are SWCNTs or MWCNTs.

CHAPTER **12**

Novel Characteristics of Nanocarbon

Madhuri Sharon

MUNAD Nanotech Pvt. Ltd., Mumbai, India

Samrat Paul

Tezpur University, Assam, India

Harish Dubey

Nanotechnology Research Centre, Birla College, Kalyan, Maharashtra, India

12.1   Introduction

Various unique nanoforms of carbon with remarkable electronic and mechanical properties have been synthesized by different procedures. Interest from the research community first focused on the exotic electronic properties of carbon nanoforms; carbon nanotubes (CNTs) may be considered as prototypes for a one-dimensional quantum wire. Over the years, other useful properties have been discovered, particularly their strength, which has created interest among scientist not only to study these properties but also to use them for various useful applications, such as in electronics to strengthen polymer materials, in diagnostics, and in medicine.

The structure and synthesis of carbon nanoforms have been discussed in detail in previous chapters. This chapter focuses on the unique properties of carbon nanomaterials (CNMs).

The morphology and measurements of the size of CNTs and other nanoforms, including carbon nanofiber (CNF) and carbon nanobeads (CNB), and the chiral angle have been made with scanning-tunneling microscopy (SEM) and transmission electron microscopy (TEM). However, it remains a major challenge to determine the thickness and surface area at the same time for measuring the physical properties such as resistivity. This is partly because CNTs are small, and they can be damaged by the electron beam in the microscope.

Because each unit cell of a CNT contains a number of hexagons, each of which contains six carbon atoms, the unit cell of a CNT contains many carbon atoms. If the unit cell of a CNT is N times larger than that of a hexagon, the unit cell of the CNT in reciprocal space is $1/N$ times smaller than that of a single hexagon.

In a scanning electron microscope, the CNT looks like a mat of carbon ropes. The ropes are between 10 and 20 nm across and up to 100 µm long. When examined in a transmission electron microscope, each rope is found to consist of a bundle of single-walled carbon CNTs (SWCNTs) aligned along a single direction. X-ray diffraction, which views many ropes at once, also shows that the diameters of the SWCNTs have a narrow distribution with a strong peak.

X-ray diffraction measurements have shown that bundles of SWCNTs form a two-dimensional triangular lattice. The lattice constant is 1.7 nm, and the tubes are separated by 0.315 nm at the closest approach. Various studies have shown that the width and peak of the diameter distribution depend on the composition of the catalyst, the growth temperature, and various other CNM synthesis parameters.

Before reviewing the various properties of CNTs, it may be useful to examine the structural aspects of CNTs. Rolling up of a honeycomb graphite sheet of specified size gives the middle cylindrical part of CNTs with both ends capped by half a fullerene molecule (Fig. 12-1).

The structure of each CNTs is defined by a unit cell, which is the smallest group of atoms that constitutes it. Mathematically, it is

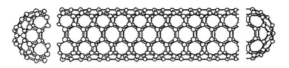

FIGURE 12-1 Model of single-walled carbon nanotube. (The left and right end is half of a fullerene.)

defined by a vector C_h. The vector C_h, called the *chirality* or *chiral vector*, is expressed as:

$$C_h = na_1 + ma_2 \qquad (12.1)$$

where a_1 and a_2 are the unit cell vector of the graphite sheet that rolls up to form nanotube and n and m are the integers. It has been seen that for all the zigzag tubes, $m = 0$, and for all the armchair tubes, $m = n$. For any other combination of m and n, the tubes are chiral types (Fig. 12-2).

Not only integers, but also the chiral angle may determine the type of carbon CNTs, whether they are armchair, zigzag, or chiral tubes. For armchair tubes, $n = m$ and $\theta = 30$ degrees; for zigzag tubes, $m = 0$ and $\theta = 0$ degrees; and chiral tubes, θ ranges from 0 to 30 degrees. All SWNTs are normally described by two integers (n, m), as described above, that specify the tube diameter (Dresselhaus, 2001), which is:

$$d_t = \sqrt{3}\, a_{C-C}\, (n^2 + mn + m^2)^{1/2} / \pi \qquad (12.2)$$

where a_{C-C} is the nearest-neighbor C–C distance (0.142nm), and the chiral angle ϕ is given by:

$$\phi = \tan^{-1} [\sqrt{3}m / (m + 2n)] \qquad (12.3)$$

This gives the orientation of the carbon hexagons with respect to the tube axis.

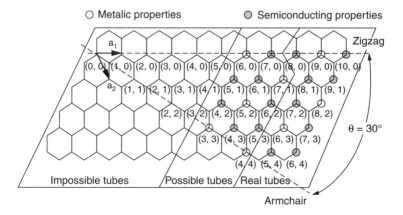

FIGURE 12-2 A self-defined graphene layer with atoms labeled using (n, m) notation and their possible properties.

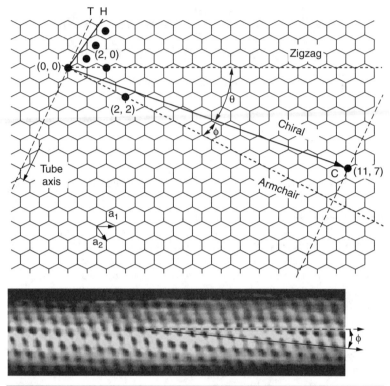

FIGURE 12-3   A (11, 7) graphene sheet and its scanning tunneling microscope (STM) image (Wildoer et al., 1998).

Figure 12-3 shows a graphene sheet and an automatically resolved scanning tunneling microscope (STM) image of (11, 7) CNT. In the STM image, the black dot is attributed to the center of hexagon, which can be used to determine the chiral angle (i.e., the angle between the hexagon rows [H] and the tube axis [T]).

It has been predicted that all of the armchair ($n = m$) tubes are metallic (Dresselhaus et al., 1996). But in the case of zigzag and chiral tubes, two possibilities exist. When $n - m = 3l$ (where l is an integer), the tubes are expected to be metallic, but when $n - m \neq 3l$, the tubes are predicted to be semiconducting with an energy gap of about 0.5 eV. This gap depends on the diameter of the tube:

$$E_{gap} = 2\gamma_0 a_{c-c} / d_t \qquad (12.4)$$

where γ_0 = C–C tight-binding overlap energy

a_{C-C} = nearest distance between two neighboring carbon atoms (0.142 nm)

d_t = tube diameter (Odom et al., 1998; Wildoer et al., 1998).

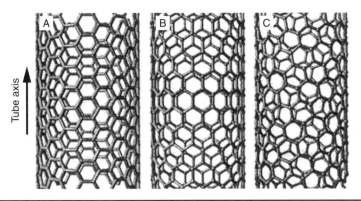

Tube axis

FIGURE 12-4 Types of carbon nanotubes: armchair (A), zigzag (B), and chiral (C) (Terrones et al., 1999).

This energy gap is due to curvature-related effects, which have been discussed by Ouyang et al. (2001) and Collins et al. (2001) in detail. These three types of CNTs are shown in Fig. 12-4.

All CNTs are formed by repeating a unit structure called a *unit cell* of the nanotube. Fig. 12-5 shows the unit cell diagram of a (6, 6) armchair tube and a (12, 0) zigzag tube. Each unit cell of a specified nanotube (n, m) consists of $2N$ number of carbon atoms. N can be calculated with the following formulas:

$$N = 2(n^2 + m^2 + nm)/d_H \qquad \text{if } n - m \neq 3rd_H \qquad (12.5)$$

and

$$N = 2(n^2 + m^2 + nm)/3d_H \qquad \text{if } n - m = 3rd_H \qquad (12.6)$$

where d_H is the highest common divisor of n and m, and r is an integer. A tube denoted by (80, 67) with a 10-nm diameter has unit cell of length 54.3 nm and contains 64,996 carbon atoms (Harris, 1999).

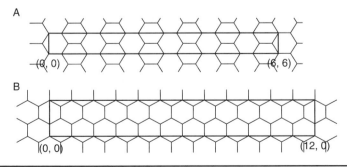

A

(0, 0)   (6, 6)

B

(0, 0)   (12, 0)

FIGURE 12-5 Unit cell of (6, 6) armchair (A) and (12, 0) zigzag tubes (B).

180 Characterization of Carbon Nanomaterials

These three types of structures and the magnitude of N govern the various physical properties of CNTs. It is important to realize, however, that although this type of information is useful in carrying out any theoretical analysis to predict the physical and chemical properties of CNTs, experimentally it is extremely difficult to observe which forms of CNT have been produced during the synthesis. Moreover, it may also be extremely difficult to synthesize one type of CNT (i.e., zigzag, armchair or chiral type). To confirm the type of CNT produced, high resolution transmission electrom microscopy (HRTEM) is needed, which is not available in every lab. Therefore, instead of classifying CNT under these forms, one should determine the concentration of carbon atoms of configuration of sp^2 and sp^3 (which can be determined experimentally with nuclear magnetic resonance [NMR] and Fourier transform infrared spectroscopy [FTIR]) and the band gap of the CNT (which can be experimentally determined with the optical absorption method). From these two pieces of information, it is possible to predict the physical properties of CNT.

12.2 Electrons in Carbon Nanotubes

The unique electronic properties of CNTs are due to the quantum confinement of electrons normal to the CNT axis. In the radial direction, electrons are confined by the monolayer thickness of the graphene sheet. Around the circumference of the CNT, periodic boundary conditions come into play. For example, if a zigzag or armchair CNT has 10 hexagons around its circumference, the eleventh hexagonal coincides with the first (see Fig. 12-2).

Because of this quantum confinement, electrons can only propagate along the CNT axis, so their wave vectors point in this direction (see Fig. 12-4). The resulting number of one-dimensional conduction and valence bands effectively depends on the standing waves that are set up around the circumference of the CNT. These simple ideas can be used to calculate the dispersion relations of the one-dimensional bands, which link the wave vector to energy from the well-known dispersion relation in a graphene sheet.

Although the choice of n and m determines whether the CNT is metallic or semiconducting, the chemical bonding between the carbon atoms is exactly the same in both cases. This is because of the special electronic structure of a two-dimensional graphene sheet, which is a semiconductor with almost 0 band gap. In this case, the top of the valence band has the same energy as the bottom of the conduction band, and this energy equals the Fermi energy. Because the band gap in semiconducting CNTs is inversely proportional to the tube diameter, the band gap approaches 0 at large diameters, as for a graphene sheet.

Calculations have shown that concentric pairs of metal–semiconductor and semiconductor–metal CNTs are stable.

Nanometer-scale devices could therefore be based on two concentric CNTs or the junction between CNTs. For example, a metallic inner tube surrounded by a larger semiconducting (or insulating) CNT would form a shielded cable at the nanometer scale. One might then envision nanoscale electronic devices made completely from carbon that would combine the properties of metals and semiconductors, without the need for doping.

12.3   Properties of Carbon Nanomaterial

The special nature of carbon combines with the molecular perfection of SWCNTs to endow them with exceptional material properties, such as very high electrical and thermal conductivity, strength, stiffness, and toughness. No other element in the periodic table bonds to itself in an extended network with the strength of the C–C bond. The delocalized π-electron donated by each atom is free to move about the entire structure rather than remain with its donor atom, giving rise to the first known molecule with metallic-type electrical conductivity. Furthermore, the high-frequency C–C bond vibrations provide an intrinsic thermal conductivity higher than even diamond. The properties of CNMs can be categorized as follows:

- High electrical conductivity
- Very high tensile strength
- High flexibility (can be bent considerably without damage)
- High elasticity (~18% elongation to failure)
- High thermal conductivity
- Low thermal expansion coefficient
- Good field emission of electrons
- High aspect ratio (length ~1000/diameter)

12.3.1   Electrical Conductivity

There has been considerable practical interest in the conductivity of CNTs. CNTs with particular combinations of (n, m) (i.e., structural parameters indicating how much the CNT is twisted) can be highly conducting and hence can be said to be metallic. Their conductivity has been shown to be a function of their chirality (degree of twist) as well as their diameter. CNTs can be either metallic or semiconducting in their electrical behavior.

Conductivity in MWCNTs is quite complex. Some types of "armchair"-structured CNTs appear to conduct better than other metallic CNTs. Furthermore, interwall reactions within MWCNTs have been found to redistribute the current over individual tubes non-uniformly, but there is no change in current across different parts of metallic SWCNTs. However, the behavior of ropes of semiconducting SWCNTs is different in that the transport current changes abruptly at various positions on the CNTs.

The conductivity and resistivity of ropes of SWCNTs have been measured by placing electrodes at different parts of the CNTs. The resistivity of the SWCNT ropes was in the order of 10^{-4} ohm-cm at 27°C. This means that SWCNT ropes are the most conductive carbon fibers known. The current density that was possible to achieve was 10^7 A/cm^2; however, in theory, the SWCNT ropes should be able to sustain much higher stable current densities, as high as 10^{13} A/cm^2.

It has been reported that individual SWCNTs may contain defects. Fortuitously, these defects allow the SWCNTs to act as transistors. Likewise, joining CNTs together may form transistor-like devices. A CNT with a natural junction (where a straight metallic section is joined to a chiral semiconducting section) behaves as a rectifying diode (i.e., a half-transistor in a single molecule). It has also recently been reported that SWCNTs can route electrical signals at high speeds (≤ 10 GHz) when used as interconnectors on semiconducting devices.

12.3.2   Strength and Elasticity

The carbon atoms of a single (graphene) sheet of graphite form a planar honeycomb lattice in which each atom is connected via a strong chemical bond to three neighboring atoms. Because of these strong bonds, the basal-plane elastic modulus of graphite is one of the largest of any known material. For this reason, CNTs are expected to be the ultimate high-strength fibers. SWCNTs are stiffer than steel and are very resistant to damage from physical forces. Pressing on the tip of a CNT will cause it to bend but without damage to the tip. When the force is removed, the tip returns to its original state. This property makes CNTs very useful as probe tips for very high-resolution scanning probe microscopy.

Quantifying these properties has been rather difficult, and an exact numerical value has not been agreed upon. However, using atomic force microscopy (AFM), the unanchored ends of a freestanding CNT can be pushed out of their equilibrium position, and the force required to push the CNT can be measured. The current Young's modulus value of SWCNTs is about 1 tera pascal (Tpa), but this value has been disputed, and a value as high as 1.8 Tpa has been reported. Other values significantly higher than this have also been reported (Ye et al. (2009). The differences probably arise through different experimental measurement techniques. Yu Wang et al. (2005) have shown that the Young's modulus theoretically depends on the size and chirality of the SWCNTs, ranging from 1.22 to 1.26 Tpa. They have calculated a value of 1.09 Tpa for a generic CNT. However, when working with different MWCNTs, others have noted that the modulus measurements of MWCNTs using AFM techniques do not strongly depend on the diameter (Javey et al. (2004)). Instead, they argue that the modulus of the MWCNTs correlates to the amount of disorder in the CNT walls. Not surprisingly, when MWCNTs break, the outer-most layers break first.

In most materials, however, the actual observed material properties (e.g., strength, electrical conductivity) are degraded very substantially by the occurrence of defects in their structure. For example, high-strength steel typically fails at about 1% of its theoretical breaking strength. CNTs, however, achieve values very close to their theoretical limits because of their molecular perfect structure.

12.3.3   Thermal Conductivity

CNTs have extraordinary electrical conductivity, heat conductivity, and mechanical properties. Research indicates that CNTs may be the best heat-conducting material ever known. Ultra-small SWCNTs have even been shown to exhibit superconductivity below 20°K (Mohammadizadeh.2006) Research suggests that these exotic strands, already heralded for their unparalleled strength and unique ability to adopt the electrical properties of either semiconductors or perfect metals, may someday also find applications as miniature heat conducts in a host of devices and materials. The strong in-plane graphitic C–C bonds make them exceptionally strong and stiff against axial strains. The almost 0 in-plane thermal expansion, but large inter-plane expansion of SWCNTs implies strong in-plane coupling and high flexibility against non-axial strains. Many applications of CNTs, such as in nanoscale molecular electronics, sensing and actuating devices, or as reinforcing additive fibers in functional composite materials, have been proposed.

Reports of several recent experiments on the preparation and mechanical characterization of CNT–polymer composites (Ajayan, 1969) have also appeared. These measurements suggest modest enhancements in the strength characteristics of CNT-embedded matrixes compared with bare polymer matrixes. Preliminary experiments and simulation studies on the thermal properties of CNTs show very high thermal conductivity. It is expected, therefore, that CNT reinforcements in polymeric materials may also significantly improve the thermal and thermomechanical properties of the composites.

12.3.4   Electron Field Emission

CNTs are probably the best electron field emitters. Field emission results from the tunneling of electrons from a metal tip into vacuum under application of a strong electric field. The small diameter and high aspect ratio of CNTs is very favorable for field emission. Even for moderate voltages, a strong electric field develops at the free end of supported CNTs because of their sharpness. This was observed by

Walt A de Heer (1995). He also immediately realized that these field emitters must be superior to conventional electron sources and might find their way into all kind of applications, most importantly flat-panel displays. It is remarkable that after only five years, Samsung actually realized a very bright color display, which will be shortly commercialized using this technology.

Studying the field emission properties of MWCNTs, emission of light was also observed. This luminescence is induced by the electron field emission because it is not detected without applied potential. This light emission occurs in the visible part of the spectrum and can sometimes be seen with the naked eye.

CNTs are the best-known field emitters of any material because of the incredible sharpness of their tips. The sharpness of the tips also means that they emit at especially low voltage, an important fact for building low-power electrical devices. CNTs can carry an astonishingly high current density, possibly as high as 10^{13} A/cm². Furthermore, the current is extremely stable. An immediate application of this behavior receiving considerable interest is in field emission flat-panel displays. Instead of a single electron gun, as in a traditional cathode-ray tube (CRT) display, in CNT-based displays, there is a separate electron gun (or even many of them) for each individual pixel in the display. Their high current density; low turn-on and operating voltages; and steady, long-lived behavior make CNTs very attractive field emitters in this application. Other applications using the field emission characteristics of CNTs include general types of low-voltage cold-cathode lighting sources, lightning arrestors, and electron microscope sources.

12.3.5   High Aspect Ratio

CNTs represent a very small, high aspect ratio conductive additive for plastics of all types. Their high aspect ratio means that a lower loading (concentration) of CNTs is needed compared with other conductive additives to achieve the same electrical conductivity. This low loading preserves more of the polymer resins' toughness, especially at low temperatures, while also maintaining other key performance properties of the matrix resin. CNTs have proven to be an excellent additive to impart electrical conductivity in plastics. Their high aspect ratio (~1000:1) imparts electrical conductivity at lower loadings compared with conventional additive materials such as carbon black, chopped carbon fiber, and stainless steel fiber.

12.3.6   Light Emission by Carbon Nanotubes

Light-emitting diode (LED) operates in a reverse direction to photovoltaic cells. In the cells, light energy is converted into electrical energy; in LED, electrical energy is converted into light energy. LEDs are special diodes that emit light when connected in a circuit. They are frequently used as "pilot" lights in electronic appliances to indicate

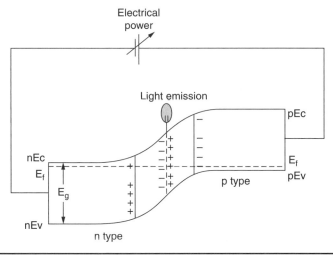

FIGURE 12-6   Schematic diagram of the p:n junction (light-emitting diode) connected to a battery such that the p type is positively charged. On application of potential, a light is emitted from the junction.

whether the circuit is closed or not. LEDs operate at relative low voltages between about 1 and 4 V and draw currents between about 10 and 40 mA. The most important part of a LED is the semiconductor diode (Fig. 12-6). The diode has two regions separated by a junction. The p region is dominated by positive electric charges, and the n region is dominated by negative electric charges. The junction acts as a barrier to the flow of electrons between the p and n regions. When sufficient voltage is applied to the semiconductor diode, current can flow, and the electrons can cross the junction into the p region, producing light.

When sufficient voltage is applied to the diode across the diode, electrons can move easily in only one direction across the junction (i.e., from p- to n-type regions). The p region contains more positive charges than negative; the n region contains more electrons than positive charges. When a voltage is applied and the current starts to flow, the electrons in the n region have sufficient energy to move across the junction into the p region. After they are in the p region, the electrons are immediately attracted to the positive charges because of the mutual Coulomb forces of attraction between opposite electric charges. When an electron moves sufficiently close to a positive charge in the p region, the two charges "recombine."

Each time an electron recombines with a positive charge, electric potential energy gets converted into electromagnetic energy. For each recombination of a negative and a positive charge, a quantum of electromagnetic energy is emitted in the form of a photon of light with a frequency characteristic of the semiconductor material. Only photons in a very narrow frequency range can be emitted by a material. LEDs that emit different colors are made of different semiconductor materials and require different energies to produce light.

12.4 Carbon Nanomaterial as a Light Source

It has been observed that an individual MWCNT may carry a current density as high as 10^{13} A/cm², which is about 1000 times larger than that of copper. It has also been reported that CNTs may emit fluorescent light when they are excited by a laser light source. These effects are not surprising. Thomas Edison used carbon filament in a high vacuum to show its incandescent properties. However, the carbon bulb filaments were very fragile, and the bulbs darkened rapidly as a result of the deposition of carbon on the glass envelop and suffered easily from premature burnout.

Recently, Jinquan (2004) demonstrated lighting from macroscopic SWCNTs and DWCNTs (double walled carbon nanotubes). SWCNTs were first immersed in alcohol and then assembled into long filaments under the surface tension when the alcohol evaporated. Tungsten filament of safelight of approximately 36 V and 40 W was then replaced by the CNT filament. The CNT filament was connected to the electrodes using silver solder and sealed in a glass bulb after a high vacuum was achieved inside the enclosure.

Sharon (unpublished work) synthesized a carbon rod by pyrolysing a shell of a coconut with a length of 2 cm and a diameter of 0.5 mm. Two ends were connected with a conducting wire. The entire system was then kept in a glass tube and sealed after evacuating it. An AC 5 V was allowed to pass through the carbon rod. It illuminated, giving almost white light (Fig. 12-7).

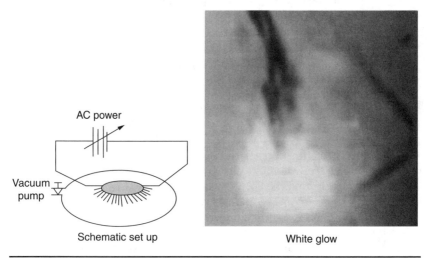

Schematic set up   White glow

Figure 12-7 An illuminated carbon rod obtained by pyrolysing a coconut shell showing light emission at an applied voltage of 5 V.

The carbon filament emits incandescent light evenly along its entire length and has a resistance of about 16 Ω at room temperature.

12.5 Summary

Structural aspects and various physical properties of CNTs (electrical, thermal, mechanical strength, electron field emission, LED, light emission) have been discussed in detail in this chapter.

Depending on their chiral angle, CNTs exhibit three prominent types of structures: armchair, zigzag, and chiral. The folding of graphene sheets is responsible for inducting various properties in CNTs. It is also suggested that identifying the presence of such structures is not an easy task, so information such as the presence of carbon of configuration sp^2 and sp^3 (with NMR or FTIR), the band gap (with the optical absorption method), and the diameter (with TEM) may reveal the properties of CNTs.

The unique electronic properties of CNTs are due to the quantum confinement of electrons normal to the CNT axis. Because of this quantum confinement, electrons can only propagate along the CNT axis, so their wave vectors point in this direction.

The electrical conductivity of CNTs is complex because their conductivity is a function of their chirality (degree of twist) and hence their diameter. Moreover, CNTs can be either metallic or semiconducting in their electrical behavior. Diamond, although it is highly resistive, shows a very high thermal conductivity.

CNTs are very strong and have a high tensile strength. Being the best electron emitter, CNTs have a lot of potential applications. Moreover, they exhibit high light-emitting properties.

PART IV

Applications of Carbon Nanomaterials

CHAPTER 13

Electron Field Emission from Carbon Nanomaterials

Debabrata Pradhan and I. Nan Lin

Department of Physics, Tamkang University Taiwan, Republic of China

13.1 Introduction

Electron field emission is an old theory that got renewed and became the center of prime attention after the discovery of carbon nanotubes (CNTs) (Iijima, 1991). In the past few years, extensive research has

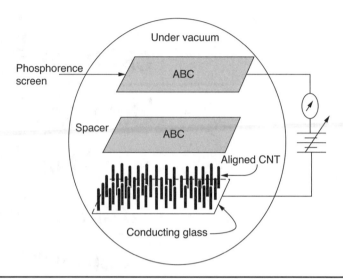

FIGURE 13-1 Schematic diagram of an electron field emission setup.

been carried out using different carbonaceous materials as cathodes for electron field emission. Field emission is a quantum mechanical tunneling phenomenon in which electrons escape from a solid surface into a vacuum.

In contrast to the commonly used thermionic emission from hot filaments, field emission occurs at room temperature from unheated "cold" cathodes under an electric field, so it is also called *cold cathode emission*.

An oversimplified schematic diagram of an electron field emitter is shown in Fig. 13-1, where CNTs acting as cathode release the electrons, which travel toward the anode through a separator. When electrons strike the anode, which is made of phosphorescence material, the image is produced (ABC as shown in Fig. 13-1). The entire setup is kept under vacuum to make large, mean free path of electrons and to stop any corrosion of electrodes.

Field emission display (FED) is one of the main applications that has been attracting technologists to make electron field emission practical for market-oriented applications.

13.2   Field Emission Display

Over the past few decades, various flat-panel display technologies have emerged, including liquid crystal display (LCD) and plasma displays. LCD currently dominates the market of display technology. LCD technology provides a flat screen with the drawbacks of low image quality, a restricted field of view, and a high cost. Some of these drawbacks were partially resolved in active matrix LCD and plasma displays, which have much higher refresh times and an increased image quality. However, none of the display technology matches the brightness and resolution of the cathode-ray tube (CRT), which was discovered in the 1920s, in which a heated filament ejects a stream of electrons. This stream of electrons is then accelerated, deflected, and sent through a vacuum tube to finally hit the opposite phosphor-coated glass screen.

In display technology, FED has many advantages compared with other display technologies (Tables 13-1 and 13-2). FED promises slimmer, more energy-efficient, and portable display (Chalamala, 1998; Mann, 2004).Traditionally, sharp-pointed cathodes made of silicon, tungsten, or molybdenum have been used for electron field emission. These tips tended to break down quickly in the powerful electric fields, which compelled the researchers to focus their efforts on diamond film as cathode material. However, diamond film has proven tough to fabricate into the sharply pointed shapes needed for electron emission. While researchers were struggling with diamond emitters, the discovery of CNTs and superior field emission behavior of CNTs stole attention from all other materials.

Electron emission is an important phenomenon of ejecting electrons from the surface of a solid material by application of either heat or voltage. This phenomenon occurs in photoemission, thermal emission, cold cathode emission, and secondary electron emission.

13.3   Types of Electron Emission

13.3.1   Cold Cathode Field Electron Emission

Unlike thermionic emission, electron field emission occurs at room temperature, which is why it is termed *cold cathode emission*. In this case, a high electric field is applied to pull the electrons from the metal surface. The applied field should be enough to eject the electrons from the emitter surface by overcoming the potential barrier across the metal–vacuum interface. This barrier is called *work function* and is the difference between the Fermi level of the metal and the vacuum level.

Basic requirements for electron emission for FED or other applications are:

1. Low threshold field and therefore low operating voltage

2. Uniform electron emission from the whole area of the cathode

3. Large emission site density (ESD; ~10^6 cm^{-2}) to meet resolution requirements

4. Stable emission and long life of the emitter

	Cathode ray tube	Liquid crystal display	Plasma display	Filed emission display
Working principle	An electron beam steered by magnetic fields strikes phosphors on a glass screen	Polarized light shines through liquid-crystal "gates" that control pixel color and intensity	An electric pulse sets off a burst of ionized gas in each pixel as though it were a tiny neon sign	CNTs glued to a substrate shoot electron at phosphors on a glass screen
Advantages	Reliable	Reliable	Pixels switch quickly	Pixels switch quickly
	No burn-in	No burn-in	Sharp, bright images	No burn-in
	Viewable from any angle	Thin	Thin	Thin
	Inexpensive	Light	Viewable from any angle	Light
	Phosphors can display fast motion			Viewable from any angle
				Low power consumption
Disadvantages	The electron gun must be placed far behind the screen, making the tubes bulky and heavy	The viewer must be positioned directly in front of the screen	High power consumption	Unsolved technical problems, such as maintaining a vacuum between the substrate and the glass
		The pixel switch slowly, smearing fast-moving images	Burn-in (motionless images displayed for too long become seared into the screen)	Currently cannot be manufactured affordably
		Expensive	Expensive	

CNT = carbon nanotube.

Table 13-1 Basic Working Principles of Various Display Techniques and their Pros and Cons

Characteristics	Thin-film transistor liquid crystal display	Electroluminescent display	Plasma display panel	Field emission display
Emission type		Thin-film phosphor	Photoluminescence	Low- and high-voltage phosphors
Brightness (cd/m^2)	200	100	300	150 (low V) >600 (high V)
Viewing angle (degrees)	±40	±80	±80	±80
Emission efficiency (lm/W)	3–4	0.5–2.0	1.0	10–15
Response time	30–60 ms	<1 ms	1–10 ms	10–30 μs
Contrast ratio	>100:1	50:1	100:1	100:1
Number of colors	16 million	16 million	16 million	16 million
Number of pixels	1024 × 768	640 × 480	852 × 480	640 × 480
Resolution (μm pitch)	0.31	0.31	1.08	0.31
Power consumption, W (size)	3 (26.4 cm)	6 (26.4 cm)	200 (106.7 cm)	2 (26.4 cm)
Panel thickness (mm)	8	10	75–100	10
Operating temperature range (°C)	0–50	–5–85	–20–55	–5–85

TABLE 13-2   Characteristics of Some Flat Panel Displays

The five physical ways of increasing the emission current are:

1. Increasing the voltage
2. Decreasing the inter-electrode distance
3. Decreasing the material work function of the emitter material by increasing the dopant concentration or by chemiadsorption of surface adsorbates
4. Increasing the effective field at the tip through sharpening
5. Increasing or decreasing the number of tips (depending on their height) in a given area (i.e., tip density)

However, all of the above physical ways are not applicable to all cases. The next sections discuss the effects of all the parameters that crucially control the emission properties.

Effect of Voltage

The lower voltage (~100 V) exposes the emitters to a lower risk of damage from the ionization of ambient gas, creating a positive ion that then impacts the tip in the vacuum (Brodie, 1975). Emitters can be used at a higher pressure with a longer operating lifetime at lower operating voltage. Therefore, the emitters (cathodes of the devices) must be capable of emitting electrons at low macroscopic electric fields (typically in the range of 1 to 20 V/μm) with sufficient current density (in the range of 10 to 100 mA/cm^2) to generate bright fluorescence from the associated phosphor on the anode. There are two ways to reduce the operating voltage:

1. Emitters should be chosen from a low work function material, which allows less energy to eject electrons from its surface.
2. The shape of emitter tip should be very sharp so that local field strength becomes high enough even at low gate voltage.

Effect of Inter-electrode Distance

The inter-electrode distance plays a significant role in flat panel display. A smaller distance between the cathode and anode is not only easier for achieving a higher current at lower applied voltage, but it also increases the compactness of devices. However, care should be taken to maintain a well-defined gap between electrodes to avoid charging and short circuit.

Effect of Work Function

Work function is one of the major properties of cathode material that determines the emission voltage (Fig. 13-2). Materials with a low

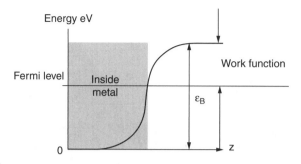

FIGURE 13-2   Band diagram of a solid showing work function.

work function are suitable for ejecting electrons from a metal surface at a lower threshold voltage.

It has been shown that emitters with a lower work function show lower current fluctuation as well. However, most materials with low work function are not stable at room temperature and undergo decomposition at the operating emission voltage. This restricts the choice in selecting the emitter metals. Table 13-3 depicts the work functions of various metals and their melting points. Out of these materials, tungsten, molybdenum, and carbon have been studied widely for electron field emission because of their moderate work functions and high melting points.

Effect of Emission Tip
A sharp tip of radius r in the range of 0 to 100 nm is necessary to build a higher electric field at the tip of emitter, which allows field emission

Metal	ϕ (eV)	Melting point (°C)
Al	3.7	660
Au	4.6	1064
Ba	2.3	725
C	4.4	3550
Cu	4.5	1083
Fe	4.4	1535
Ir	5.2	2410
Mo	4.3	2620
Os	5.4	3045
W	4.5	3410

TABLE 13-3   Work Function and Melting Points of Selective Metals

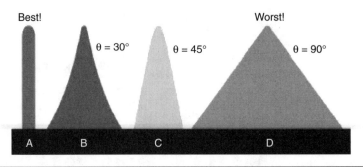

Best!

$\theta = 30°$ $\theta = 45°$ Worst! $\theta = 90°$

A   B   C   D

FIGURE 13-3   Effectiveness of four emitter tips (A and D) (Utsumi, 1991).

of electrons to occur at a lower applied voltage. A high field enhancement is possible with a one-dimensional structure (having a high aspect ratio) followed by a cone with a tiny radius. Fig. 13-3 shows the best to worst field emitters. The emission is observed at a lower field from emitter tips with smaller radiuses of curvature (Stepanova, 1998).

Tip Density

Electron emission increases manifold by creating sharp tips or protrusions on the flat cathode surface. This is normally called the *field enhancement factor*. The higher the field enhancement factor, the better emission is at a lower threshold voltage. The field enhancement factor directly depends on the aspect ratio (i.e., the ratio of the length to the diameter) of the emitter tip. At the same time, the tip density of the emitter plays a significant role in electron emission. A reduction in emission properties occurs when sharp peaks are densely packed. This is due to a screening effect, which prevents the field from concentrating on the tip of the emitters.

Pressure Inside the Chamber

Because the FED is a vacuum device, pressure in between the cathode and anode plays a significant role in the stability of the tips and electron emission. The presence of oxygen, water vapor, or other residual gases becomes a severe problem for large-area panels. A high vacuum or even ultra-high vacuum is needed for better emitter stability and higher emission. Because field emission is more sensitive to surface properties of cathode materials, tips get contaminated in a lower vacuum, degrading emission performance. However, selective contamination or deposition of certain metals as adsorbates on emitter tips has been found to enhance emission (e.g., depositing alkali metal such as Na, K, or Cs on emitter tips). This is either because of a lowering of the work function or a change in the band structure of the emitter material (Chhowalla et al., 2001; Jeong et al., 2001; Tanemura et al., 2005).

13.4   Electron Field Emission from Carbonaceous Material

Carbon is one of the most fascinating elements known because of the variety of materials it forms and because it exists in many allotropic forms. Researchers have focused their eyes on diamond film, diamondlike carbon film, and CNTs as cathode material. The insulating nature of diamond and the difficulty in doping to make it conducting disfavors its use as the best emitter; instead, the discovery of CNTs and their superior field emission behavior became the main attraction for researchers. Table 13-4 shows the threshold fields of widely used emitter materials. It is obvious from Table 13-4 that CNTs and nanodiamond have remarkably high electron emission characteristics compared with any other emitter material.

Many different mechanisms are involved as the electrons travel from the negative end of the power supply through the various interfacial contacts, through the bulk of the film itself and to the film surface, tunnel through the potential barrier, and propagate through the vacuum gap before finally reaching the anode.

Cathode material	Threshold field (V/µm) for a current density of 10 mA/cm²	Reference
Si tips	50–100	Brodie and Spindt, 1992
Mo tips	5–100	Brodie and Spindt, 1992
p-Type diamond	160	Zhu et al., 1996
Defective CVD diamond	30–120	Zhu et al., 1996
Amorphic diamond	20–40	Kumar et al., 1995
Cesium-coated diamond	20–30	Zhu et al., 2001
Nanodiamond	3–5	Zhu et al., 2001
Carbon nanotube	1–4	Saito and Uemura, 2000 Bonard et al., 1999

CVD = carbon vapor deposition.

TABLE **13-4**   Threshold Field of Electron Emission from Various Cathode Materials

13.4.1   Emission from Diamond

Bare diamond surface has a very small electron affinity of +0.37 eV. However, hydrogenated diamond film exhibit negative electron affinity (NEA), which is one of the excellent properties that diamond possesses for electron field emission. The electron affinity χ is the energy difference between the vacuum level (E_{vac}) and the conduction band minimum (E_c) in a semiconductor:

$$\chi = E_{vac} - E_c \qquad (13.1)$$

In most cases, the electron affinity is more than 0. However, chemisorption of hydrogen onto the diamond surface lowers the vacuum level, and the χ value decreases to –1.27 eV for a fully hydrogenated surface (Cui et al., 1998). Thus, the valence band lies between 5.9 and 4.2 eV below the vacuum level. Because of this excellent NEA property of diamond, the field emission properties of diamond film deposited by chemical vapor deposition (CVD) are significantly higher in hydrogenated films compared with unhydrogenated CVD diamond, as shown in Fig. 13-4. Hydrogenation has two salutary effects for enhancing field emission. It increases the conductivity of diamond

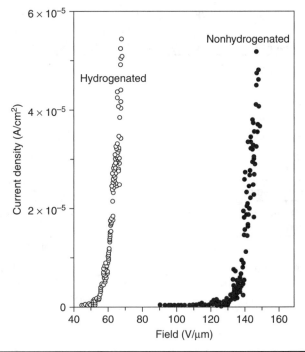

Figure 13-4   Field emission characteristics of the same chemical vapor deposition diamond film before and after plasma hydrogenation (Cui et al., 2000).

films, which provides a conducting path for electrons to the emission sites. It also lowers the effective emission threshold of graphite in contact with diamond that exhibits negative electron affinity after hydrogenation.

13.4.2   Emission from Carbon Nanotubes

Because of the many excellent physical and chemical properties of CNTs, it has been the most widely studied material in the past few years. Electron field emission application is one of the foremost and very promising applications of CNTs. There are broadly two types of CNTS: multiwalled carbon CNTs (MWCNTs), in which the tubes each have more than one wall of graphitic layer, and single-walled CNTs (SWCNTs), in which the tubes each have only one wall of graphene layer.

The characteristics of CNT that have made it most attractive and promising emitter material include:

1. **High aspect ratio:** Because of its nanoscale nature, the field enhancement factor of CNT emitters increases manifold and is far higher than any other emitter material available so far.

2. **Small radius of curvature at their tips:** CNTs have an inherent nature of having sharp tips, which prohibits any special fabrication for sharpening the tips. A single layer of carbon atoms is found to be sufficient to shield most of the electric field except at the tip, where strong field penetration occurs. The penetration leads to a nonlinear decrease of potential barrier for emission, which is equally responsible for the low threshold voltage in addition to the well-known geometrical field enhancement factor (Zheng et al., 2004).

3. **High conductivity:** Because CNTs are constituted from graphitic layers, they are either semiconducting or metallic depending on their diameter and chirality.

4. **High chemical and thermal stability:** For durability and longevity, these two properties are very important. CNTs have been found to be inert and stable up to 2000 K.

5. **High mechanical strength:** This property contributes to the stability of the emitter material.

It is controversial whether emission is higher from open-tip or closed-tip CNTs (Fig. 13-5) and the reason behind it. Closed CNTs produce significant emission currents at much lower applied voltages than open CNTs (Bonard et al., 1999). However, emission current depends on the field enhancement factor. In the case of open MWCNTs, field emission takes place at a lower applied field. Recent density functional calculation shows that in the case of the open-ended CNTs,

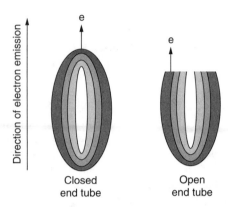

Direction of electron emission

Closed
end tube

Open
end tube

FIGURE **13-5**   Schematic diagram showing the direction of electron flow under an applied potential with closed-end (*A*) and open-end (*B*) multiwalled carbon nanotubes.

the field emission originates primarily from the dangling-bond states localized at the edge; the pentagonal defects are the main source of the field emission in the capped tubes.

Vertically aligned CNTs grown with the CVD process are directly used as emitter material without any postfabrication. In such cases, CNTs are either densely packed like a film (Fig. 13-6*A*) or grown in a controlled fashion as an array to reduce the screening effect (Fig. 13-6*B*).

SWCNT films have shown higher emission uniformity and lower turn-on voltage than MWCNT films. The low turn-on field of SWCNTs is due to a stronger local electric field on the small tip radius. However,

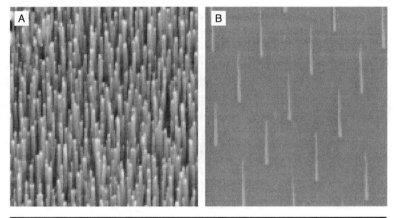

FIGURE **13-6**   (*A*) Nanotube films on a nickel catalyst layer (Chhowalla et al., 2001). (*B*) Array of multiwalled carbon nanotubes on nickel nanodots (Teo et al., 2003).

SWCNT films show less emission stability than MWCNT films because of tip degradation. The degradation of SWCNT is probably due to ion bombardment (by gas phase electron ionization or by ion desorption from the anode, both induced by the emitted electrons). SWCNT degrades faster than MWCNT (a factor >10).

Moreover, design is also important to get uniform brightness at the screen of FED. Figure 13-7 shows the calculated electrical field distribution and trajectories of emitted electrons. There is a strong enhancement of the field from the edge of electrodes, leading to less brightness at the central part of the electrodes. This problem was overcome by designing special FED (Fig. 13-8). In the new design, CNTs with a width of 120 μm were printed on one edge of the 390-μm wide cathode electrode. Figure 13-9 shows magnified emission images of pixels. Very uniformly emitting pixels of green, blue, and red were observed over the 9-inch panel, without cross-talk between pixels. The non-uniformity of electron field distribution has been also overcome by using a triode-type FED structure. A triode-type FED has shown a high-gray scale and a high brightness at a video speed operation. In the triode structure, it is observed that electron emission from CNT emitters is controlled by modulation of gate voltages. Mukul et al. (2001) have reported the field emission of electron using thin film of carbon synthesized from kerosene.

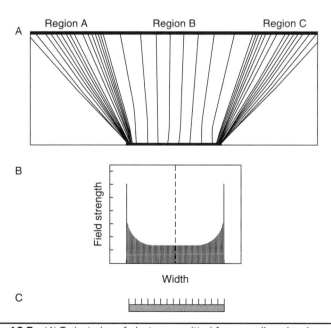

FIGURE 13-7   (A) Trajectories of electrons emitted from an aligned carbon nanotube (CNT) cathode to an anode. (B) Electric field distribution. (C) Schematic of aligned CNTs on the cathode (Choi, 2001).

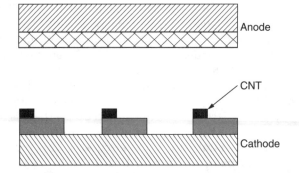

FIGURE 13-8   Cross-sectional view of field emission display to reduce the edge effect (Lee et al., 2001). CNT = carbon nanotube.

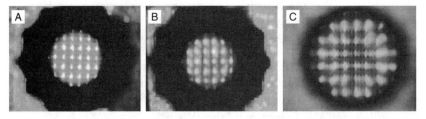

FIGURE 13-9   Magnified emission images of a field emission display where (A) green, (B) blue, and (C) red are lit up (Lee et al., 2001).

13.5   Application of Electron Field Emission

13.5.1   Field Emission Display

Recently, a number of prototypes of FED using diamond (Fink, 1998) and CNT (Choi et al., 2001) emitters have been demonstrated. In 1998, CRT using MWCNTs was reported with a high current density $10\,mA/cm^2$ on the cathode surface (effective area, ~2 mm²) at an average field strength of 1.5 V/μm. The cold cathode CRT lighting elements made from MWCNTs of 20 mm in diameter and 74 mm in length were reported to be twice as bright as conventional CRTs and had a lifetime of more than 8000 hours. A number of FEDs of various sizes (4 to 9 inch) have been successfully fabricated using CNTs. Very recently, a 38-inch full-color television has been demonstrated by Samsung using a CNT emitter (Mann, 2004) (Fig. 13-10). These studies establish promising applications for electron field emission in FED.

13.5.2   Gas Discharge Tube

A gas discharge tube (GDT) consists of two electrodes parallel to each other in a sealed ceramic case filled with a mixture of noble gases. In case of transient overvoltage due to lightning or alternating current

FIGURE 13-10   A 38-inch full-color television made from a carbon nanotube emitter (Mann, 2004).

power cross faults, a plasma is generated in between the electrodes, making the system short circuit, thus protecting the entire components (Standler, 1989).

GDTs are normally used in telecom network interface device boxes and central office switching gears. Compared with solid-state protectors, GDTs can carry much higher currents. The lower breakdown voltage and a factor of 4 to 20 reduction in breakdown voltage fluctuation make CNT-based gas discharge tubes a better option than voltage protection units for advanced telecommunication networks such as asymmetric digital subscriber lines and high-bit-rate digital subscriber line, in which the tolerance is narrower than what can be afforded by the current commercial GDTs (Rosen et al., 2000). The breakdown voltage of the GDT with SWCNT–Fe–Mo electrodes is 448.5 V, with a standard deviation of 4.8 V over 100 surges. A lower breakdown voltage and its low fluctuation during surges make SWCNT-based GDTs a better candidate over other commercially available GDTs.

13.5.3   X-Ray Tube

Normally, thermionic emission from hot filament is used for X-ray tube devices. For effective thermionic electron emission, metal filament has to be heated to at least 1000°C, which is a basic drawback. Therefore, for room temperature operation, field emission devices are found to be an acceptable feature for fabricating X-ray tubes. Most diagnostic applications require tube current in the order of 10 to 100 mA

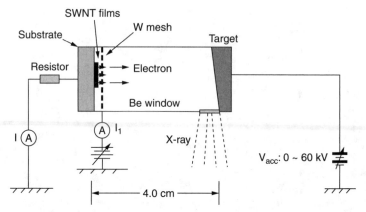

Note: SWNCNT = Single-walled carbon nanotube.

Figure 13-11   Schematic drawing of carbon nanotube–based field emission X-ray sources. The gate electrode is a W mesh 50 to 200 μm away from the cathode. Electron emission is triggered by the voltage applied between the gate and the cathode. An X-ray is produced when the emitted electrons are accelerated and bombarded on the copper or molybdenum target (Cheng, 2003).

and an operating voltage in the range of 30 to 150 kV, which was difficult to accomplish for the field emission X-ray tubes (Cheng and Zhou, 2003). X-ray tubes using CNTs as cathode have been successfully fabricated (Fig. 13-11).

This device has readily produced both continuous and pulsed X-rays. The X-ray intensity obtained from CNT-based X-ray tubes was sufficient to image a human extremity at 14 kVp and 180 mA (Fig. 13-12).

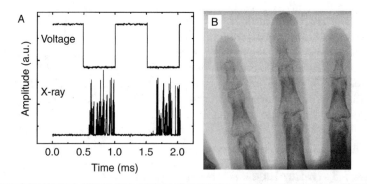

Figure 13-12   X-ray generation by a carbon nanotube (CNT)–based field emission source. (A) 1-KHz and 50% duty cycle pulsed X-ray signals recorded using an oscilloscope by applying pulsed gate voltage with the same wave form. (The height of the signal is proportional to the photon energy.) (B) X-ray image of a human hand taken using the carbon nanotube source on a Polaroid film. The X-ray energy is 40 KV, and the image was taken 40 cm away from the X-ray source without using an image intensifier (Cheng and Zhou, 2003).

13.6 Summary

In this chapter, electron field emission behavior from various CNMs has been reviewed. CNTs are robust and deliver high-emission current with a current density routinely exceeding 1 A/cm^2. However, greater effort is needed toward large-scale fabrication, for stability at a high current density, and for uniformity of emission.

The high electron emission of CNTs is due to its high aspect ratio and very small tip curvature. Despite the graphitic nature, CNTs have high work function. In contrast, the negative electron affinity property of hydrogenated nano crystalline diamond (NCD) surfaces shows high emission at lower threshold voltage.

Therefore, the collective favorable properties of CNTs (i.e., a high aspect ratio of CNTs and negative electron affinity of hydrogenated diamond surface) can be combined in a single cathode material for further enhancing field emission characteristics. To make it feasible, CNT tips can either be hydrogen terminated or converted to diamond. However, more studies on the surface adsorbates at CNT tips or diamond films surface to enhance their emission properties are needed. The stability of tips and the emission noise of emitters have to be controlled more precisely.

CNTs have shown a promising future in display technology at the laboratory scale. In the near future, FED using CNTs as an emitter is expected to come to commercial-level production.

CHAPTER 14

Carbon Nanomaterial as an Electrode in Fuel Cells

K.D. Barhate and Maheshwar Sharon

Nanotechnology Research Centre, Birla College, Kalyan, Maharashtra, India

14.1   Introduction

A fuel cell is an energy conversion device that converts the chemical energy of the electrochemical reaction between a fuel (i.e., which burns to produce energy, such as hydrogen or methanol) and an oxidant (oxygen) into a low-voltage DC electricity. Fuel and oxidant are supplied to separate electrodes, which are supported on opposite sides of a thin layer of an electrolyte. A fuel cell thus converts chemical energy directly into electrical energy without involving a thermal process. The efficiency of fuel cells is, thus, not controlled by the Carnot's cycle limitation. It operates without combustion and is thus virtually pollution free. Moreover, it has quieter operation because there are no moving parts. Unlike with batteries, the electrical energy from fuel cells can be generated as and when fuel is injected into the system.

Despite various advantages and a higher efficiency of fuel cells, some technical and engineering challenges are still to be solved; for example, the cost of fuel cells is too high to make them commercially viable. The materials used for the electrodes of fuel cells are expensive and require some special precautions to handle them.

In search for a cheaper electrode, scientists have embarked upon using a novel material, carbon nanoforms. Carbon nanomaterials (CNMs) are known to be electrochemically stable and could be produced cheaply. Therefore, scientists are trying to develop electrodes for fuel cells using CNMs.

14.2   Historical Overview and Development of Fuel Cells

Although the fuel cell has received much attention in recent years, it is not a new technology. William Grove made the first working fuel cell in 1839 while investigating the electrolysis of water. He noticed that when the current was switched off, a small amount of current began to flow through the circuit in the opposite direction, which was due to the reaction of hydrogen with oxygen. In 1889, Ludwig Mond and Charles Langer called the apparatus used by William Grove a *fuel cell*. The first major practical alkaline fuel cell was developed by Francis Thomas Bacon in 1932, which was used on the *Gemini* earth-orbit space program by NASA.

Elements of Cell	Characteristic
No. of cells in a stack	33
Rated power (Kw)	2
Rated voltage (V)	24
Current density (mA/cm^2)	300
Anode material	Raney nickel (with titanium 110 mg/cm^2)
Cathode catalyst	Nickel doped with silver (60 mg/cm^2)
Electrolyte	6N KOH
Working temperature (°C)	80
Fuel	Hydrogen and oxygen
Dimension (cm^3)	44 × 55 × 32
Total weight (kg)	53

TABLE 14-1   Fuel Cell Developed for the *Apollo* Mission to the Moon

United Aircraft Corporation made a fuel cell for the *Apollo* space mission, which put the first man on the moon. Characteristics of this fuel cell are given in Table 14-1.

In 1993, Ballard made a fuel cell for running a bus and launched 250-Kw fuel cell as a stationary power generator in 2000. In 2002, Siemens built a solid oxide fuel cell power plant of 220 Kw.

14.3   Principles of Fuel Cells

A simple design of a fuel cell is shown in Fig. 14-1. A fuel cell consists of two electrodes (Fig. 14-1*B* and *D*), one for oxidation of fuel (oxidation of hydrogen), and the other for the reduction of the oxidant (reduction of oxygen). These electrodes are separated by a membrane soaked with an aqueous solution of a suitable electrolyte (Fig. 14-1*C*). Electrodes are externally connected with a load. Fuels are injected into both electrodes through their respective chambers (Fig. 14-1*A* and *E*). Both electrodes are porous, containing suitable electrocatalysts to assist electrochemical reactions. The size of the pores is small enough to prevent leakage of electrolyte into chambers A and E.

Fuel injected through chamber A enters the pores of electrode B, where its oxidation occurs. The oxidized product diffuses out of the electrode to chamber C, and electrons are transferred to the other electrode through an external load. Likewise, oxidant enters through chamber E and enters the pores of electrode D, where its reduction takes place with the electrons transferred from electrode B. The oxidized product of fuel diffuses to chamber C, where it combines with the oxidized ions that came through electrode B to give the final product.

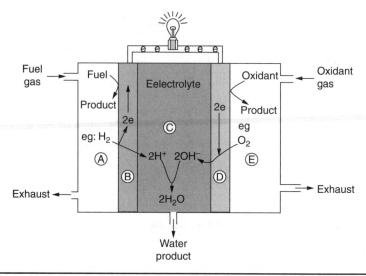

FIGURE 14-1 Schematic diagram of the basic components of a fuel cell.

Unreacted gases are allowed to escape from their respective chambers through the exhaust. Assuming hydrogen as one fuel and oxygen as an oxidant gas, the following reactions take place at the respective electrodes.

If operating the cell under acidic media:

$$2H_{2\ (gas)} \rightarrow 4H^+ + 4e^- + 0.0 \text{ V (at the anode)} \tag{14.1}$$

$$O_{2\ (gas)} + 4\ H^+ + 4e^- \rightarrow 2H_2O + 1.23 \text{ V (at the cathode)} \tag{14.2}$$

$$2H_{2\ (gas)} + O_{2\ (gas)} \rightarrow 2H_2O + 1.23 \text{ V (total reactions)} \tag{14.3}$$

If operating the cell under alkaline media:

$$2H_{2\ (gas)} + 4OH^- \rightarrow 4H_2O + 4e^- + 0.35 \text{ V (at the anode)} \tag{14.4}$$

$$O_{2(gas)} + 2H_2O + 4e^- \rightarrow 4OH^- + 0.88 \text{ V (at the cathode)} \tag{14.5}$$

$$2H_{2(gas)} + O_{2(gas)} \rightarrow 2H_2O + 1.23 \text{ V (overall reaction)} \tag{14.6}$$

Thus, a fuel cell operating with hydrogen and oxygen gas will produce water as the final product and will have a potential of 1.23 V irrespective of whether it works under acidic or alkaline media. However, it would not be possible to get 1.23 V because of various electrochemical factors. Most of the fuel cells operating with hydrogen or oxygen produce a potential of about 1.0 V. However, the current produced by the cell depends on the electrode surface area, the efficiency of electrochemical reactions occurring at the two electrodes, and the rate at which gas is injected into the fuel cell.

14.4   Thermodynamics of Fuel Cells

Various steps are involved (Fig. 14-2) in getting electrical energy by converting fuel and an oxidant. Electrical energy can be produced either with the help of a fuel cell (path A) or through the conventional process of directly converting chemical into heat energy (path B).

Irrespective of the path selected, at each stage of conversion of energy from one form to other, there will be a loss. However, loss of energy at stage I and stage III may be small, and the efficiency of the process depends on the design parameters. Conversion of energy at stage II includes loss factors that cannot be avoided by any improvement. The reason lies in the specific nature of heat as a form of energy. Heat is the energy of chaotically moving molecules. If there are two heat reservoirs (source and sink), then under the most favorable conditions, the degree of efficiency of heat Q into work A depends on the temperature of the sink and source:

$$\frac{A}{Q} \le \left(1 - \frac{T_1}{T_2}\right)$$ (14.7)

For example, power generation by steam has an efficiency between 4-8%, a thermal power station working on coal or gas has efficiency of about 30%, and an internal combustion engine (automobile car) has an efficiency of around 40 to 50%. Fuel cells operating with path A can, in theory, have efficiency as high as 100% because this method does not involve conversion of heat into electrical power, and its efficiency is not limited by Eq. (14.7).

Thus, calculation of conversion efficiency of a fuel cell should be considered differently. At the anode, fuel is converted to fuel oxidant products and electrons. Similarly, at the cathode the oxidant is reduced by the electrons to produce an oxidant reduction product. Electrons pass through the external circuit from the anode to the cathode to perform useful work (A_{max}). Thus:

$$A_{max} = ZFE = - \Delta G$$ (14.8)

where ΔG is the free energy for the reaction.

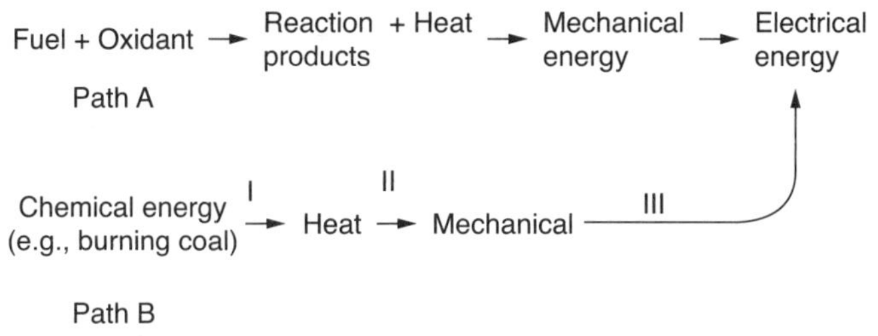

Fuel + Oxidant → Reaction + Heat → Mechanical → Electrical
products   energy   energy

Path A

Chemical energy ⟶ Heat → Mechanical ⟶
(e.g., burning coal)
         I   II   III

Path B

FIGURE 14-2   Schematic representation of various steps involved in converting chemical energy into electrical energy.

The total changes of chemical energy stored in the reaction system corresponds to the enthalpy change:

i.e.,
$$Q = -\Delta H \qquad (14.9)$$

Thus, the thermodynamic efficiency of a fuel cell is:

$$\eta = \frac{ZFE}{Q} = \frac{-\Delta G}{-\Delta H} = 1 - \frac{T\Delta S}{\Delta H} \qquad (14.10)$$

From this equation, it is clear that η is 1 or greater if $T\Delta S$ has a negative value. To make this quantity negative, the change in entropy of the reaction should be a negative value, which is the case when hydrogen and oxygen (both being gas) are converted to water (which is a liquid):

$$2H_2 + O_2 \rightarrow 2H_2O \qquad (14.11)$$

The net change in the number of molecules is negative (e.g., 2 molecules (i.e. product) –3 molecules (i.e. reactant) = –1). Therefore, the theoretical efficiency of a fuel cell can be more than 100%, which can never be achieved by any other process.

14.5 Advantages of Fuel Cells

Fuel cells have a distinct advantage over other clean energy generators, such as wind turbines and photovoltaic cells, because they can produce continuous power as long as there is a constant supply of hydrogen. The ability to produce continuous power makes fuel cells well suited for supporting critical loads for security applications. Another advantage of fuel cells is that the powerhouse of the system, the fuel cell stack, does not contain any moving parts as in traditional generators, which may lead to mechanical breakdowns. With the further development of technology, fuel cells are bound to become more reliable than conventional engines. In addition to low noxious air emissions, fuel cells can also produce significant amounts of power (electrical and thermal) with much less noise than standard generators. These factors are viewed favorably when setting fuel cell systems in populated areas and even inside facilities. The advantages of fuel cells can be summarized as follows:

- Power can be drawn continuously by feeding a reactant.
- Reduction occurs only when the cell is connected to the load (i.e., when the cell is short circuited). Hence, the cells' life cycle is very long.

- There are no moving parts. Hence, fuel cells can be used as silent electric power generator, causing fewer operational hazards.

- The final product (water) does not create any environmental hazards.

Environmental Benefits

Even though fossil fuels are consumed in the electrochemical reaction inside a fuel cell, fuel cells do not produce the same unhealthy air pollution emissions that are generated by burning gasoline in cars or burning coal and other fossil fuels in power plants. With fuel cells, there is no combustion, so fewer gases are released into the environment. For example, almost no sulfur oxides (SO_2) or nitrogen oxides (NO_x) are emitted, and emissions do not include any particulate matter. Also, the high electrical efficiency of fuel cells provides much more electricity per unit of product like water released in hydrogen or oxygen fuel cell than conventional generators of similar size.

14.6   Disadvantages of Fuel Cells

Fuel cells are not the perfect solution to the world's energy needs. Several obstacles need to be overcome before widespread use of fuel cells is possible.

14.6.1   Cost

The biggest hurdle for fuel cells is their cost. Although some fuel cell systems are in use today, very few are currently cost effective. For stationary fuel cells, typical capital costs for installed systems exceed $5000 per kilowatt, well above the target capital cost of $1000 to $1500 used by most energy generation developers.

14.6.2   Early Technology Risks

Fuel cells are still in a relatively early stage of development, and even the few commercially available models have limited fleet operating experience. This emerging technology requires risk-taking early adopters (as end users) to ultimately expose more consumers to the benefits of fuel cells.

14.6.3   Reliability and Durability

To become widely accepted as a clean distributed generator, fuel cells must prove their adaptability for a variety of applications. Also, certain fuel cell system components—such as the cell stack, which can require a costly replacement every 1 to 5 years, depending on the model—must be developed to have a longer lifespan or be easily and cheaply replaced.

14.7   Chemicals Used in Fuel Cells

The type of chemical used for a fuel cell depends on the working temperature of the fuel cell. Normally, solid fuels are not used. Liquid fuels, such as methyl and ethyl alcohol and formaldehyde, have been used. Gaseous fuels, such as ethylene, butane, hydrocarbon, CO, and hydrogen gases, have also been used. As the oxidant, either oxygen or air has been used.

The electrodes are separated by a suitable electrolyte, which has a high ionic conductivity, is stable, and does not cause corrosion to the cell. Phosphoric acid and KOH are the most common electrolytes used for fuel cells.

Most of the fuel cells developed have been using Ni–Ag alloy containing 10 to 50% Ni or Raney nickel doped with titanium. Recently, scientists have started using CNM as an electrode for the fuel cell. Electrocatalysts such as nickel or silver are deposited over the CNMs, and this is used as an electrode in the fuel cell.

14.8   Various Applications of Fuel Cells

All fuel cells operate on the principle of combining oxygen and hydrogen to generate electricity, heat, and water. They have many potential shapes and sizes suited for diverse applications. Three important applications of fuel cells are:

- They are used in automobiles.
- They are portable and hence are used for many other purposes.
- Stationary installation of fuel cells is also possible.

Many vehicle manufacturers, including Daimler Chrysler, Ford, General Motors, Honda, and Toyota, are actively researching and developing transportation fuel cells for future use in cars, trucks, and buses. These fuel cells will have to be relatively small, light, and durable. The manufacturers will also need to overcome the real but solvable challenges associated with storing hydrogen gas onboard in vehicles.

Because the electrochemical reaction in a fuel cell is much more efficient than the burning process in a conventional internal combustion engine, fuel cell vehicles are much more efficient than conventionally powered vehicles. They also have advantages over battery operated electric vehicles because they do not need to be slowly recharged, but can instead be replaced by charged battery at a filling station similar to refueling conventional cars by petrol.

Portable fuel cells are small and are designed for a variety of purposes. These units can provide backup electricity generation for military, temporary, or special needs applications. With the advancement in technology, even smaller fuel cells can be constructed. The applications for portable fuel cells are limitless. Laptop computers, cellular phones, video recorders, and hearing aids could be powered by portable fuel cells.

Stationary fuel cells are the largest, most powerful fuel cells. They are designed for installation in permanent settings such as hospitals, banks, airports, military bases, universities, and homes. The market advantage of stationary fuel cell systems is that they can provide a clean source of onsite power to a variety of end users. These fuel cell systems have the potential to provide end users with these added values.

Assured power can be readily obtained from fuel cells for applications requiring a high availability of power to maintain the operation of critical equipment, even when the electric grid fails.

14.9   Classification of Fuel Cells

Fuel cells are classified according to the electrolytes, operating temperature, and fuel used in them. Various characteristics of different types of fuel cells are briefly given in Table 14-2 and are discussed in detail later.

14.9.1   Alkaline Fuel Cells

Alkaline fuel cells differ from other types of fuel cells in the chemical reaction and the operating temperature. The basic schematic representation of an alkaline fuel cell is given in Fig. 14-3.

The chemical reaction that occurs at the anode is:

$$2H_2 + 4OH^- \rightarrow 4H_2O + 4e^- \qquad (14.12)$$

The reaction at the cathode occurs when the electrons pass around an external circuit and react to form hydroxide ions, OH^-:

$$O_2 + 4e^- + 2H_2O \rightarrow 4OH^- \qquad (14.13)$$

In the 1940s, F. T. Bacon at Cambridge University proved that alkaline fuel cells are a viable source of power. Alkaline fuel cells were later used in the *Apollo* space shuttle, which took the first man to the moon. Because of the success of the alkaline fuel cell in the space shuttle, Bacon was able to continue his research toward further advancement in alkaline fuel cells. They were tested in many different applications, including agricultural tractors and power cars and in providing power to offshore navigation equipment and boats.

The use of alkaline fuel cells encountered many problems, including their cost, reliability, ease of use, and safety, which were not easily solved. Attempts at solving these problems proved to be uneconomical. Proton exchange membrane fuel cells became very successful, so alkaline fuel cells were given less developmental importance.

Type of fuel cell	Electrolyte	Mobile ion	Fuel	Operating temperature (°C)	Reactions on different electrodes	
					Anode	Cathode
Alkaline	KOH	OH^-	H_2	100	$H_2 \rightarrow 2H^+ + 2e^-$	$\frac{1}{2}O_2 + H_2O + 2e^- \rightarrow 2OH^-$
Phosphoric acid	H_3PO_4	H^+	H_2 (tolerates CO_2)	180–240	$H_2 \rightarrow 2H^+ + 2e^-$	$\frac{1}{2}O_2 + 2H^+ + 2e^- \rightarrow H_2O$
Methanol	Polymer membrane	H^+	CH_3OH	50–100	$CH_3OH + H_2O \rightarrow CO_2 + 6H^+ + 6e^-$	$6H^+ + 6e^- \rightarrow 3/2O_2 + 3H_2O$
Solid oxide	ZrO_2, Y_2O_3, Lu_2O_3, Dy_2O_3	O^{2-}	H_2, CO, propane, natural gas	900–1000	$H_2 + O^{2-} \rightarrow H_2O + 2e^-$	$1/2O_2 + 2e^- \rightarrow O^{2-}$
Molten carbonate	Li_2CO_3, K_2CO_3, Na_2CO_3	CO_3^{2-}	H_2, CO	600–700	$H_2 + Co_3^{2-} \rightarrow H_2O + CO_2 + 2e^-$	$1/2O_2 + CO_2 + 2e^- \rightarrow CO_3^{2-}$

TABLE **14-2** Characteristics of Different Types of Fuel Cells and Their Electrode Reactions

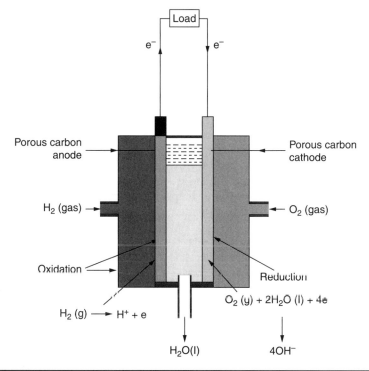

FIGURE 14-3 A basic alkaline fuel cell.

However, improvement in alkaline fuels cell remained an important project for space research programs.

Alkaline fuel cells have some major advantages over other types of fuel cells. The first is that the activation over voltage at the cathode is usually less than with an acid electrolyte fuel cell. The second advantage is that the electrodes do not have to be made of precious metals.

Types of Alkaline Electrolyte Fuel Cells

Alkaline fuel cells are categorized depending on their pressure, temperature, and electrode structure, which vary widely between the different designs. One major similarity between all alkaline fuel cells is the use of potassium hydroxide solution as the electrolyte.

Mobile Electrolyte

The mobile electrolyte fuel cell uses pure hydrogen, H_2, as the fuel at the anode and air for the reaction at the cathode. The electrolyte is pumped around an external circuit. One of the major problems that the mobile electrolyte fuel cell faces is the chemical reaction between the potassium hydroxide electrolyte, KOH, and carbon dioxide

(CO$_2$ that is present in the supplied air). The problem is shown in the following reaction:

$$2KOH + CO_2 \rightarrow K_2CO_3 + H_2O \qquad (14.14)$$

The potassium hydroxide is slowly converted to potassium carbonate in the presence of CO$_2$. This is unfavorable because the efficiency and performance of the fuel cell depend on keeping the potassium hydroxide in its pure form. To avoid this problem, a CO$_2$ scrubber is used to remove as much CO$_2$ as possible from the supplied air. In addition, potassium hydroxide is circulated to remove water which is produced during the operation of the fuel cell. The circulated potassium hydroxide helps to prevent the produced water from diluting potassium hydroxide.

Dissolved Fuel Alkaline Fuel Cells
The dissolved fuel alkaline fuel cell is the simplest alkaline fuel cell to manufacture. But it does not work well for large power generation applications. Potassium hydroxide is used as the electrolyte along with a fuel such as hydrazine or ammonia combined with it. Hydrazine (H$_2$NNH$_2$) dissociates into hydrogen and nitrogen on the surface of a fuel cell electrode. The resulting hydrogen is then used as the fuel. However, it is not a preferred fuel because hydrazine is toxic, carcinogenic, and explosive.

Electrode for Alkaline Electrolyte Fuel Cells
Alkaline electrode fuel cells operate at wide range of temperatures and pressures using different types of electrodes. Some of electrodes used in alkaline electrolyte fuel cells are explained here.

Sintered Nickel Powder
Sintered nickel powder was used by F. T. Bacon in his first fuel cell because of the low cost and simplicity of the material. The powder form of the nickel makes it much more porous and therefore more advantageous for fuel cells because the porosity increases the surface area for the chemical reactions. The sintering is used to make the powder a rigid structure. Two different sizes of nickel powder are used to give the optimum porosity for the liquid and gas fuel. The liquid is better with a smaller pore size, and a larger pore size is better for the gas fuel.

Raney Metal
Raney metals are a good materials for achieving the activity and porosity needed in an electrode. A Raney metal is formed by mixing an active metal, for example, Ni (needed for the electrode) and an inactive metal such as aluminum. The inactive metal is then removed from the mixture by dissolving the metal using a strong alkali. The

remaining structure is a highly porous structure made entirely of active metal. An advantage of this process is that the pore size can be easily changed for the desired application by simply altering the mixture ratio of active to inactive metal. Raney metals are often used for the anode i.e., negative side of the fuel cell, and silver is often used for the cathode i.e., positive side of the fuel cell.

Rolled Electrodes

In rolled electrodes, nanocarbon coated with metal catalysts is mixed with polytetraflouroethelene (PTFE) used as an electrode. The purpose of the PTFE is to act as a binder and to control the mixture's porosity. This mixture is then rolled onto a sheet of nickel. Because of the strength-to-weight ratio and the conductivity, carbon nanofiber is often used to increase the strength, conductivity, and roughness of the mixture. Rolled electrode manufacturing can be done on an paper machine, which makes it easy to manufacture at a relatively low cost. Also, the use of nonplatinum electrodes greatly reduces the cost of producing the electrodes.

However, some problems are associated with rolled electrodes. The electrode has a layer of PTFE that is nonconductive, which it increases the resistance of the cell. Joong-Pyo (1998) used an anode based on granules of Raney nickel mixed with PTFE. But in either methods, PTFE increases the resistance of the electrode. In addition, localized generation of water at the site of PTFE takes place during the operation. This causes localized explosion and spoils the electrode. Hence, efforts are being made to avoid using PTFE as well to reduce the quanity of nickel by mixing the nickel metal with some other electrochemically active metals. This is done to reduce the cost of the electrode without losing the electrochemical efficiency of the cell.

Methanol Fuel Cells

To combat the problems arising from use of hydrazine or PTFE, an alternative fuel such as methanol can be used, creating a new type of fuel cell, the methanol fuel cell. The methanol reaction at the anode is:

$$CH_3OH + 6OH^- \rightarrow 5H_2O + CO_2 + 6e^- \qquad (14.15)$$

But as it can be seen, the produced CO_2 will react with the KOH, producing carbonate, which is unfavorable. Because the CO_2 cannot be easily removed from the system, the use of methanol is impractical.

Operating Pressure and Temperature

Alkaline electrolyte fuel cells generally operate at pressure and temperature that are much higher than the environment in which they operate. The open circuit voltage of a fuel cell depends on the temperature and pressure and increases with increasing pressure and temperature. The increase in voltage depends on the pressure of the gas used. Because of the high flammability of pure hydrogen and oxygen, it is essential in this type of storage device to ensure that no leak occurs.

14.10   Molten Carbonate Fuel Cells

The molten carbonate fuel cell (MCFC) uses molten mixture of alkali metal carbonates. The electrolyte is usually a binary mixture of lithium and potassium or lithium and sodium carbonates, which is held in a ceramic matrix of $LiAlO_2$. A highly conductive molten salt is formed by the carbonates at very high temperatures (~600° to 700°C).

Unlike with alkaline fuel cells, CO_2 must be supplied to the cathode instead of being extracted from the supply. The CO_2 and oxygen are essential to react and form the carbonate ions by which the electrons are carried between the cathode and anode. Note that two moles of electrons and one mole of CO_2 are transferred from the cathode to the anode. This is given by:

$$H_2 + \frac{1}{2}O_2 + CO_2(\text{cathode}) \rightarrow H_2O + CO_2(\text{anode}) \qquad (14.16)$$

The Nernst reversible potential for an MCFC is:

$$E = E° + \frac{RT}{2F}\ln\left(\frac{P_{H_2}P_{O_2}^{\frac{1}{2}}}{P_{H_2O}}\right) + \frac{RT}{2F}\ln\left(\frac{P_{CO_{2c}}}{P_{CO_{2a}}}\right) \qquad (14.17)$$

This type of fuel cells operates at relatively high temperatures, which allows them to attain high efficiencies. The CO_2 produced at the anode is commonly recycled and used by the cathode. The reactant air is preheated by burning unused fuel.

Another advantage of this type of fuel cell is that this does not require noble metal for the electrode. Instead of using hydrogen gas, carbon monoxide can also be supplied to the anode as the fuel. This causes production of twice the amount of CO_2 at the anode.

14.10.1   Molten Carbonate Fuel Cell Components

Electrolytes

Molten electrolyte fuel cells contain approximately 60 wt% carbonate in a matrix of 40 wt% $LiOAlO_2$. The $LiOAlO_2$ is made of fibers smaller than about 1 cm in diameter. The matrix is produced using tape casting similar to those used in the ceramics and electronics industries. The ceramic materials are put into a "solvent" during the manufacturing process. A thin film is formed on a smooth surface by use of an adjustable blade device. The material is then heated, and organic binding agents are burned out. The thin sheets are then put on top of each other to form a stack. The operating voltage is largely related to the ohmic resistance of the electrolyte. The most significance factor of ohmic losses is the thickness of the electrolyte described by:

$$\Delta V = 0.533t \tag{14.18}$$

where t is the thickness of the electrolyte in centimeters. Using tape casting, the electrolyte thickness can be reduced to approximately 0.25 to 0.50 mm.

Anode

MCFCs are made of porous sintered Ni–Cr/Ni–Al alloy. The anodes can be made to a thickness of 0.4 to 0.8 mm. The anodes are manufactured by hot pressing the fine powder or tape casting as used for the electrolytes. Chromium is commonly added to reduce the sintering of the nickel, but it causes problems such as increased pore size, loss of surface area, and mechanical deformation under a compressive load in the stack. These problems can be reduced by adding aluminum to the anode. The reactions that take place at the face of the anode are relatively fast at high temperatures; therefore, a large surface area is not needed. This reduces the amount of porosity needed in the structure and allows partial flooding of the anode with molten carbonate without problems. Tape casting allows the manufacture of electrodes with various sizes of porosity.

Cathode

The cathode of an MCFC is made of nickel oxide. The problem with nickel oxide is its solubility in molten carbonates. Nickel ions diffuse into the electrolyte towards the anode. Metallic nickel then precipitates out in the electrolyte. This precipitation causes internal shortages within the fuel cell, which cause electrical problems. The formation of nickel ions is described by:

$$NiO + CO_2 \rightarrow Ni^{2+} + CO_3^{2-} \tag{14.19}$$

This can be reduced using a more basic carbonate in the electrolyte. Nickel dissolution can be reduced by:

- Using a basic carbonate
- Reducing the pressure in the cathode and operating at atmospheric pressure
- Increasing the thickness of the electrolyte to increase the amount of time and distance it takes for Ni^{2+} to reach the anode

14.11 Polymer Electrolyte Fuel Cells

The main advantage of polymer electrolyte fuel cells (PEFC) is that they have the ability to operate at very low temperatures. Because they have the ability to deliver such high power densities at this temperature, they can be made smaller, which reduces their overall weight, the cost of production, and the specific volume of the electrode. Because the polymer electrolyte membrane fuel cell (PEMFC) has an immobilized electrolyte membrane, there is simplification in the production process that in turn reduces corrosion, providing for a longer stack life. This immobilized proton membrane is a solid-state cation transfer medium. Some groups call this type of cell a solid polymer fuel cell (SPFC). Similar to all fuel cells, PEFCs consist of three basic parts: the anode, the cathode, and the membrane. PEFCs are being actively pursued for use in automobiles and portable applications.

14.11.1 Polymer Membrane

Although many different types of membranes are used, the most common is Nafion, produced by DuPont. Other types of membranes being researched are (1) polymer–zeolite nanocomposite proton-exchange-membrane sulfonated polyphosphazene-based membranes and (2) phosphoric acid-doped poly (bisbenzoxazole) high-temperature ion-conducting membranes. The Nafion layer is essentially made of carbon chain that has a fluorine atom attached to it. The Nafion membrane possesses the following useful properties:

- Highly chemically resistant
- Mechanically strong (possible to make them as small as 50 μm)
- Acidic
- Highly absorptive to water
- A good proton (H) conductor if well hydrated

14.12 Direct Methanol Fuel Cells

The operation of the whole DMFC (direct methanol fuel cells) system is similar to the operation of the PEMFC in terms of the physical manufacturing of the cell. The major difference is in the fuel cell supply. The fuel is a mixture of water and methanol. Fuel reacts directly at the anode according to:

$$CH_3OH + H_2O \rightarrow 6H^+ + 6e^- + CO_2 \tag{14.20}$$

and at the cathode:

$$6H^+ + 6e^- \rightarrow 3/2O_2 + 3H_2O \tag{14.21}$$

Because the boiling point of methanol at atmospheric pressure is 65°C, the cells require an operating temperature of around 70°C. The reaction mechanism is much more complex. The overall electrochemical reactions at both the cathode and the anode are as follows:

$$CH_3OH + 1.5O_2 \rightarrow 2H_2O + CO_2 \qquad (14.22)$$

This corresponds to a theoretical voltage of 1.21 V at standard temperature and pressure (STP).

The voltage losses of a DMFC are similar to those associated to a hydrogen fuel cell. It has activation losses, ohmic losses, mass transport losses, and fuel crossover losses. During the electrochemical reaction, it has been observed that products such as formic acid (HCOOH) and formaldehyde (HCOH) are formed, which inactivate the electrode.

DMFCs are less attractive than the pure hydrogen fuel cells for several reasons. These problems are associated with the inability to get the full potential out of the anode and cathode.

14.13   Phosphoric Acid Fuel Cells

Phosphoric acid fuel cells are very similar to PEMFCs. They use a proton-conducting electrolyte. The chemical reactions use highly dispersed electrocatalyst particles within carbon black. The electrode material is generally platinum. Concentrated phosphoric acid is used as the electrolyte, hence the name of the fuel cell. This electrolyte conducts protons. The phosphoric acid fuel cell operates at approximately 180° to 200°C.

14.13.1   Electrolyte

Phosphoric acid is used as the electrolyte because it is the only inorganic acid that exhibits the required thermal stability, chemical and electrochemical stability, and low enough volatility to be effectively used. Phosphoric acid does not react with CO_2 to form carbonate ions such as the case with alkaline fuel cells; therefore, carbonate formation is not a problem with phosphoric acid fuel cells. Phosphoric acid has a freezing point of 42°C, which is high compared with the electrolyte materials used in other fuel cells. Small amounts of the acid electrolyte are lost during operation; therefore, the excess acid should be initially put into the fuel cell or the acid should be replenished.

14.13.2   Electrodes and Catalysts

The phosphoric acid fuel cell uses gas diffusion electrodes. The primary choice for the catalyst is platinum on carbon. Carbon is bonded with PTFE to create the catalyst structure. Carbon plays following major functions:

- It disperses the Pt catalyst to ensure good utilization of the catalytic metal.
- It provides micropores in the electrode for maximum gas diffusion to the catalyst and electrode–electrolyte interface.
- It increases the electrical conductivity of the catalyst.
- The quantity of platinum is reduced by using carbon materials.
- Platinum loadings are approximately 0.10 mg Pt/cm^2 in the anode and about 0.50 mg Pt/cm^2 in the cathode.

14.14   Use of Carbon in Fuel Cells

Phosphoric acid fuel cells and alkaline fuel cells have been developed, and working models of these fuel cells are available. However, none of these fuel cells can find its use for terrestrial application because of the high cost involved in making suitable electrodes for these cells. Scientists are searching for economical electrodes to replace the expensive materials such as Raney nickel (Dicks, 2006).

Carbon, on the other hand, can be synthesized with properties similar to metal as well as semiconductors from the same precursor by altering the temperature of pyrolysis and the catalyst. Hence, the cost of the electrode can be reduced tremendously by using CNMs for developing electrodes for fuel cells.

14.15   Fuel Cells Using Carbon Nanomaterials

Sharon et al. (2003a&b), have been trying to develop CNMs for fuel cells. Their approach to reduce the cost of the electrodes is by (1) synthesis of CNMs from plant-derived precursors so that cost of CNTs could be reduced and (2) use inexpensive pumice porous stone as the substrate for developing electrodes for alkaline fuel cells. Because though pumice stone is made from ceramic oxides, it is highly nonconducting but it can be made conducting by depositing conducting carbon over it, and it is inexpensive.

With chemical vapor deposition (CVD), a thin layer of conducting carbon is deposited over the porous pumice stone. Over this electrocatalyst, nickel alloy with Sn, Ag, or Mg of desired quantity is deposited. To protect the corrosion of this electrocatalyst, a thin layer of semiconducting carbon is deposited by CVD. This electrode can be used either as hydrogen electrode (i.e., those deposited with Ni–Sn alloy) or oxygen electrode (nickel deposited with Mg or Ag). This type of cell has given potential of about 0.9 V per cell and current in the range of 200 mAcm^{-2}.

14.16   Summary

This chapter has dealt with various aspects of different types of fuel cells. The basic design and principles of fuel cells have been discussed to explain why the generation of electrical power through an electrochemical process using fuel cells is more efficient than the conventional process of converting chemical energy to electrical power via the thermal route. The characteristics of various types of fuel cells have been discussed. Moreover, the possibility of using CNMs as electrodes has also been discussed.

CHAPTER 15

Electric Double-Layer Capacitors and Carbon Nanomaterials

Maheshwar Sharon and Vilas Khairnar

Nanotechnology Research Centre, Birla College, Kalyan, Maharastra, India

Chi-Chang Hu

Department of Chemical Engineering, National Chung Chen University, Taiwan

15.1 What Is A Supercapacitor?

Supercapacitors are electrochemical double-layered capacitors with moderate energy and high-power densities. A supercapacitor is a device that uses the storage of charge in the electric double layer between an electrolyte and an electrochemically inert electrode. To be more precise, an electric double-layer capacitor (EDLC) is as a device that uses induced ions between an electronic conductor, such as activated carbon, and an ionic conductor, such as an organic or aqueous electrolyte. Supercapacitors are an intermediate system between electrochemical batteries that can store high energy associated with low power and dielectric capacitors, which can deliver high power in few milliseconds. Recently, there has been much interest in EDLCs because of their practical applications as high-power devices for having large capacitance and long cycling lives and because they are free from toxic materials. EDLCs have been developed as energy storage devices applicable in consumer electronic products and electric vehicles for the purpose of memory backup or pulse current supply. They may be coupled with batteries to provide pulses of peak power during acceleration, especially during climbing up hill gradients. They overcome a capacitor's limited ability to store charge and a battery's limited ability for rapid energy discharge. Hence, they are able to store energy like a battery, which generate energy via chemical reactions.

The discovery of the possibility of storing an electrical charge on the surface arose from phenomena associated with the rubbing of amber in ancient times. This led to the discovery of the conventional capacitor. The conventional (electrostatic) capacitor is made up of two electrodes separated by a dielectric media (Fig. 15-1). The extent of capacitance held is given by equation 15.1. This can store charge to

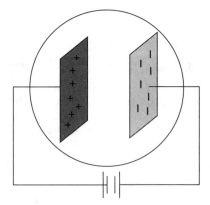

FIGURE 15-1   A conventional capacitator. Parallel plates are separated by a high dielectric media such as a vacuum, forming a conventional capacitor. This can store charge to the extent that there is no leakage of charge across the two plates through the dielectric media.

the extent the charge can be prevented from its discharge through the dielectric mediator.

$$\frac{1}{C_{total}} = \frac{1}{C1} + \frac{1}{C2}$$

(15.1)

$$C_{total} = \frac{C1C2}{C1+C2} \ldots\ldots \quad \ldots\ldots \quad \ldots\ldots$$

EDLCs differ from conventional capacitors. EDLCs have the same basic structure as conventional capacitors except that instead of high dielectric media, EDLCs use electrolyte either of low pH (acidic media) or high pH (alkaline media) or organic liquid. The charges are held in the same fashion at the electrode except that the opposite charge is held near the electrode separated by a distance of few nanometers.

The amount of energy a capacitor can hold is measured in micro-farads or µF (1 µF = 0.000,001 farad). Whereas small capacitors are rated in nanofarads (1000 times smaller than 1 µF) and picofarads (1 million times smaller than 1 µF), supercapacitors come in farads. The gravimetric energy density of a supercapacitor is 1 to 10 Wh/kg. This energy density is high compared with a regular capacitor. The electrochemical battery, similar to a lead acid battery, delivers a fairly steady voltage in the usable energy spectrum; the variation in voltage of the supercapacitor is linear and drops evenly from full voltage to 0 V. Because of this, the supercapacitor is unable to deliver the full charge. For example, if a 6-V battery is allowed to discharge to 4.5 V before the equipment cuts off, the supercapacitor reaches that threshold within the first quarter of the discharge cycle. The remaining energy slips into an unusable voltage range.

Supercapacitors are most commonly used as memory backups to bridge short power interruptions. Another application is in improving the current handling of a battery. The supercapacitor is placed parallel to the battery terminal and provides a current boost on high load demands. The supercapacitor will also find a ready market for portable fuel cells to enhance peak-load performance. Because of their ability to rapidly charge, large supercapacitors are used for regenerative braking on vehicles.

The charge time of a supercapacitor is about 10 seconds. The ability to absorb energy is, to a large extent, limited by the size of the charger. The charge characteristics are similar to those of an electrochemical battery. The initial charge is very rapid; the topping charge takes extra time. Provision must be made to limit the current when charging an empty supercapacitor.

In terms of the charging method, supercapacitors resemble lead-acid batteries. Full charge occurs when a set voltage limit is reached. Unlike with an electrochemical battery, a supercapacitor does not require a full-charge detection circuit. Supercapacitors take as much energy as needed. When full, they stop accepting charge; there is no danger of overcharge.

Supercapacitors can be recharged and discharged virtually an unlimited number of times. Unlike with electrochemical batteries, there is very little wear and tear induced by cycling, and age does not have much of an effect on supercapacitors. In normal use, a supercapacitor deteriorates to about 80% after 10 years. The self-discharge of supercapacitors is substantially higher than that of electrochemical batteries. Supercapacitors with an organic electrolyte are affected the most. In 30 to 40 days, the capacity decreases from full charge to 50%. In comparison, a nickel-based battery discharges about 10% during that time.

Supercapacitors are relatively expensive in terms of their cost per watt. Supercapacitors and chemical batteries are not necessarily in competition. Rather, they enhance each other. They have virtually unlimited cycle life. They have low impedance and enhance load handling when put in parallel with a battery. It is charged within few seconds. There is no need to use full-charge detection; no danger of overcharge. Linear discharge voltage prevents use of the full energy spectrum.

However, supercapacitors have a low energy density, typically holding one fifth to one tenth the energy of an electrochemical battery. Cells in supercapacitors have low voltages; serial connections are needed to obtain higher voltages. Voltage balancing is required if more than three capacitors are connected in series.

15.2   Historical Development of Supercapacitors

The discovery of the possibility of storing an electrical charge on the surface arose from phenomena associated with the rubbing of amber in ancient times; however, the origin of such effects was not understood until the mid-eighteenth century in the period when the physics of so-called "static electricity" was established. In relation to such historic investigations, the development of Leydenjar and the discovery of the principle of charge separation and charge storage on the two surfaces in the Leyden jar, separated by a layer of glass, were of major significance for the physics of electricity and later for electrical technology, electronics, and electrochemical engineering. Utilization of this principle to store the electrical energy for practical purposes, as in a cell or battery in an aqueous solution, was first proposed and claimed as an original development in the patent granted to Becker in 1957.

Commercial double-layer capacitors originated at the Standard Oil and Ohio Research Center (Cleveland) in 1961 to 1962. Sohio used the double-layer capacitance of high surface area carbon materials in a nonaqueous solvent containing a dissolved tetra alkyl ammonium salt electrolyte. Because of a lack of sales, however, Sohio halted its development efforts in 1971 and later licensed the technology to Nippon Electric Company, which developed and marketed commercial double-layer products. During the 1980s, Matsushida Electric Industrial Co. of Japan patented methods for producing double-layer capacitors containing improved electrodes. One, made from activated carbon fibers woven into a fabric, was the initial basis for the development of the EDLC called the Gold capacitor.

15.3   What Is An Electrical Double Layer?

Before discussing EDLCs, it is useful to digress a little from the subject and discuss the electric double layer. The electrolyte in an aqueous media is normally associated with water molecules known as *hydrated ions* (Fig. 15-2).

Each water molecule has a dipole, and when it comes in contact with cation, its charges are directed toward the positively charged cations. Thus, cations in the presence of water molecules get hydrated, as shown in Fig. 15-2. The number of water molecules attached to cations depends on the charges of the cation and its ionic radii. In aqueous solution, the distribution of hydrated cations is such that there is no accumulation of charge anywhere in the solution; the solution remains neutral. However, when a metal (which has large number of mobile electrons) is dipped into the electrolyte, the neutrality of the solution is disturbed. Hydrated cations accumulate near the interface of the metal and the electrolyte because of electrostatic attraction between the hydrated cations and the mobile electrons of the metal. Accumulation of electrons of the metal occurs near its surface. The number of hydrated cations (equal number of electrons accumulated at the metal interface) that gets accumulated near the surface of the metal depends on the nature of the metal (see Fig. 15-3). Because of accumulation of electrons and the hydrated ions near the interface of metal and electrolyte, an electrostatic potential develops. The magnitude of potential depends on the nature of electrolyte and

Cation surrounded with water molecules directing the dipole towards cation

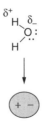

Water molecule with dipole

FIGURE 15-2   The water molecule possesses a dipole due to loan pairs of oxygen. When cation interacts with water molecules, negative charge gets directed toward the cation, making a hydrated ion.

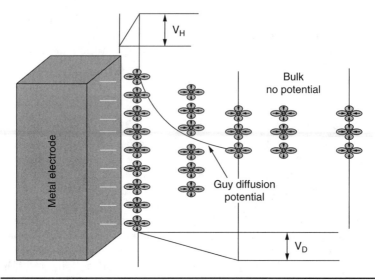

Figure 15-3   Schematic representation of double layers formed when a metal electrode is immersed in an electrolyte. V_H is the Helmholtz potential formed between electrons present at the surface of metal electrode and hydrated positive ions present at the interface. V_D is the Guoy diffusion potential formed due to the difference in the concentration of cations between the Helmholtz plane and the plane from where the bulk concentration starts.

the metal. It is worth noting that the distance of separation of the cation from the surface of the metal is almost equal to the diameter of the water molecule. In other words, two types of charges are formed at the interface of the metal and the electrolyte, and these charges behave as if they are present in two parallel planes separated by a water molecule.

The potential developed from these charges is known as the *Helmholtz potential* (V_H). Because the hydrated plane near the interface is separated by a distance equal to the diameter of the water molecule, all cations lie in one plane parallel to the metal surface. Hence, the plane occupying these hydrated cations is referred to as the *Helmholtz plane*. This condition resembles that of a parallel plate capacitor; one plane is the surface of the metal, and other is the Helmholtz plane that contains the hydrated cations. The magnitude the capacitance created due to formation of this double layer of charges is about 12 to 14 $\mu F/cm^2$.

As a result of the accumulation of hydrated cations at the interface of metal, a concentration gradient of cations between the bulk and the interface of the metal and electrolyte is created. The potential developed due to such change of concentration is know as the *Gouy Chapman potential* (V_D), and the capacitance created due to this variation in charges is known as the *Gouy-Chapman capacitance*. Because there are two types of capacitances formed in series, the total capacitance is given by equation 15.2:

$$\frac{1}{C_{total}} = \frac{1}{C_H} + \frac{1}{C_D}$$ (15.2)

Whereas the Helmholtz plane is separated by 2.0 to 3.0 nm, the Gouy-Chapman capacitance is separated by a distance in the range of few hundred nanometers. If capacitance is measured for this type of system (i.e., capacitance due to Helmholtz $[C_H]$ and Gouy-chapman $[C_D]$), the total capacitance would be equal to the Helmholtz capacitance only, because $1/C_D$ would become a very small quantity compared with $1/C_H$.

The magnitude of the capacitance of such type of double layers is given by equation 15.3:

$$C = \frac{\varepsilon}{4\pi\delta} \int dS$$ (15.3)

where C is the capacitance of the double layer, ε is the dielectric constant of the electrolyte, δ is the distance from the electrode interface to the center of the ion, and S is the surface area of the electrode interface. This shows that the capacitance is a constant quantity for a given area of the electrode. Thus, the capacitance for unit area (i.e., 1 cm² of electrode) is around 12 to 14 µF. However, if the area of the electrode could be increased to 1000 m²/g, the total capacitance becomes almost 12×10^{-4} Fcm$^{-2} \times 1000 \times 10^2 = 1200$ F/g (Fig. 15-4).

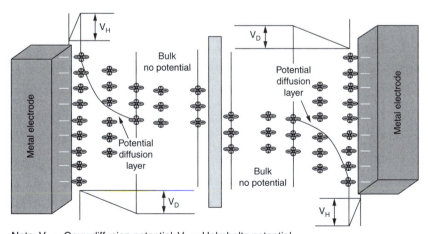

Note: V_D = Gouy diffusion potential; V_H = Helmholtz potential.

FIGURE 15-4 Two similar capacitors formed by immersing two electrodes in one solution. Two electrodes are separated by a membrane. Helmholtz and Gouy diffusion layers are formed on both electrodes.

15.3.1 Double-Layer Capacitor (Nonfaradaic Capacitor)

It is worth noting that an electrochemical cell forming this kind of double-layer capacitor has two separate electrodes separated by a membrane. Hence, at each electrode, a double layer of the type, as discussed earlier, is formed (see Fig. 15-4).

Both of these electrodes are arranged in series (Fig. 15-5) and hence the total capacitance would be given by equation 15.1.

Under these conditions, if both electrodes contain identical double layers, then the total capacitance would depend on the magnitude of individual capacitance governed by equation 15.1.

This type of double-layer capacitance is also known as *nonfaradaic capacitance* because there is no net charge transfer across the electrode, and current passing through the two electrodes is due to exchange current or equilibrium current established at the interface of each electrode.

15.3.2 Pseudo Capacitor (Faradaic Capacitor)

However, it is possible to store some additional charge at the electrode by performing a redox reaction with adsorbed ions on the electrode (i.e., by performing the following reactions):

$$\text{Adsorbed species} + ne^- \leftrightarrow \text{Reduced species} \qquad (15.4)$$

Alternatively, electrodes can be loaded with some metal or ions, which under specific electrode potential get reduced:

$$M^{n+} + S + ne^- \leftrightarrow SM \qquad (15.5)$$

where M^{n+} and S are cations and the metal electrode at which the reaction is occurring or any adsorbed species on the electrode, and n is the number of electrons involved in the reaction.

In both cases, the electrode gets accumulated with additional electrons. This process is equivalent to charging the electrode with a

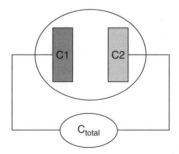

Figure 15-5 Two capacitance formed at two electrodes of a cell (see Fig. 15-4) will act as if they are connected in series; hence, total capacitance will be given by equation 15.3. C1 and C2 represent the Helmholtz capacitances present at the two electrodes of a cell.

Device	Energy density (Wh/dm³)	Power density (W/dm³)	Cycle life
Battery	50–250	150	<104
Electrochemical capacitor	5	>105	>105
Conventional capacitor	0.05	>108	>106

TABLE 15-1 Comparison of Properties of Electrical Energy Stored by Batteries and Capacitors

negative charge. In the Helmholtz capacitance, there is no flow of current (i.e., this is a nonfaradaic capacitance). However, under the reactions represented by equations 15.4 and 15.5, the metal electrode gets additional negative charge because of the flow of current. The capacitance formed due to accumulation of this additional charge is known as the *Faradaic capacitance*. The only difference is that this type of charging follows Faradaic laws and the amount of charge deposited on the electrode depends on the number of species reduced. In other words, charge generated on the electrode is equivalent to the amount of current passed through the system. Capacity generated due to such a Faradaic reaction is known as *pseudo capacitance*. One of the necessities for this kind of capacitance is that the charging process must be fully reversible and there should be no degradation of either the reactant or the product. This type of capacitance is almost equivalent to charge generated in a battery. In Table 15-1, the properties of electrical energy stored by the battery and capacitor are given for comparison.

Apart from the power density or energy density consideration, an electrochemical capacitor is extremely useful for providing its full power to the load in few seconds, which is not possible with batteries. Batteries can provide a constant power over a longer period, but electrochemical capacitors (i.e., supercapacitors) provide surges of power. Therefore, the utility of batteries and supercapacitors differs, and these technologies are complementary to each other.

15.4   Design of a Supercapacitor

There are different supercapacitor's designs, including cylindrical, prismatic, button, or coin types, with some larger embodiments being of cake-tin sizes or larger and some multicell series for higher voltage with bipolar electrodes having edge seals. However, to explain their characteristic properties, a basic design of supercapacitors is discussed here.

Supercapacitor electrodes are normally composed of a current collector and a highly porous, nonreactive material. Powdered nonreactive

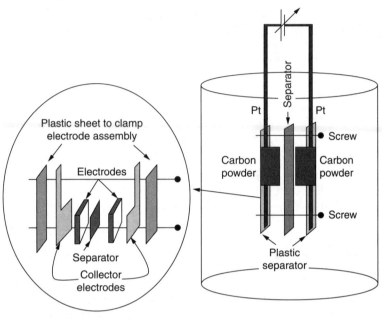

Note: Pt stands for platinum.

FIGURE 15-6   A typical electrode assembly of a double-layer capacitor.

material is mixed with a suitable binder and sometimes with an electronic conductor. It is then laminated on each side of the current collector. The current collector should be electrochemically inactive in the potential window in which the system works. A schematic diagram of a double-layer capacitor is shown in Fig. 15-6.

Powder carbon nanomaterial (CNM) or any other material to be used for the electrode of a known weight (in the range of 5 to 10 mg) is spread over a platinum plate (or any conducting inert material) covering an area of 1 cm² (or any desired size). A Whatmann filter paper (or any other separator that can allow transfer of ions from one side of the half cell to other side of the half cell) is placed over the powder, over which a platinum plate (or the conducting materials) is placed. This entire assembly is then sandwiched between two plastic plates and screwed to make it tight. The length of the platinum plates should be such that small portion of both could project out of the plastic plate for making electrical connections.

The electrode assembly is dipped in a suitable electrolyte (e.g., 1M KOH or 0.1M H_2SO_4) solution overnight so the carbon powder and separator get fully soaked with the electrolyte. Finally, this assembly is kept in a beaker (or any container) containing the same electrolyte (see Fig. 15-6). The end of the platinum plates is connected to a potentiostat or galvanostat (or a DC power source). A saturated calomel electrode with 1M KOH solution or any other reference electrode is used to measure the potential applied to the electrode (this is needed only if potential developed to electrode is to be measured). For normal charging purposes, the two platinum electrodes can be connected to a DC power supply.

15.4.1   Electrodes for Supercapacitors

The requirements of electrode for electrochemical capacitors are substantially different from those for battery electrodes. The electrodes for the capacitors must possess the following properties:

- Capable of multiple cycling (cycle life $>10^5$ times)
- Long-term stability related to the cycle life
- Resistance to electrochemical reduction or oxidization
- Maximum operating potential range so that it can operate at a higher potential at which the electrolyte does not decompose
- Optimized pore-size distribution for maximum specific area in the range of 1000 to 2000 m^2g^{-1}
- Minimized internal electrolyte resistance
- Good wettability and hence a favorable electrode–solution interface contact angle
- Minimized ohmic resistance of the electrode material and other electrical contacts
- Minimum self-discharge on open circuit

15.4.2   Electrolytes for Supercapacitors

When the electrolyte used in an EDLC possesses a high coefficient of viscosity or a low degree of conductivity, the internal resistance of the EDLC as a whole increases, thereby decreasing the voltage of the capacitor during charging and discharging. Therefore, an electrolyte used in an EDLC is required to possess a low coefficient of viscosity and a high degree of conductivity. Also, because an EDLC is used in a sealed state for an extended period of time and is charged and discharged repeatedly, the electrolyte used in the capacitor must possess long-term stability. Conventionally, in view of ensuring long-term stability, electrolytes for use in EDLCs consist of a tetra alkyl ammonium salt, such as triethylmethylammonium tetrafluoroborate, dissolved in a cyclic carbonate, such as propylene carbonate.

15.4.3   Measurement of the Capacitance of Supercapacitors

Double-layer capacitance can be measured by either the potentiostatic method (in which variable potential is applied and the corresponding current is measured) or galvanostaticaly (in which a constant current is applied and the change in potential with time is measured).

Potentiostatic Method

Cyclic voltammetry is a kind of potentiodynamic electrochemical measurement. The three-electrode method is the most widely used because the electrical potential of reference does not change easily during the measurement. This method uses a reference electrode, working electrode, and counter electrode (also called the secondary or auxiliary electrode). Electrolyte is usually added to the test solution to ensure sufficient conductivity. The combination of the solvent, electrolyte, and specific working electrode material determines the range of the potential.

The potential is measured between the reference electrode and the working electrode, and the current is measured between the working electrode and the counter electrode. This data are then plotted as current (i) versus potential (E). Some time symbool for current is also expressed as "C" and for potential "V". As the waveform shows (Fig. 15-7), the forward scan produces a current peak at certain high potential (in the vicinity of 4V). For measuring the capacitance, potential range is selected only where magnitude of current shows almost independent of applied potential. Normally (as shown in Fig. 15.7) the current increases as the potential reaches the reduction potential of the analyte. If potential mantained at the reduction potential, current falls off as the concentration of the analyte is depleted. As the applied potential is reversed and reaches a potential that reoxidizes the product formed in the first reduction, reaction current of reverse polarity is observed. This oxidation peak usually has a shape that is similar to the reduction peak. A typical graph obtained by measuring current versus applied potential to the electrodes of the cell (see Fig. 15-6) is shown in Fig. 15-7.

The potential range is shown from about -1 to 4 V. The potential range, in general, may be from a negative potential to a positive

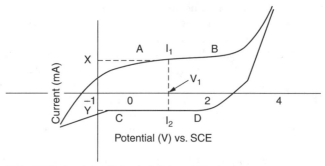

Note: SCE stands for Saturated Calomel Electrode.

***

Figure 15-7 A typical cyclic voltammogram of a double-layer capacitor.

potential, or the entire potential range may be in a positive potential. This range depends on the type of materials used. It is necessary that for certain range of potential, the current variation should be minimum (e.g., AB or CD range, as shown in Fig. 15-7). Under such conditions, the middle of the current range (i.e., I_1 of AB or I_2 of CD) is taken as the magnitude of the current at potential V_1. If there appears to be a peak during the scanning potential from either −1 to 4 V or from 1 to −1 V, it would mean that the electrolyte, the metal electrode, or the carbon powder is not electrochemically stable. In these cases, the material responsible to give the oxidation/reduction peaks should be eliminated from making a supercapacitor. If current-voltage curve (CV) shows a plateau (e.g., AB and CD), then the capacitance C is calculated in the following manner:

$$C = \frac{dQ}{dV} \tag{15.6}$$

where (dQ/dV) is the rate of change of surface charge density of the double layer with electrode potential. In cyclic voltammetry, the scan rate can be expressed as:

$$S = \frac{dV}{dt} \qquad or \qquad dV = Sdt \tag{15.7}$$

where S is the scan rate. Scan rate means the rate at which the potential applied to the electrode changes with time. By combining these two equations, we have:

$$C = \frac{dQ}{Sdt} \tag{15.8}$$

and (dQ/dt) is equal to current I where:

$$I = \frac{I_1 - I_2}{2} \tag{15.9}$$

where I_1 and I_2 are the anodic and cathodic current density.
Therefore:

$$C = \frac{I \frac{amp}{cm^2}}{S \frac{volt}{sec}} = \frac{(amp)(sec)}{volt\ cm^2} = F\ cm^{-2} \tag{15.10}$$

In the potentiostatic method, the current–voltage curve can be obtained with the electrochemical cell at different scan rates, and capacitance is calculated from the curve obtained with different scan rates by using equation 15.10. It is normally observed that capacitance is dependent on the scan rate. Therefore, capacitance calculated with the potentiostaic method should always mention the scan rate at which the (CV) was measured. A typical graph of the variation of scan rate with capacitance is shown in Fig. 15-8. This graph suggests

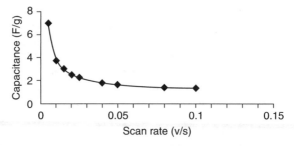

FIGURE 15-8 Variation in capacitance of carbon nanofibers obtained from rice straw with respect to the scan rate.

that as the scan rate increases the capacitance decreases. When the scan rate increases the time given to the electrode to adjust with the applied potential decreases. If the materials used for making the supercapacitor are not able to store the charge quickly (i.e., adjust the charge as potential changes with time), the dependence of capacitance with scan rate is seen. Therefore, it is important to realize that during the application of a capacitor, because in practice charging is normally done by DC voltage, the value of capacitance obtained with the slowest scan rate should be considered as a useful value of capacitance.

Galvanostatic Method
In the galvanostatic method, charging of the capacitor is done by passing a constant current, and the development of potential is measured with respect to time. Likewise, the capacitor is discharged by reversing the current, and the change in potential is again measured with respect to time. Thus, a graph is plotted between the potential and corresponding time. A typical graph of charging and discharging of a capacitor is shown in Fig. 15-9.

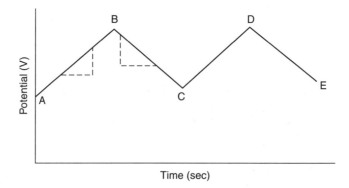

FIGURE 15-9 Typical graph of charging and discharging of a double-layer capacitor. AB and CD represents the charging curve, and BC and DE represent the discharging curve of a capacitor at a fixed constant current.

The slope of linear graph of either charging or discharging process is:

$$slope = \frac{dV}{dt} \tag{15.11}$$

The capacitance thus becomes:

$$C = \frac{I(amp)}{\frac{dV}{dt}} = \frac{I\left(\frac{Amp}{cm^2}\right)}{slope\left(\frac{volt}{sec}\right)} = \frac{F}{cm^2} \tag{15.12}$$

It is customary to check the capacitance by measuring the change in potential with time for different currents to confirm that the system is working perfectly. If the system is working perfectly, the same capacitance should be obtained for different charging and discharging currents.

Capacitance by Using an Equivalent RC Circuit

Assuming an equivalent circuit of cell of configuration "electrode/electrolyte/electrode" to behave like simple RC circuit (Fig. 15-10), the variation in current I flowing through the circuit with time will follow as per equation 15.13:

$$I = I_o e^{-\frac{t}{RC}} \tag{15.13}$$

where I_o is the saturation current (also known as exchange current), R is the resistance, C the capacitance of the cell, and t is time.

Taking the log of this equation, we have:

$$\ln(I) = \ln(I_o) - \frac{t}{RC} \tag{15.14}$$

Thus, if a constant potential is applied to the cell, charging of capacitance will take place, and current flowing through the circuit

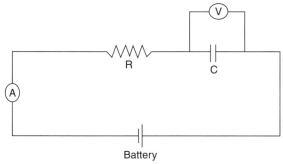

Battery

Note: A = Ammeter, R = Resistance, C = Capacitor, V = Voltmeter

Figure 15-10 Equivalent RC circuit for measuring the capacitance of a capacitor

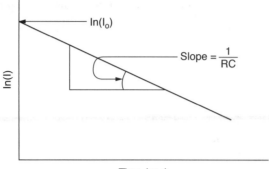

Note: RC = Resistance C = capacitance (RC is time constant of the cell).

FIGURE 15-11 A linear relationship between ln (current) versus time when a fixed potential is applied to the cell.

will also change with time. When ln I is plotted with time t, a linear graph is obtained (Fig. 15-11). The slope will be equal to 1/RC, and from the intercept, the value of I_o can be calculated. For a good capacitor, the magnitude of exchange current should be very low so it can prevent the leakage of charge. If the value of resistance R is known, the capacitance can be calculated.

15.5 Precautions in Using Supercapacitors

1. EDLCs have an inherent polarity. Before using the cell, the polarity of the electrode should be confirmed. If the capacitor has a reversed voltage, there will be an extremely large leakage current, which will lead to reduction of capacitance and an increased internal resistance. In some scenarios, the capacitor will be ruined.

2. Voltages beyond the rated voltage should never be applied to the cell. When a voltage exceeding the rated voltage is applied, there will be a large leakage of current and intense heat generation, causing a reduction in capacitance. Also, the internal resistance will increase, and in some cases, the capacitor will be ruined.

3. EDLCs should never be used for leveling of power supply (ripple attenuation). Because EDLCs have large internal resistances, using these capacitors for power supply leveling will cause excessive internal heating, causing a reduction in the capacitance and increased internal resistance. In some cases, the capacitor will be ruined.

4. EDLCs should never be used in circuits in which repetitively and rapidly charge and discharge processes are occurring. Using EDLCs in circuits where there is frequent and rapid charging and discharging will cause excessive internal heating; capacitance reduction and an increased internal resistance. In some cases, the capacitor will be ruined.

5. The ambient temperature has a major effect on the lifespan of the EDLC. The lifespan is approximately doubled when the temperature is reduced by 10°C. The capacitor should be placed as far away as possible from heat sources.

6. Voltage decreases in EDLCs when used as a backup power source. It should be noted that there will be a momentary decrease in voltage because of the internal resistance of the EDLC. The momentary decrease is caused by the operating current when the backup power supply switches to the backup mode.

7. EDLCs should not be used in environments that have water droplets or poisonous gases. Such environments may cause the lead wires and case of the EDLC to corrode, leading to open circuits.

8. EDLCs should not be stored in a high-temperature, high-humidity environment. The capacitor should be stored in a cool, dry place, with a temperature between 5° and 30°C and a relative humidity of less than 60%. Rapid changes in temperature should be avoided because this variance may cause condensation or may damage the product.

9. When EDLCs are used on double-sided circuit boards, caution is needed so that the interconnect pattern does not cross the location where the capacitor is attached. Depending on how the capacitor is mounted, there is a danger that such a layout could cause a short circuit.

10. When soldering the capacitor to the wiring board, the body of the capacitor should not be attached to the circuit board. If the body of the capacitor is attached directly to the circuit board, the flux or solder can blow through the holes in the circuit board, negatively impacting the capacitor.

11. After attaching the capacitor, the body of the capacitor should not be forcibly tilted or inverted. The added force may cause the leads to become detached. If a strong force is applied to the internal cell by the lead, the capacitor may experience a permanent loss in performance.

12. The leads are coated with solder plating. If the leads are scraped and the plating is removed, it will make the capacitor difficult to solder.

13. Excessive heating of the capacitor during solder dip will greatly diminish the lifespan of the product.

14. When EDLCs are connected in series, if the voltage balance between the capacitors is lost, then one or more capacitors may be subjected to an excessive voltage. This excessive voltage may destroy the capacitor.

15.6   Carbon Nanomaterials for Supercapacitors

Since the discovery of carbon nanotubes in 1990, there has been intensive research activity in the area, not only because of their fascinating structural features and properties, but also because of their potential technological applications. Vilas Khairnar et al. (2004 and 2008) have used carbon materials synthesized from plant derived precursors to make supercapacitor giving capacitance in the order of 40-80F/g. The main practical requirements for better capacitance are high specific areas (a few hundreds to thousands m^2/g), optimized porosities and pore size (2 to 50 nm) distributions (Vilas Khairnar et al., 2008), and good electrochemical stability on cycling. These properties are likely to be obtained with CNMs.

15.7   Summary

There is a difference between a supercapacitor and a conventional capacitor. Whereas the media between the two electrodes of a supercapacitor contains liquid, the electrodes in a conventional capacitor are separated by a high dielectric material, or the entire system is kept under vacuum. The capacitance formed due to the presence of a double layer between the electrode and the electrolyte is one of the components of a supercapacitor. This is known as nonfaradaic capacitor. In addition to this, electrode material can be loaded with some reversible ions such as Mn or Sn. This adds an additional capacitance to the system. This capacitance is known as pseudo capacitance or Faradaic capacitance. In this chapter, the theory of double-layer capacitors (supercapacitors) and techniques to measure the capacitance have been discussed in detail.

CHAPTER 16

Hydrogen Storage by Carbon Nanomaterials

Sandesh Jaybhaye and Maheshwar Sharon

Nanotechnology Research Centre, Birla College, Kalyan, Maharashtra, India

16.1   Introduction

In 1766, British scientist Henry Cavendish first identified hydrogen as a distinct element after he evolved hydrogen gas by reacting zinc metal with hydrochloric acid. In a demonstration for the Royal Society of London, Cavendish applied a spark to hydrogen gas, yielding water. This discovery led to the finding that water (H_2O) is made of hydrogen and oxygen.

Hydrogen is the simplest and lightest element and is considered to have been the primordial substance from which all other elements in the universe evolved. Stars were formed by gravitational forces from a rotating mass of hydrogen; the resultant high temperature led to the fusion reaction converting hydrogen to helium, releasing thermal energy, as in the sun, and leading to the formation of the rest of elements found on earth.

Hydrogen is lighter than air and 10 times less dense than water. One gram of hydrogen gas occupies about 11 L (2.9 agl) of space at atmospheric pressure. Hydrogen in the form of water is absolutely essential to life, and it is present in almost all organic compounds. Hydrogen is ninth in abundance by weight of the elements found in the earth's crust (and third in the number of atoms), most of it being found in water, which contains 11.2% hydrogen.

Hydrogen has great potential as an energy source. Unlike petroleum, it can be easily generated from renewable energy sources. It is also nonpolluting and forms water as a harmless byproduct during various reactions. Hydrogen has received increased attention as a renewable and environmentally friendly element to help meet today's energy needs. The rediscovery of energy imbalances and the increasing long-range concern over fuel supplies accelerated both scientific and economic interest in hydrogen as early as 1970. In the late 20th century and beginning of the 21st century, many industries worldwide began producing hydrogen, hydrogen-powered vehicles, hydrogen fuel cells, and other hydrogen products. In Table 16-1, some of the important properties of hydrogen are listed.

16.2   Hydrogen as Fuel

Hydrogen's characteristics give it many advantages over fossil fuels in terms of safety. Hydrogen's low density and ability to rapidly disperse allow hydrogen to escape to the atmosphere when a leak occurs. Propane and gasoline, with their high densities and slow dispersals, cause the fuels to congregate near the ground, increasing the risk of explosion. Hydrogen has to reach a concentration of 4% in the surrounding atmosphere before it poses a danger. On the other hand, gasoline becomes ignitable and dangerous at a concentration of only 1%.

	Property	Units
1	Melting point	13.96 k
2	Heat of fusion at 14 K	14 cal/g
3	Boiling point	20.4 K
4	Heat of vaporization at 20.4 K	107 cal/g
5	Density: Solid at 4.2 K Liquid at 20.4 K	0.089 g/cm³ 0.071 g/cm³
6	Critical temperature	33.3 K
7	Critical pressure	12.8 atm abs
8	Critical volume	65 cm³/mole
9	Critical density	0.031 g/cm³
10	Heat of transition of converting ortho hydrogen to para hydrogen at 20.4 K	168 cal/g
11	Specific heat at constant pressure Solid at 13.4 K Liquid at 17.2 K	 0.63 cal/g 1.93 cal/g
12	Specific heat at constant volume (0° to 200°C)	2046 cal/g
13	Gas density at 0°C and 1 atm	0.0899 g/l
14	Gas specific gravity (air = 1.0)	0.0695 (cal)/(s) (cm) (°C)
15	Gas thermal conductivity at 25°C	0.00044(cal)/(s) (cm) (°C)
16	Heat of combustion at 25°C	68.32 kcal/g mole
17	Energy release upon combustion	29,000 cal/g 2050 cal/cm³
18	Flame temperature	2483 K
19	Auto ignition temperature	858 K
20	Flammability limit In oxygen In air	 4% to 94% 4% to 74%
21	Heat atomization	218 kj/mol

TABLE 16-1   Properties of Hydrogen

Hydrogen offers three important benefits:

1. The use of hydrogen greatly reduces pollution. When hydrogen is combined with oxygen in a fuel cell, energy in the form of electricity is produced. This electricity can be used to power vehicles, as a heat source, and for many other uses. The advantage

of using hydrogen as an energy carrier is that when it combines with oxygen, the only byproducts are water and heat. No greenhouse gases or other particulates are produced by the use of hydrogen as fuel.

2. Hydrogen can be produced locally from numerous sources. Hydrogen can be produced centrally and then distributed, or it can be produced onsite where it is to be used. Hydrogen gas may be produced from methane, gasoline, biomass, coal, or even water.

3. To ignite, hydrogen requires a higher concentration in the atmosphere than other fuels. When concentration of hydrogen reaches a level of 4% concentration in the atmosphere, the possibility of its getting ignited increases greatly. A concentration level of 4% for hydrogen does not seem that high, but when compared with gasoline, which is flammable at 1%, hydrogen offers a significantly lower risk of explosion.

16.2.1 Properties of Hydrogen versus Other Fuel

As discussed, hydrogen's properties give it many advantages over fossil fuels in terms of safety. Table 16-2 provides a comparison of the characteristics of gasoline, methane, and hydrogen related to ignition and explosion hazards.

A comparison between three above-mentioned fuels clearly shows that hydrogen has the highest energy-to-weight ratio of all fuels. One kg of hydrogen contains the same amount of energy as 2.1 kg of natural gas and 2.8 kg of gasoline.

Property	Gasoline	Methane	Hydrogen
Density (kg/m^3)	4.40	0.65	0.084
Diffusion coefficient in air (cm^2/s)	0.05	0.16	0.610
Specific heat at constant pressure (J/Gk)	1.20	2.22	14.89
Ignition limits in air (vol%)	1.0–7.6	5.3–15.0	4.0–75.0
Ignition energy in air (Mj)	0.24	0.29	0.02
Ignition temperature (°C)	228–471	540	585
Flame temperature in air (°C)	2197	1875	2045
Explosion energy (G TNT/kj)	0.25	0.19	0.17
Flame emissivity (%)	34–43	25–33	17–25

TABLE 16-2   Characteristics Related to the Fire Hazards of Fuels

16.3 Methods of Hydrogen Storage

One of the major hurdles in using hydrogen as a fuel for energy generation is storing and transporting it. Following are the methods adopted for this purpose.

16.3.1 Storage as Liquid Hydrogen

Hydrogen can be liquefied and stored in liquid form. Liquid hydrogen has to be stored at 20 K (–253°C). This requires storing liquid hydrogen under cryogenic conditions. The container of hydrogen should prevent its evaporation. Moreover, a special type of pump is also needed to transfer liquid hydrogen to the site of its application. However, the temperature requirements for liquid hydrogen storage necessitate expending energy to compress and chill the hydrogen into its liquid state. The cooling and compressing process requires energy, resulting in a net loss of about 30% of the energy that the liquid hydrogen has stored. The margin of safety concerning liquid hydrogen storage is a function of maintaining the tank integrity and preserving the Kelvin temperatures that liquid hydrogen requires

The real energy needed to liquefy hydrogen is about 11 kWh/kg, which is about 30% of its energy content. This is one of the biggest problems concerning the use of liquid hydrogen. However, this loss of energy is compensated with a high energy density of liquid hydrogen storage.

Hence, the energy required for liquefying hydrogen as well as storing it under special conditions is very expensive compared with other storage methods. Research in the field of liquid hydrogen storage centers around the development of composite tank materials (which are lighter but stronger) and improved methods for liquefying hydrogen.

16.3.2 Storage as Gaseous Hydrogen

Hydrogen can be compressed into high-pressure tanks. This process requires energy to accomplish, and the space that the compressed gas occupies is usually quite large, resulting in a lower energy density compared with a traditional gasoline tank. A hydrogen gas tank can store 3000 times more energy than a gasoline tank.

Compressing or liquefying the gas is expensive. Hydrogen can be compressed into high-pressure tanks in which each additional cubic foot compressed into the same space requires another atmosphere of pressure of 14.7 psi; hydrogen gas can be compressed to a pressure of 6000 psi. Such tanks should be periodically tested and inspected to ensure their safety.

Considering these disadvantages of storing hydrogen in gaseous form, scientists are researching to find some materials into which hydrogen can be stored in form of adsorbed hydrogen and by suitable thermal treatment, hydrogen could desorbed from the same materials. This is done by using some suitable metal hydride or carbon nanomaterials. These are discussed here in some detailed manner.

Hydrogen Storage as Metal Hydride

Metal hydrides are specific combinations of metallic alloys that act like a sponge soaking up water. Metal hydrides posses the unique ability to absorb hydrogen and release it later, either at room temperature or through the heating of the tank. One of the main requirements for hydrogen storage is that system must be fully reversible (i.e., the metal hydride must be able to absorb hydrogen at a specific temperature and must also release hydrogen at the specific temperature). In addition, the release of hydrogen should be at a constant pressure. The total amount of hydrogen absorbed is generally 1 to 2% of the total weight of the tank. Some metal hydrides are capable of storing 5 to 7% of their own weight at 250°C or higher. The percentage of gas absorbed to volume of the metal is still relatively low, but hydrides offer a valuable solution to hydrogen storage.

Metal hydrides offer the advantages of safely delivering hydrogen at a constant pressure. The life of a metal hydride storage tank is directly related to the purity of the hydrogen it is storing. The alloys absorb not only hydrogen but also the impurities present in the hydrogen. Adsorption of impurities by the tank and retaining them there cause release of pure hydrogen only. However, the impurities are not a very welcomed ingredient because they affect the life of the storage tank.

Hydrogen forms three types of metal hydrides: ionic, metallic, and covalent. Depending on the type of metal, different types of ionic metal hydrides are formed. For example, with an alkali metal, the NaCl-type structure is formed; with an alkaline earth metal, the $BaCl_2$-type structure is formed. However, metal hydrides formed with certain metals (not with magnesium) are very stable and hence are not suitable for hydrogen storage. Similarly, covalent hydrides such as BeH, CaH_2, and AlH_3 can be in a solid, liquid, or in gaseous form and very unstable. These types of hydrides are also unsuitable for hydrogen storage. Metallic hydrides are mostly formed with transition metal and with the elements of groups IIIA to VIIIA. These hydrides are suitable for hydrogen storage.

It is useful and necessary to consider the thermodynamics of the adsorption/desorption process of hydrogen by a metal hydride.

Thermodynamics of the Adsorption/Desorption Process   The adsorption/desorption process by a metal hydride can be represented by equation 16.1:

$$M + \frac{x}{2}H_2 \Leftrightarrow MH_x + Q \qquad (16.1)$$

where M = metal
MH_x = metal hydride
Q = release of heat because this process is exothermic type.

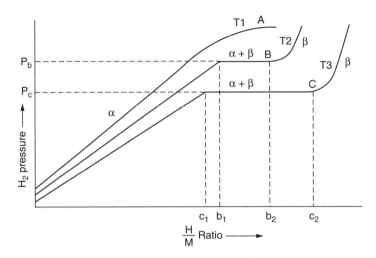

Figure 16-1 Schematic diagram showing the isotherm of metal hydride with the metal:hydrogen ratio.

The direction of this reaction depends on the hydrogen pressure, the temperature of the system, and the composition of the isotherm. In other words, if the amount of hydrogen released (i.e., its pressure) is plotted with the composition of MH_x, a graph showing three types of nature of isotherms is obtained from three different types of metal hydrides (Fig. 16-1).

Metal hydride generally has three phases: α phase, β phase, and a mixture of these two phases. The α phase corresponds to a condition in which hydrogen starts forming metal hydride. The β phase also represents a condition of formation of metal hydride, but its slope is different than that of the α phase. It is important to realize that a plateau is formed when the β phase starts. This condition continues until both phases coexist. Some metal hydrides exhibit a plateau at the mixture of the two phases (Fig. 16-1B and 16-1C), and some show no plateau at all (Fig. 16-1A). Metal hydride is of no use if one wishes to work in either of the pure α or β phase because the pressures released by these two phases are linearly dependent on the H/M ratio. However, at the phase that exhibits both of the phases (i.e., at the plateau region), there is no change in the pressure of hydrogen released for a different H/M ratio. Considering this aspect of the isotherm, one would select a metal hydride that gives larger plateau and high pressure available at the plateau and the lower temperature needed for the desorption process. Consider the phase rule:

$$F = C - P + 2 \tag{16.2}$$

Because there are two reactant constituents—hydrogen and hydride—the value of C is 2. The value of P is 3, representing three phases (gas phase, α phase, and β phase). This makes F equal 1 (i.e., degree of freedom is 1, which means that pressure will remain constant at a chosen temperature).

When hydrogen is stored in metal hydrides, the increase in the operating temperature results in a decrease of the plateau range. At some critical temperature, the miscibility gap disappears and the α and β phases continuously convert into the β phase (Fig. 16-1A). In some metal hydrides, an additional γ phase appears that gives another plateau at a higher H/M ratio (not shown Fig. 16-1).

Hysteresis in the Adsorption/Desorption Process   Hysteresis is a property of systems (usually physical systems) that do not instantly follow the forces applied to them but react slowly or do not return completely to their original state (i.e., systems whose states depend on their immediate history).

Some metal hydrides show hysteresis during isotherm for the absorption and desorption processes (Fig. 16-2). This occurs when the transition pressure in isotherm is higher for absorption than for desorption. The cause of hysteresis is not well understood, but it may be due to a lattice expansion and contraction process.

Storage Criteria for Metal Hydride   Considering these thermodynamics properties of metal hydrides, the selection of hydride should be based on the following criteria:

- The metal hydride should be capable of storing large quantities of hydrogen.
- The metal hydride should be readily formed and decomposed to release hydrogen at a lower temperature.
- The metal hydride should be safe to handle.

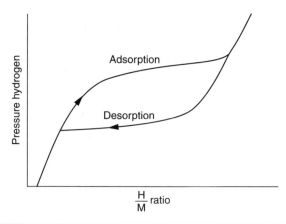

Figure 16-2   Schematic diagram showing hysteresis in hydrogen adsorption/desorption with metal hydride.

- The kinetics of adsorption/desorption should be reasonably fast.
- The number of operating cycles of adsorption/desorption should be large.
- Hysteresis should be small.
- The cost of the material should be reasonably low.
- The weight and amount of hydrogen adsorbed should be a small number.

Unfortunately, no metal hydride can meet all of these requirements. However, metal hydrides that have been tried belong to groups AB, AB_5, and A_2B. Among the various AB-type hydrides, $LaNi_5$ is the most common. Following is the reaction of this material with hydrogen:

$$LaNi_5 + 3H_2 \Leftrightarrow LaNi_5H_6 \qquad (16.3)$$

Lanthanum has strong affinity for hydrogen; nickel has less. Lanthanum has tolerance with other gases as well. The plateau at room temperature is formed giving a hydrogen pressure of 2.2 atm. Another hydride of this group is $SmCo_5$, which can give a hydrogen pressure of 3.5 atm at room temperature.

Among the AB type of hydrides, FeTi has been studied the most. This material becomes poisonous with oxygen; therefore, before using this material, degassing is essential. At around room temperature, FeTi reaches its plateau, giving about 8 atm pressure of hydrogen. A_2B- and AB_2-type metal hydrides have been studied most recently. These materials have a high storage capacity and do not require any catalyst to initiate the hydrogen adsorption process. These materials are lightweight and easy to carry. Mg_2Ni, which forms with hydrogen Mg_2NiH_4 at 200°C, has also been studied.

Hydrogen Storage in Glass Spheres

Glass spheres are small, hollow like microballoons with a diameter from 25 to 500 μm and a wall thickness of about 1 μm. The spheres are filled with hydrogen at a high pressure and a temperature of 200° to 400°C. High temperature makes the glass wall permeable, and the hydrogen is able to fill in. After the glass has cooled down to ambient temperature, the hydrogen is trapped inside the spheres. The hydrogen may be released by heating or crushing the spheres. The crushing naturally prevents the reuse of spheres and is not necessarily a favorable option. The glass spheres may also cause accidents when breaking down if not handled properly. The storage capacity of spheres is about 5 to 6 wt% (percentage by weight) at 200 to 490 bar.

Chemical Storage

Chemical compounds containing hydrogen can also be considered as a kind of hydrogen storage. These include methanol (CH_3OH), ammonia

(NH$_3$), and methyl-cyclohexane (CH$_3$C$_6$H$_5$). At standard temperature and pressure (STP), all of these compounds are in liquid form; thus, the infrastructure for gasoline could be used for transportation and storage of the compounds. This has an advantage compared with gaseous hydrogen, which demands leak-proof, preferably seamless, piping and vessels. The hydrogen storage capacity of these chemical compounds is quite good at 8.9 wt% for CH$_3$OH, 15.1 wt% for NH$_3$, and 13.2 wt% for CH$_3$C$_6$H$_5$. These figures do not include the containers in which the liquids are stored. Because the containers can be made of lightweight composites or even plastic in some cases, the effect of the container is negligible, especially with larger systems.

Chemical storage of hydrogen also has some disadvantages. The storage method is nonreversible (i.e., the compounds cannot be "charged" with hydrogen reproducibly). The compounds must be produced in a centralized plant, and the reaction products have to be recycled somehow. This is difficult, especially with ammonia, which produces highly pollutant and environmentally unfavorable nitrogen oxides. Other compounds produce carbon oxides, which are also quite unfavorable.

16.4   Hydrogen Storage in Carbon

16.4.1   Carbon Nanotubes

Carbon nanotubes (CNTs) were discovered in 1991 accidentally while fullerenes were being synthesized. A CNT is a graphite sheet rolled up in a seamless cylinder with a diameter in a scale of nanometers. CNTs have several interesting properties. For example, their modulus of elasticity is about five times the value of steel. They also have special electronic properties depending on the chirality of the CNT; some behave like metallic conductors, and the others behave like semiconductors. Thus, CNTs can be used for micro- and nanoscale electronic devices.

Hydrogen can be stored into CNTs by chemisorptions or physisorption. The methods of trapping hydrogen are not known very accurately, but density functional calculations have shown some insights into the mechanisms. Calculations indicate that hydrogen can be adsorbed at the exterior of the tube wall by H–C bonds with a H/C coverage of 1.0 or inside the tube by H–H bonds with a coverage up to 2.4. The hydrogen relaxes inside the CNT, forming H–H bonds. The adsorption into the interior wall of the CNT is also possible, but reproducibility is a problem.

Multiwalled CNTs (MWCNTs), in which two or more single tubes are entangled because of van der Waals attraction, can adsorb hydrogen between the layers of wall of the CNTs. The hydrogen causes the radius of the tubes to increase and thus makes MWCNTs less stable (Lee and Lee, 2000). In CNTs, bundles hydrogen can also be adsorbed in between the two layers of the graphene sheets (Bae et al., 2000).

Density functional calculations have shown that theoretically, in proper conditions, a single-walled CNT (SWCNT) can adsorb up to 14 wt% and a MWCNT about 7.7 wt% of hydrogen. Dillon et al. (1997) reported their first experimental result of high hydrogen uptake by CNTs. They estimated that hydrogen could achieve a density of 5 to 10 wt%. Chen et al. (1999) have reported that alkali-doped CNTs are able to store even 20 wt% under ambient pressure, but are unstable and require elevated temperatures. The result has shown to be in a great disagreement with other results and has been thought to be incorrect.

Recent results on hydrogen uptake of SWCNTs are promising. At 0.67 bar and 600 K, about 7 wt% of hydrogen has been adsorbed and desorbed with a good cycling stability (Allemann et al., 2000). Another result at ambient temperature and pressure shows that 3.3 wt% can be adsorbed and desorbed reproducibly and 4.2 wt% can be adsorbed and desorbed reproducibly with slight heating (Cheng et al., 1999).

Although the price of commercial CNTs is quite high, they have good potential for storing hydrogen. Hence, when the manufacturing techniques are improved and some engineering problems solved, CNTs may become highly competitive against other hydrogen storage technologies.

16.4.2 Graphitic Nanofibers

Graphite nanofibers (GNFs) are graphite sheets that are perfectly arranged in a parallel ("platelet" structure), perpendicular ("tubular" structure), or angle orientation ("herringbone" structure) with respect to the fiber axis. These fibers are also referred as carbon nanofibers (CNF). These fibers could be hollow as carbon nanotubes, but the surface of fibers is made of broken graphene sheet whereas carbon nanotubes are made of unbroken graphene sheet. A schematic of the structure of a GNFs with some hydrogen adsorbed between the sheets is represented in Fig. 16-3.

The most critical factor affecting the hydrogen adsorption by carbon nanofibers (CNFs) is the demand for a high surface area because the hydrogen is adsorbed in between the graphite sheets. It has been reported that some CNFs can adsorb more than 40 to 65 wt% of hydrogen. However, these results have not been able to be reproduced.

Studies have shown that only 0.7 to 1.5 wt% of hydrogen adsorption in a CNF under ambient temperature and pressures slightly above 100 bar is possible (see Fig. 16-3). Some other studies claim that about 10 to 15 wt% of hydrogen has been adsorbed in graphitic and nongraphitic CNFs. The cyclic stability and other properties of CNFs have not yet been studied in detail; thus, it is difficult to say whether the CNFs will be competitive against other hydrogen storage technologies.

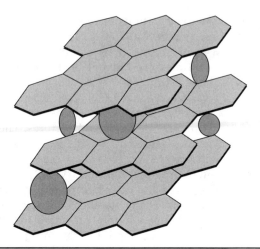

FIGURE **16-3** Schematic diagram of graphite carbon nanofibers showing adsorbed hydrogen. (Modified from Baker et al., 1998.)

Recently, Sharon (2007), and Jayabhaye (2006, 2007) have synthesized carbon fibers from plant based precursors (baggas) and reported the hydrogen adsorption to the tune of 4.5 wt%. It is believed that this type of fibers after some modifications of synthesis procedures and activation with suitable metal catalyst, hydrogen adsorption can be enhanced to about 9 wt%. If this happens then cost of production of the fibers would become very economical and may be useful to use it for hydrogen storage devices.

16.4.3 Fullerenes

Fullerenes are synthesized carbon molecules usually shaped like a football, such as C_{60} and C_{70}. Fullerenes are able to hydrogenate through the reaction:

$$C_{60} + xH_2O + Xe^- \leftrightarrow C_{60}H_x + xOH^- \tag{16.4}$$

An experimental study by Chen et al. (1999) shows that more than 6 wt% of hydrogen can be adsorbed on fullerenes at 180°C and at about 25 bar. Usually, the bonds between the C and H atoms are so strong that temperatures over 400°C are needed to desorb the hydrogen, but Chen et al. (1999b) were able to do this at a temperature below 225°C. Despite quite a high hydrogen storing ability, the cyclic tests of fullerenes have shown poor properties for storing hydrogen.

16.4.4 Activated Carbon

Bulky carbon with a high surface area, so-called *activated carbon*, is able to adsorb hydrogen in its macroscopic pores. The main problem is that the pores are very small to catch the hydrogen atom, so high pressures must be applied to get the hydrogen into the pore. About 5.2 wt% of hydrogen adsorbed into the activated carbon has been achieved at cryogenic temperatures and pressures in the range of 45 to 60 bar. In ambient temperature and a pressure of 60 bar, the amount adsorbed has been only approximately 0.5 wt%. Some studies show that carbon-adsorbent in a pressure vessel can adsorb little more hydrogen than what would fit into an empty vessel as gas. This is true for pressures below 150 bar, after which an empty vessel can store more hydrogen. The poor variation in pressure P and temperature T relationship for hydrogen absorption of activated carbon prevent them from being suitable for hydrogen storage in practical applications.

16.5   Hydrogen Storage Measurement

To determine the hydrogen storage capacity of a carbon material, generally three methods have been used: volumetric, gravimetric, and temperature programmed desorption (TPD).

16.5.1   Volumetric Method

The basis for the volumetric method is that when a degassed sample of carbon in a container of known volume is exposed to a known amount of hydrogen at a high pressure, the carbon adsorbs some of the hydrogen and causes a reduction in pressure. From the reduction in pressure, the volume of adsorbed hydrogen is calculated. To achieve this, two volumetric methods have been used, direct pressure measurement and differential pressure measurement. Volumetric method of determining hydrogen storage is the nearest to simulating the conditions of a storage tank.

Direct Pressure Measurement

The direct pressure method was originally used to determine the hydrogen storage capacity of metal hydrides at pressures between 1 and 40 bar using Sievert's-type apparatus, as shown in Fig. 16-4 (Sandrock and Huston 1981; Sivakumar et al., 1999). A known amount of hydrogen is exposed to the storage material, and the resulting

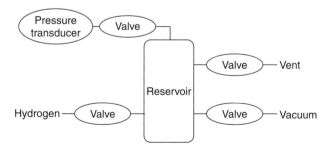

FIGURE 16-4   Simple schematic representation of a Sievert's-type apparatus.

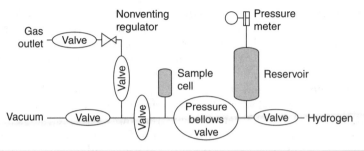

FIGURE 16-5 Schematic representations of the basic apparatus used for the measurement of hydrogen uptake in various materials.

decrease in pressure is monitored using a pressure transducer from which the hydrogen storage capacity of the material can be determined.

In 1998, Chamber et al. (1998) used a Sievert's-type apparatus to determine the hydrogen storage of some CNFs grown with carbon vapor deposition. Their apparatus consisted of a conventional high-pressure stainless steel sample cell (~20 cm³), connected to a high-pressure hydrogen reservoir container (~75 cm³) via a high-pressure bellows valve (Fig. 16-5). The system is calibrated with several blank runs to take into account the pressure decrease on exposing the evacuated sample cell to the reservoir. For each run, about 0.2 g of sample is used. The system is then evacuated to 0.1 Pa while being heated to 150°C to remove all physisorbed species. Hydrogen is then introduced into the reservoir to the desired pre-expansion pressure with the target postexpansion pressure being 120 bar. The bellows valve between the reservoir and the sample cell is then opened, exposing the sample to the hydrogen. The pressure is then monitored for 24 hours. After the system has reached equilibrium, the gas is vented through the regulator, allowing the desorbed gas to be measured by water displacement (not shown in Fig. 16-5) and analyzed by gas chromatography–mass spectrometry to confirm that only hydrogen was present as adsorbed gas. Many different variations of this apparatus have been built and used to study the potential for carbons as a hydrogen storage medium.

Differential Pressure Measurement

A novel method using differential pressure measurement for the determination of the hydrogen storage capacities of carbon materials (Fig. 16-6) was proposed by Browning et al. (2002). The hydrogen adsorption capacity of CNF was determined at 120 bar using differential pressure measurements between four volumetrically balanced chambers. The hydrogen adsorption was measured as a differential pressure between the sample and reference limbs. Before each experiment, the sample (~50 to 100 mg) is heated to 150°C at 100 Pa. The lower linked valves are closed, and the reservoirs are pressurized

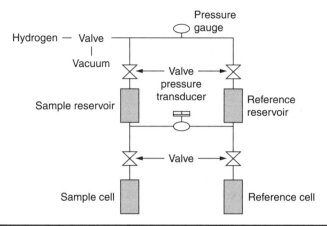

FIGURE **16-6** Schematic diagram of the experimental setup of differential volumetric hydrogen adsorption apparatus.

with hydrogen. The upper linked valves are then closed, and the lower linked valves are opened to expand the hydrogen into the sample limbs to begin the experiment. The differential pressure is recorded at regular time intervals. To prove that the rig is leak free, blank runs are carried out. In addition, the rig is calibrated using a commercial metal hydride ($MmNi_{4.5}Al_{0.5}$, where Mm denotes misch metal of lanthanides), which was found to observe to adsorb 1.3 wt%, coinciding well with the manufacturer's data.

This technique is claimed to possess several advantages over traditional direct pressure measurement. First, high-pressure transducers are far less precise than differential pressure transducers operating at high pressure; thus, the accuracy of the pressure monitoring system is higher. The method also attempts to eradicate many of the problems associated with the expansion of non-ideal hydrogen gas, which may introduce large errors when performing this type of measurement by using simultaneous expansion of the sample and reference cells. Browning et al. (2002) claimed that the apparatus is capable of reproducing hydrogen uptakes as low as 0.1 wt% when using 100 mg of material.

16.5.2   Gravimetric Method

The basis for the gravimetric method is that when a degassed sample of carbon kept in a container of known volume is exposed to hydrogen, the carbon will adsorb some of the hydrogen and cause an increase in the weight of the sample. To achieve this, two methods have been used by various teams: high-pressure and ambient pressure systems. This method is sensitive to all types of gases adsorbed because it is purely based on weight. Thus, although this method gives very accurate results, it gives information about all types of gas that have been adsorbed, not only hydrogen. Hence, for this method, pure hydrogen gas must be used.

High-Pressure Isothermal Method

In high-pressure Thermogravimetric Analysis (TGA) experiments, the quantity of hydrogen adsorbed is calculated at the equilibrium condition (Chen et al., 1997). In this method, a degassed sample is exposed to high-pressure hydrogen, and then the weight change is monitored as a function of time.

At the equilibrium condition, there would be no weight change. The hydrogen storage capacity is calculated at this equilibrium condition. Using a series of such measurements, an adsorption–desorption isotherm can also be constructed. For taking a measurement at pressures of over 100 bar, Strobel et al. (1999) developed a high-pressure system using a super-micro S3D-P microbalance (Fig. 16-7). The balance chamber can be pressurized to 150 bar while operating at ambient temperature. The reference crucible R is loaded with quartz crucible and the sample crucible S containing carbon material. Buoyancy is numerically corrected. Before the high-pressure experiment is begun, the chamber is purged with hydrogen and helium. The chamber is then evacuated until the pressure and mass remain constant for 30 minutes. To record a hydrogen adsorption isotherm for each sample, the experiment is repeated at pressures from 0 to 130 bar with high-pressure balance.

More recently, high-pressure gravimetric experiments have been performed in a direct gravimetric analyzer with a high-pressure, temperature-controlled balance chamber (Badzian et al., 2001).

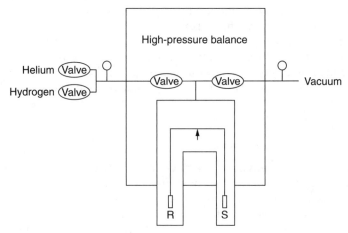

Notes: R = Reference material S = Sample whose storage capacity is needed

FIGURE 16-7   Schematic representation of experimental setup of a gravimetric hydrogen adsorption apparatus.

These systems commonly operate between 1 and 60 bars and at temperatures between −196 and 1000°C (Pradhan et al., 2002). In a typical run, a sample of carbon is degassed under high vacuum at 150° and 500°C (de la Casa-Lillo, 2002). The sample is then exposed to hydrogen at the desired pressure, and the change in weight is monitored and used to calculate the hydrogen storage capacity of the material.

16.5.3   Dynamic Temperature or Ambient Pressure Method

Ambient pressure systems use direct TGA analysis and are usually dynamic measurements recording the hydrogen storage capacity of a material as a function of temperature. In a typical run, a sample (~10 mg) is loaded onto the balance and then degassed. Bai et al. (2001) have published a detailed protocol that is representative of many of the other methods used (Chen et al., 1997). The sample is purged with hydrogen and then preheated to 400°C. The adsorption cycle begins by decreasing the temperature to 25°C. The desorption cycle is then recorded by heating it to 400°C. Some methods also include a correction for buoyancy to compensate for the difference of taking the measurement in hydrogen and air (Badzian et al., 2001). Another feature of this method is that a mass spectrometer can be attached to identify which gases are being desorbed to prove that only hydrogen is present (Bai et al., 2001).

16.5.4   Temperature Programmed Desorption

TPD, sometimes referred to as *temperature desorption spectroscopy*, measures hydrogen desorption only in a high vacuum using mass spectrometry. In 1999, Dillon et al. reported that they had used TPD to determine the hydrogen storage capacity of SWCNT soot. Their method was used later by several other groups. The experiments were carried out in an ultra-high vacuum chamber equipped with a cryostat and a mass spectrometer. In these experiments, samples of carbon material (~1 mg) are kept in platinum foil packets with pinholes for gas diffusion. The packets' temperature is monitored by a thermocouple and controlled by resistive heating. Standard hydrogen exposures are carried out at 40 kPa for 10 minutes at 0°C followed by 3 minutes at −140°C. The samples are then cooled to −183°C under a high vacuum (1×10^{-5} Pa) and then heated to 700°C at a rate of 1°C s⁻¹ under a vacuum. Throughout the process, the presence of hydrogen is monitored by mass spectrometry. From this, the amount of hydrogen desorbed is quantitatively calculated (Chen et al., 1997). Deuterium rather than hydrogen (Hirscher et al., 2002) has also been used for such measurements, greatly enhancing the sensitivity.

16.5.5   Electrochemical Method

The electrochemical method investigates electrochemical charge–discharge cycles of ionic hydrogen (Frackowiak and Beguin, 2002).

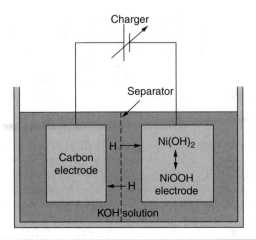

Figure 16-8   Schematic diagram of an electrochemical hydrogen storage apparatus.

Carbon to be analyzed is grinded and mixed with conducting transition metals and polymer binders to create a carbon composite electrode. This electrode and a counter electrode, often salt of a first series transition metal, are submersed in an electrolyte solution separated by a membrane (Fig. 16-8). The voltage across the two electrodes is measured as a function of time while a constant current is maintained until equilibrium is achieved (Dai et al., 2002). The accuracy of this method may be increased by using a reference electrode to take the voltage measurements between the working electrode and the reference electrode. The voltage–time profile can then be used to calculate the hydrogen uptake achieved by the sample materials.

16.6   Mechanism of Hydrogen Storage

It is accepted that the adsorption of a gas in microporous and mesoporous carbon materials may occur via the physisorption mechanism. The carbon atoms exert van der Waals forces on the molecules. Physisorption is reversible by varying the physical conditions of a system; thus, at a given temperature, the amount of hydrogen gas adsorbed is a function of the pressure and can be desorbed by reduction of the pressure in the system. A mechanism of hydrogen storage on CNMs that explains the much-higher-than reported uptake remains unclear. Chambers et al. (1998) proposed that the interlayer spacing of CNTs and CNFs produces an array of nanoropes accessible to hydrogen directly from the edge of the material. The hydrogen penetrates the nanopores formed by the layers of CNF and the interior of CNT, forming an intercalated layer of hydrogen (Chen et al., 1997; Dillon et al., 1997).

In addition, the nanopores could undergo expansion to accommodate hydrogen in a multiplayer configuration. A slight variation to this theory was proposed by Browning et al. (2002), who hypothesized that the exposed edge sites of the graphene sheets, which constitute the bulk of the CNF surface, catalyze the dissociation of hydrogen, followed by intercalation of the graphene layers. Others have proposed that the hydrogen condenses in the hollow cores of the CNTs and CNFs because of a capillary effect (Fan et al., 2001).

Another theory is that hydrogen dissociates on the metal catalyst with the CNF and CNT present in the tube as particles left during the preparation (Lueking et al., 2002). The presence of functional groups has also been attributed to the enhancement of hydrogen storage capacities by facilitating stronger bonding (Badzian et al., 2001; Bai et al., 2001; Zhu et al., 2003). Many authors attribute the large hydrogen storage capacities of CNTs and CNFs to a novel, yet undefined uptake mechanism (Liu et al., 1999; Strobel et al., 1999). However, until reliable and repeatable results are obtained, the question of the mechanism of hydrogen storage cannot be conclusively answered.

16.7   Future Directions

Large numbers of reproducible evidence suggests that CNTs have the potential to be used as a hydrogen storage medium. However, it is still not very practical because of the limitations on mass production and utilization of CNTs. And the following obstacles still needed to be solved by scientists:

1. Mass production of carbon nanostructure materials with a controlled microstructure at a reasonable cost

2. Purification of CNTs and the development and optimization of pretreatment methods for opening the caps at the tube ends to improve their hydrogen storage capacity

3. Elucidation of the microstructure of CNTs, especially pore structure and surface microstructure, in the viewpoint of hydrogen adsorption/desorption

4. Elucidation of volume storage capacity and how to improve it

5. Further investigation of the adsorption/desorption process, thermodynamics, kinetics, and the cycling behaviors of CNTs

6. A more practical hydrogen adsorption model to design a CNT-based hydrogen storage medium so that it could be confirmed whether the surface area, pore size, or both are factors that are important in controlling hydrogen adsorption

16.8 Summary

It can be concluded that hydrogen fuel is clean, versatile, efficient, and safe, and it will play an important role in the future world's energy structure.

Preliminary experimental results and some of the theoretical predictions indicate that carbon nanostructures (CNTs and CNFs) can be promising candidates for hydrogen storage, which may be the solution for hydrogen fuel cell–driven vehicles.

Nevertheless, many efforts still have to be made to reproduce and verify the hydrogen storage capacity of CNTs, both theoretically and experimentally, to investigate their storage capacity and absorption and desorption behaviors as well as to clarify the feasibility of carbon nanostructures as a practical hydrogen storage medium. It is suggested that carbon fibers synthesized from plant precursors like baggas can become economical for its use as hydrogen storage materials.

CHAPTER 17

Application of Carbon Nanotubes in Lithium-Ion Batteries

Sunil Bhardwaj

Nanotechnology Research Centre, Birla College, Kalyan, Maharashtra, India

17.1 Introduction

Rechargeable batteries play an important role as energy carriers in our modern society, being present in devices for everyday use such as cellular phones, video cameras, and laptop computers. The demand for batteries rapidly increased at the end of the 20th century because of the large interest in wireless devices. Today, the battery industry is a large-scale industry producing several million batteries per month.

Another important drive for technological development in the field of batteries was the introduction of hybrid electric vehicles, which significantly reduce fuel consumption and gas emissions.

Combustion engines emit greenhouse gases, which will have a serious influence on the future climate. A rechargeable battery is used to buffer the electricity produced by a traditional combustion engine and power generated from the electrical engine. For this application, batteries optimized for high power, low cost, and a long service life are essential. There is a demand to shift from combustion engine vehicles to zero-emission vehicles (ZEVs) to reduce exhaust gases.

Battery development is a major task for both industry and academic research; hence, the effort to develop powerful, cheap, and reliable rechargeable batteries continues. The battery technology dominating the market today is the lithium-ion (Li-ion) battery (Maurin et al., 2000). In portable devices, these batteries have rapidly replaced the less energetic and less environmental friendly nickel–cadmium (Ni–Cd) batteries, as well as the bulkier nickel–metal hydride (Ni–MH) cells. However, in large-scale batteries where cost is the key issue, the older battery types are still prominent.

The idea to use lithium in batteries was first proposed in 1958 and has been used for a long time in primary (nonrechargeable) batteries. Rechargeable batteries were commercialized by Sony in 1991 because of the realization that the battery was a key technology, making Sony's consumer products competitive.

Furthermore, today battery technology is recognized as a strategic key technology for many devices. As a consequence, an extraordinary amount of work has been done on all aspects of the Li-ion battery's design, manufacture, and application, and the technology is still improving significantly.

The Li-ion battery fulfills many of the demands made within the areas of portable electronics and electric vehicles (hybrid electric vehicles) and is superior in many ways to the more common Ni–Cd and Ni–MH batteries. Its superiority lies in the use of lithium with its large negative electrode potential 3.04 V vs. SHE (*standard hydrogen electrode*) and high energy density and in the development of intercalation electrodes that can repeatedly accept and release Li$^+$ ions on charge and discharge.

17.2   Fundamental Concepts of Batteries

A rechargeable battery (e.g., Li-ion battery) is an example of a system in which it is possible to convert chemical energy to electrical energy and then reconvert the electrical energy to chemical energy. A battery is composed of an anode, a cathode, and an electrolyte. The anode (the negative electrode in a galvanic cell) is the electrode in which an oxidation process occurs. The cathode is consequently the electrode in which the reduction process occurs. The electrolyte has to be an electronic insulator (to avoid short circuiting) but a good ionic conductor (to transport electrochemically active species). The free energy change, ΔG, of a cell reaction is related to its electrochemical voltage, E, by:

$$\Delta G = -nFE \qquad (17.1)$$

where n = number of electrons involved in the reaction
$\quad\quad F$ = Faraday's constant (96,487 C/mol).

The significant performance characteristics of a rechargeable battery are expressed by one of the following methods:

1. The **gravimetric energy density** (GED) of a battery is based on weight and expressed as mWh/g.

2. The **gravimetric capacity** is expressed as mAh/g. Its calculations are based on weight.

3. The **volumetric capacity** (mAh/cm³) is the ability to store energy. Capacity is expressed in Amp hours for a given voltage or as energy in Watt hours. Capacity is a function of the number of Li^+ that can be intercalated in the cathode. Specific capacity or energy is used to express capacity or energy as a function of the mass of the cathode's active material, and density denotes normalization to volume. Higher energy density gives either a longer battery life for a fixed volume or the ability to reduce the battery's size.

4. The **rate capability** is the ability to deliver a given power within a given voltage range over the life of a charge. Discharge rates are determined by the speed at which the Li^+ ion can be extracted from the anode and intercalated into the cathode. The converse is true for charging rates.

5. **Cycleability or cycling performance** is the ability of a battery to be charged and discharged repeatedly without losing capacity on successive cycles. It is quantified in terms of the number of cycles that a battery will undergo before it will no longer hold a minimum capacity (e.g., 80% of initial capacity). Cycling performance is related to the stability of the electrodes.

6. **Self-discharge characteristics** is a phenomenon where a battery though not in use, looses its capacity because of initiation of discharge reaction. This is because even under open circuit condition, some positive quantity of electricity (the magnitude of such current is extremely small; may be in the order few microamp cm^{-2}) passes through the battery which causes the battery to loose its power. This is known as self-discharge process.

The GED of a battery in mWh/g is given as:

$$GED = E \times GC \tag{17.2}$$

where E = operating voltage (in volts)
GC = gravimetric capacity (in mAh/g).

The latter indicates the total quantity of charge involved in the cell reaction. A reaction of compound A with n number of lithium ions and the corresponding number of electrons can be represented as:

$$A + n\text{Li}^+ + n\text{e}^- \cdots\!\!> \text{Li}_n A \tag{17.3}$$

The theoretical gravimetric capacity, GC (in mAh/g), for compound A is given by:

$$GC = (1/M) \times nF \tag{17.4}$$

where M = molar mass of A
F = Faraday's constant.

These factors are all influenced by the chemistry of the system, thus making the choice of battery material crucial in designing a battery.

17.3 Lithium-Ion Batteries

In Li-ion batteries, the reactions are lithium insertion in the positive electrode, and the extraction is in the negative electrode. The total discharge reaction of a Li-ion battery, resulting in to the passage of one electron between the two poles (i.e., electrodes) of the battery, is given by:

$$\text{Li-Host A } (-) + \text{Host B } (+) \cdots\!\!> \text{Host A } (-) + \text{Li-Host B } (+)$$
$$\text{Discharge reaction} \tag{17.5}$$

Host A represents the negative electrode and is generally based on a carbon material (e.g., graphite). Host B, the positive electrode, is typically based on a lithiated metal oxide such as LiCoO_2 or LiNiO_2. The electrode materials determine the battery's voltage and energy density. The high voltage of Li-ion batteries (4 V) is one major advantage; another advantage is the low weight of the materials.

The ultimate negative electrode material, in terms of energy density and voltage, is lithium metal (theoretical energy density, 3862 mAh/g). Application of pure lithium was tested commercially during the 1980s, but the poor surface properties of the material caused dendrites to grow during charging, eventually short circuiting the cell internally and causing an explosion or a fire. The carbon-based materials have poor energy densities (e.g., graphite, 372 mAh/g), giving approximately the same voltage, but are safer (Fig. 17-1).

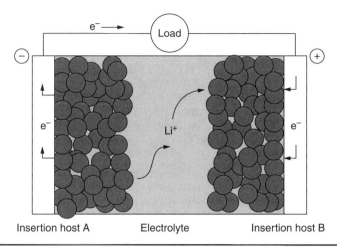

FIGURE 17-1 Schematic diagram of a lithium-ion battery showing movement of Li ions toward the electrodes.

The energy density of multiwalled carbon nanotubes (MWCNTs) has been reported to be as high as 500 mAh/g by Frackowiak et al. (1999). Similarly, with single-walled CNTs (SWCNTs), a current density of 450 mAh/g has also been observed (Gao et al., 1999). A variety of metal oxides are presently used, and all result in fairly high battery voltages (3 to 5 V) (Wartena et al., 2004). However, the poor energy density of these materials (~150 mAh/g) generally limits the overall energy density of the battery. Presently, there are two types of electrolytes in use, liquids and gels. Both are based on non-aqueous organic solvents, similar to acetone, and contain special lithium salts. The gels are solid-like because the liquid component is incorporated into a polymer matrix. There is also a great interest in electrolytes based on only polymer and salt; however, they are presently not used in commercial batteries because of their low conductivity.

17.4 Basic Structure of Carbon Materials

Numerous modifications of carbon can be used as an anode material (e.g., natural and synthetic graphite, coke, carbon fibers, CNTs). The ability to accommodate lithium depends on the morphology and structure of the carbon material.

The basic structure of graphitic carbon materials is of extended sheets of sp^2-hybridized carbon atoms arranged in hexagonal rings extended in two dimensions, sometimes referred to as *graphene sheets*. These honeycomb layers are arranged in an ABAB stacking for hexagonal graphite (referred to as 2H).

A second polymorph of graphite also exists, rhombohedral graphite, which has an ABCABC stacking (referred to as 3R). The ideal structure of graphite is never obtained in practice because of the ever-present high density of stacking faults and structural defects. The graphites are, therefore, usually characterized by the size or extension of isolated, perfectly stacked regions, so-called *crystallites*. The transformation energy between these stacking modes is very small, so most graphite materials contain both phases. The extension of the crystallites in the crystallographic directions varies from nanometers to several micrometers. The crystallites are separated by more disordered carbon regions, which dominate structures referred to as *nongraphitic carbons*.

Graphitic carbon materials have been known to intercalate lithium. Recently, Frackowiak et al. (1999) discovered that the insertion could be made electrochemically at a very low potential versus Li/Li^+, and that carbon, therefore, could be used for battery application as a replacement for the hazardous lithium metal. Carbon exhibits both electronic and ionic conductivity and can incorporate a large number of lithium ions. Their low cost, low intercalation potential, good cycling properties, and wide availability have made them the most attractive anode choice so far for practical Li-ion cells. Carbons show an almost infinitely large amount of structural modifications, ranging from highly crystalline graphites to highly disordered amorphous carbons. They all exhibit different electrochemical properties. The extent of lithium intercalation and the reversibility of the intercalation process both depend on the structure, morphology, texture, grain size, grain shape, and crystallinity of the carbonaceous host material.

17.5 Lithium Intercalation

At ambient pressure, one lithium atom can be intercalated to six carbons atoms, forming a LiC_6 compound. This reaction corresponds to a gravimetric capacity of 372 mAh/g. Intercalation occurs primarily through the edge planes (the prismatic surface); any intercalation through the basal planes occurs at defect sites only. During intercalation of lithium into graphite, the stacking order changes from ABAB stacking for hexagonal graphite to AA stacking for the graphene sheets surrounding the intercalate layers. The rhombohedral stacking is shifted in a similar manner. The intercalated lithium is accommodated in the van der Waals gaps between two honeycomb carbon rings. During the intercalation process, a number of discrete Li_xC_6 phases are formed with lower lithium content. This phenomenon is generally referred to as *staging*, and it is a consequence of the energetically favorable situation of having few highly occupied van der Waals gaps rather than a random distribution. The phases formed are generally referred to as stages I to IV, in which the Roman numerals indicate the number of graphene layers between each lithium layer. Stage I consequently corresponds to the LiC_6 phase. Lithium intercalation has an

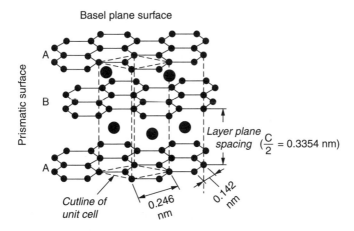

Figure 17-2   A sketch showing the layer structure of graphite of ABA type. Lithium atoms get intercalated in between the two layers of graphite. Large spherical black ball represents lithium atoms which are intercalated into the layers.

effect on the graphite interlayer spacing. The distance between the graphene layers increases by approximately 10% as the LiC_6 phase is forme (Fig. 17.2); this rather moderate volume expansion is highly advantageous when used in a Li-ion battery because electrode stability is important.

The first discharge and charge of a graphite/lithium "half cell" can be seen in Fig. 17-3. Lithium starts to intercalate at around 0.2 V versus Li/Li^+; the staging process results in distinct plateaus for the two-phase regions and vertical regions when a single phase is present. During the first discharge, a reaction due to electrolyte reduction

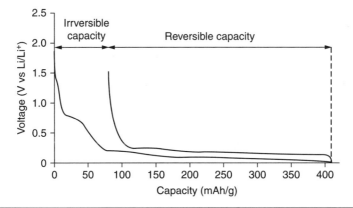

Figure 17-3   The first and second cycles of a graphite/lithium cell cycled in a 1-M $LiPF_6$ EC/DEC (2:1) electrolyte. The capacity obtained after the first cycle is smaller than the initial first cycle. Efforts are being made to get the same capacity as the second cycle after various charging/discharging cycles.

takes place at a voltage around 0.8 V versus Li/Li$^+$ as a solid electrolyte interphase (SEI) layer is formed on the graphite surface. This concept was first introduced while metallic lithium electrodes in organic electrolyte systems were being studied. The SEI constituents have shown to be very similar on lithium and graphite electrodes. On graphite, the SEI acts as a passivating layer, preventing solvent co-intercalation but allowing Li-ion transport; solvent co-intercalation with lithium into graphite causes exfoliation of the graphite sheets and has a detrimental effect on its ability to further intercalate lithium. The layer is also an electronic insulator, preventing further reduction of electrolyte as cycling continues. SEI formation is an irreversible charge-consuming reaction. The irreversible capacity (marked in Fig. 17-2) associated with this type of reaction is about 20%, depending on both the carbon material and the electrolyte used. The morphology and composition of the SEI depends strongly on the electrolyte. It is composed of a mixture of salt reduction products (e.g., LiF) and solvent reduction products (e.g., Li_2CO_3) as well as polymeric species.

17.6 Advantages of Nanomaterials

Nanocrystalline materials possess a high surface-to-volume ratio. This translates into a relatively large interfacial area between the two phases in comparison to bulk systems. Here the term *bulk* is used to denote traditionally sized particles ($\geq 1\mu m$). Equation (17.6) illustrates how interfacial energy, $\gamma^{\alpha/\beta}$, is a barrier to nucleation.

$$\Delta G^{\alpha/\beta} = -\Delta G_v^{\alpha/\beta} \times v + \gamma^{\alpha/\beta} \times A \qquad (17.6)$$

Thermodynamically, phase α will transform into the β phase when $\Delta G^{\alpha/\beta}$ is less than 0. A high interfacial surface area A to volume V cause $\Delta G^{\alpha/\beta}$ to become more positive, increasing the resistance to phase transformation. In magnetic systems, the high surface-to-volume ratio found in nanoparticles has been shown to have a marked effect on the magnetic properties arising from the surface disorder. The disorder results in an increased degree of magnetic relaxation within the particle. First-principles calculations of structural stability in $LiMO_2$ compounds, in which M is a transition metal, show that relaxation has a beneficial effect on phase stability.

Another barrier to achieving high specific capacity in many materials is the number of extracted or inserted Li$^+$ ion. The specific capacity for Li-insertion materials is calculated as:

$$\frac{mAh}{g} = \frac{Li_n M'M''X + 26.8}{\sum MW} \qquad (17.7)$$

The representation $Li_n M'M''X$ depicts the structure as a combination of a guest (Li$^+$) and a host (M'M''X). The constituents of the host determine the denominator in the equation. If more than one Li^+ per transition metal can be intercalated, the specific capacity will increase substantially.

One approach to obtaining a combination of more than one extractable Li^+ and low molecular weight is to use a light nontransition metal as M″ in combination with a transition metal for M′ that undergoes a valence state change of two or greater. An example of multiple oxidation changes on a transition metal site is found in the spinel $LiMn_{1.5}Ni_{0.5}O_4$, where removal of Li^+ is concomitant with oxidation of Ni^{2+} to Ni^{4+}.

These considerations illustrate that the high surface-to-volume ratio unique to nanocrystalline materials improves the desirable phase formation and retention in lithium battery applications. The examples suggest that, in addition to small size, narrow particle size distribution is important.

High-capacity cells require large volumes of electrolyte that must be accommodated between the electrodes. This has a double effect in reducing these cells' power-handling capability. The electrodes must be smaller and farther apart to make space for the extra electrolyte; hence, they can carry less current. Increased volume of the electrolyte means it takes longer for the chemical actions associated with charging and discharging to propagate completely through the electrolyte to complete the chemical conversion process (Table 17-1).

Decreasing the particle size may considerably improve the performance of some electrode material. One advantage of using a material

Features of nanocrystalline electrode materials	Design benefits of lithium rechargeable battery systems	Performance benefits
Nanocrystalline particle size	Consistent, small particles	Storage and cycling stability
Porous electrode characteristics	Better pore size distribution	No hot spots; maximum utilization
Ultra-smooth electrode surface	Very thin separators	High surface area; high power
Intercalation rate	Fast diffusion (more power)	Improved rate; improved pulse
Crystal structure	Stable single-particle crystals	Longer calendar life
Lower impedance to lithium insertion	More lithium intercalation to a cut-off voltage	Higher accessible capacity

TABLE 17-1   Characteristics of Nanomaterials and Their Benefits

with a large surface area is that higher charge–discharge rates can be facilitated. Particle dimensions in the nano range also lead to shorter lithium diffusion lengths, which can provide a higher power output. Particle cracking, leading to pulverization and a corresponding capacity loss, can be prevented if the particles are small enough. The enhanced surface electrochemical reactivity is generally expected to improve the performance of Li-ion batteries. However, a large surface area may also increase the solvent decomposition occurring at both the anode and cathode during charge and discharge, resulting in a large, irreversible capacity. Also, poor packing of the nanoparticles may lead to a low volumetric energy density. Therefore, it is important to optimize the particle size and morphology to maximize the performance of the electrode material.

Several attempts have been made to use nanotubes as electrode material in Li-ion batteries. CNTs were the first to attract interest because graphite is a commonly used anode material. There have been several reports on lithium intercalation into both SWCNTs and MWCNTs. However, the material normally shows a high irreversible capacity during the first cycle. CNTs have also been used in nanocomposites.

Lithium-intercalated graphite and other carbonaceous materials are commercially used in Li-ion batteries. In these cases, the specific energy capacity is partially limited by the thermodynamically determined equilibrium saturation composition of LiC_6. CNTs are interesting intercalation hosts because of their structure and chemical bonding. Nanotubes might have a higher saturation composition than graphite because guest species can intercalate in the interstitial sites and between the nanotubes. Therefore, CNTs are expected to be suitable high-energy density anode materials for rechargeable Li-ion batteries.

SWCNTs spontaneously form bundles that are called *nanoropes*. These bundles are kept together by van der Waals forces. Reversible electrochemical intercalation of SWCNT bundles with lithium has been demonstrated in the past few years. Purified SWCNTs show a reversible saturation composition of $Li_{1.7}C_6$ (632 mAh/g). In any case, this is higher than LiC_6, which is the ideal value for graphite corresponding to a capacity of 372 mAh/g. This LiC_6 value also holds for MWCNTs. Moreover, ball milling may further increase the reversible saturation Li composition of SWCNTs. This process induces disorder within the SWCNT bundles and fractures the individual nanotubes. After ball milling, the saturation composition may be as high as $Li_{2.7}C_6$, which corresponds to a capacity of 1000 mAh/g.

Gao et al. (2000) used SWCNT bundles synthesized by laser ablation. The crude materials were purified by filtering off the impurities over a micropore membrane while keeping the nanotubes in suspension. The purified material existed for 80% of SWCNT bundles with a bundle diameter varying between 10 and 40 nm. The individual nanotube diameter in the bundles was between 1.3 and 1.6 nm.

A composition of $Li_{5.4}C_6$ was obtained after the first discharge with the purified SWCNTs.

SWCNT bundles synthesized by laser ablation (90% SWCNT bundles with a length of 10 µm and a bundle diameter of 30 to 50 nm) were electrochemically reacted with lithium. Then the reversible capacity of the samples was analyzed. This capacity turned out to be Li_2C_6 (744 mAh/g). It can be concluded that SWCNTs have a better reversible capacity concerning lithium intercalation with regard to simple purified SWCNTs. Therefore, SWCNT bundles seem to be attractive host materials for energy storage. MWCNTs could also be low-cost, high-performance anode materials for rechargeable Li-ion batteries because they show an excellent reversible capacity and cycle ability during lithium insertion and extraction. Sharon and his group have been studying intercalation of lithium using carbon nanomaterials (CNMs) synthesized by chemical vapor deposition of plant-derived precursors (Kichambre, 2000; Mukul Kumar 2000, Sharon et al., 2000, 2001, 2002; Bhadwaj, 2007). It is hoped that some of these carbon materials may find its application in making an economical Li-ion battery.

For the application of CNTs in Li-rechargeable batteries, it can be concluded that the barrier height of the intercalation process is a crucial factor in battery activity. Insertion of lithium ions through the sidewall of the nanotubes seems energetically unfavorable unless structural defects are present. Release of the ion during the discharge process has to cross a very high barrier depending on the size of the rings. The electronic binding energies of the lithium ion at its equilibrium distance also decrease as the ring size increases. Thus, it seems that ions outside the tubes may intercalate easily. For the situation of two lithium ions, the binding energy strongly depends on the position of the lithium ions compared with each other. The most stable configuration is the one in which both the ions are situated outside the tube.

17.7 Summary

Compared with other types of batteries, lithium batteries are very different because their potential can be as high as 3.5 V per cell, which is the highest of all batteries. This chapter has discussed the basic principles of lithium batteries with special emphasis on the development of anodes. The specific role of CNMs in lithium batteries has also been discussed in detail.

CHAPTER 18

Carbon Solar Cells

Maheshwar Sharon

Nanotechnology Research Centre, Birla College, Kalyan, Maharashtra, India

18.1 Introduction

Application of solar energy to generate either thermal or electrical power was thought of as early as the 15th century B.C., when solar radiation was used to distill liquid and dry agricultural products. In 212 B.C., Archimedes used the sun's rays to set fire to the ships of the invading Roman fleet.

In the 17th century, the most practical application of solar energy was developed by Ehrenfried Von Tschirnhaus (1651–1700), a member of the French National Academy of Sciences. He used lenses up to 76 cm in diameter to melt ceramic materials, gold, silver, iron, copper, tin, and Hg from their ores. The first experiment relating to ovens for food preservation are described by Nicholas de Saussure (1740–1799). A temperature of 88°C was achieved. In 1908, W. Zerassky built a solar thermoelectric device in which the thermoelectric junction was formed with the wires of a zinc antimony alloy and silver-plated alloys. Before 1908, many developments were made to run engines with heat generated by focusing solar radiation with mirror and lenses.

In 1954, Bell Telephone Laboratories announced the development of a solar battery made of a photovoltaic cell. This effect, however, was discovered in selenium by Becqurel in 1839. The excitement in 1954 was due to a high conversion efficiency of 6%, which increased to 11% in 1 year. Theoretical work predicted 22%. In 1957, the first silicon solar cells were sent aloft on rockets, demonstrating that they could survive the journey and produce useful power. In 1959, the first successful Vanguard satellite carried 108 solar cells to power its radio.

The cost of solar cells has been drastically reduced in view of the volume required from an early cost of more than $1000/W to about $100/W in 1970; now it is around $5/W.

Research in the 1960s resulted in the discovery of other photovoltaic materials, including GaAs (high temperature), CdSe, and $CuInSe_2$. GaAs, CdSe, and $CuInSe_2$ cannot be used for terrestrial applications because Ga is not available in plenty, and Cd is toxic. Silicon solar cells are also not economical for general purpose use. On the other hand, because carbon has a chemistry that is similar to that of silicon, it is being considered for developing carbon solar cells. Sharon et al. (1995) were the first group to have developed homojunction solar cells, and they have predicted its cost to be around 50 cents per peak watt compared with silicon solar cells, which cost around $3 per peak watt.

Before discussing the principle of photovoltaic solar cells, we will digress a little and get familiar with certain basic properties and theories of semiconductors. In the next sections, the theory and application of semiconductor materials for the fabrication of photovoltaic solar cells are discussed in detail.

18.2   Formation of a Semiconducting Material

How a material is formed and what the role of electrons is in the formation of a material are few topics that need to be addressed to understand the physics of bond formation. Bond formation can be dealt with by using either very complex quantum chemistry or by considering simplified quantum chemistry. Because the object is to understand the role of electrons in the formation of a material, the latter approach is considered here.

Let us imagine a hypothetical a case in which two isolated atoms are sitting very far apart from each other. Electrons of each of these two atoms experience only one kind of field, which is due to the positively charged nucleus. Electrons of both of these atoms would perhaps be living with a belief that the energy of the electron is controlled by its positive charge only and would thus have acquired some specific energy values. Now let us imagine that these two atoms accidentally meet each other. Naturally, the electrons would now experience a different type of atmosphere, an attractive force due to its own nucleus (with which it was familiar earlier) and an additional experience of a repulsive force due to its negative charge (a new phenomenon the electrons did not experience earlier). As a result of this additional force, the electrons of the two atoms would no longer be in the same energy state as they were before coming across the other atoms. This behavior could be analogous to a condition in which a beautiful girl enters a classroom populated with boys. The equilibrium conditions the boys had before the girl entered the classroom would be disturbed, and a new type of equilibrium would be established. Just as the boys would be perturbed by seeing the girl in the classroom, the electrons are also perturbed by seeing another electron of another atom. The energy of the perturbed electrons of the two atoms would appear to be in new types of energy levels: one would be at some lower energy than what it had before, and other would be at a higher energy than what it had before. In terms of quantum chemistry, the electrons would be said to be hybridized, creating two bonding levels (lower bonding energy level) and an antibonding level (upper energy level). Moreover, in terms of the atom–atom interaction, the distance between them would be in the vicinity of few Angstrom only.

This interaction of two atoms, with each having one electron, has created two new energy levels. As per Hund's rule, no more than two electrons can occupy one energy level. Because each level of each electron has created two distinct levels and each of these new levels can occupy two electrons, it can be concluded that in general, if there were N number of electrons coming together, there would be 2N number of energy levels generated such that 1N energy levels would lie below the energy levels they had before coming in contact with each other and 1N energy levels would lie higher than the energy levels they had before coming in contact with each other. Because each energy level can occupy two electrons, 1N lower new levels will be able to occupy 2N number of electrons. This means that the newly created lower levels will be filled. Because no additional electrons are left, the newly formed upper levels will be empty. It is important to realize that such interactions would take place from all three directions, resulting in generation of 2N levels in space, N levels in upper levels, and other N levels in lower energy levels.

The extent of separation of energy levels would depend on how close the electrons come in contact with each other. Therefore, if an atom has electrons situated in different energy levels with different orbital shapes, then the resulting new energy levels would also be accordingly separated differently. This is analogous to the case when a beautiful girl in our earlier example entered the classroom: the level of the boys' perturbation would be proportional to the distance of separation from the girl. The boy sitting next to the girl would be perturbed to the highest level compared with a boy sitting at a far corner of the classroom. The extent of separation of the two newly formed energy levels would accordingly depend on the distance between the interacting electrons. In an atom, electrons are known to occupy different energy levels, such as 1s, 2s, 2p, 3s, 3p, 3d, and so on. Therefore, depending on the distance from which these orbitals' electrons interact, the extent of separation of new energy levels will follow accordingly.

Because the strength (i.e., the energy of formation) of a chemical bond is of the order of few eV (e.g., 5 eV), it is reasonable to assume that the depth of newly formed 1N energy levels should be equal to about 5.0 eV. Thus, the difference between each energy level that is newly formed due to the interactions should be around $5 \text{ eV}/6.023 \times 10^{23}$, which is approximately of the order of 10^{-23} eV. This quantity is very small to be differentiated from each other. Therefore, these new energy levels are designated by band instead of level. In other words, when a compound is formed, electrons of each individual atom interact with each other in three dimensions, and the net result of all interactions is the formation of two energy bands, one that is lower than what the individual electrons had before interactions, called the *valence band* and given a nomenclature E_v, and the other called the *conduction band*, which is higher than the energy levels of electrons had before the interactions took place and is designated as E_c. The splitting of energy levels and generation of new energy levels is depicted in Fig 18-1.

Electrons that are of the highest energy are of interest in understanding the chemical nature of a material and of least interest in dealing with other lower energy levels that are created by other electrons present in the atoms, although their participation in the formation of the compound plays a very important role. The behavior of

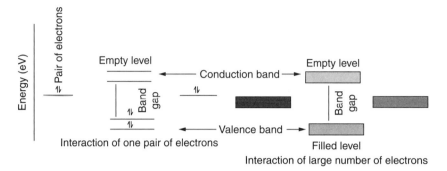

FIGURE 18-1 Schematic diagram showing the splitting of energy levels of individual electrons and the formation of two bands.

the material can be predicted by examining the energy levels of the electrons present in the outermost orbitals of the concerned atoms rather than the behaviors of entire electrons present in the material. Therefore, the discussion would be limited to the electrons occupying the uppermost levels of the valence band.

18.3 Wave Nature of Electrons

18.3.1 In A Fixed Potential

The validity of such a model can also be evaluated through a quantum mechanical approach.

Schrödinger's equation for an electron in a box of fixed potential is a well-known approach to show how an electron appears in an atom. The most general formula is given by:

$$\frac{d\psi^2}{dx^2} + \frac{8\pi^2 m}{h}(E - V)\psi = 0 \tag{18.1}$$

whose solution is given by:

$$\psi = C \exp \pm \left(2\pi \sqrt{\frac{2m(E - V)}{h}} x\right) \tag{18.2}$$

The plus or minus sign preceding the exponent in Eq. (18.2) denotes two possible solutions, one for one electron moving in the positive x direction and the other for an electron moving in the negative x direction. $E - V$ is the kinetic energy of the electron moving freely in a fixed potential V. Thus:

$$E - V = \frac{1}{2}(mv^2) = \frac{p^2}{2m} \tag{18.3}$$

where p is the momentum of the free electron. This is related to the de Broglie wavelength by:

$$p = \frac{h}{\lambda} \tag{18.4}$$

where λ is the wavelength of the electron.

It is convenient to express the wavelength in terms of wave number k, which corresponds to the number of wavelengths contained in one full period:

$$k = \frac{2\pi}{\lambda} \tag{18.5}$$

Therefore, for a free electron:

$$k = \frac{2\pi}{h}p \tag{18.6}$$

or

$$p = \frac{kh}{2\pi} \tag{18.7}$$

Equation (18.6) suggests that the wave number k is proportional to the momentum of the electron that is freely moving in the fixed potential V. The kinetic energy $(E - V)$ would then be equal to:

$$(E - V) = \frac{h^2 k^2}{8\pi^2 m} \tag{18.8}$$

Therefore, by substituting the value of $E - V$ in Eq. (18.2), we have a general solution of the Schrödinger equation for an electron moving in a constant potential field:

$$\psi = C \exp \pm ikx \tag{18.9}$$

18.3.2   Wave Nature of Electrons Under Periodically Changing Potential

The solution expressed by Eq. (18.9) cannot be used in crystals because the potential in the crystal is no longer fixed (each atom in the crystal is expected to be periodically arranged in some sort of symmetry). In addition, each atom of the structure would show a potential V, but in between the two atoms in the space, there would be a gradual variation of potential such that at one place in space it is maximum (V) and then gradually decreases to 0 at some point in the space and then increases again to value V. In other words, there is a periodicity of potential V in all three directions in space, and the electrons would be moving under these periodically changing potentials. This condition will thus also have an impact on the momentum of the electrons. Let us observe the effect of this condition on Schrödinger's equation for at least one direction (i.e., x direction). The potential may thus be expressed as a function of x, that is, $f\infty$ (V_x). Under this condition, Schrödinger's equation becomes:

$$\frac{d\psi^2}{dx^2} + \frac{8\pi^2 m}{h}(E - V_x)\psi = 0 \qquad (18.10)$$

The solution to this equation was given using Bloch's solution to the momentum vector of electrons:

$$\psi_{V_x} = U_{kx} \exp\pm ikx \qquad (18.11)$$

In substitution for ψ_x in equation 18.9, we get:

$$\psi_x = U_{kx} \exp\pm ikx \qquad (18.12)$$

where U_{kx} is the potential changing periodically in the lattice. This means that the nature of the wave would be the same irrespective of the point of origin—that is, whether $x = x$ or $x = (x + a)$—where a is a constant number, normally the lattice distance of the material.

$$U_{k(x+a)} = U_{kx} \qquad (18.13)$$

Similarly, it can also be shown that:

$$\psi_{(x+a)} = U_{k(x+a)} \exp\{ik(x+a)\} \qquad (18.14)$$

which is equivalent to:

$$\psi_{x+a} = \psi_x \exp(\pm ika) \qquad (18.15)$$

because:

$$\exp ik(x+a) = \exp(ikx)\exp(ika) \qquad (18.16)$$

and

$$\psi_{(x+a)} = U_{k(x+a)} \exp\{\pm ik(x+a)\} = U_{kx} \exp(ikx) X \, U_{ka} \exp(ika) \quad (18.17)$$

which would have same form as $\psi_x \exp(ika)$. This exercise suggests that U_{kx} has the following form:

$$\psi = 0 = C \sin\left(\frac{2\pi(\sqrt{2mE})}{h}\right) L \qquad (18.18)$$

A simple model of this solution can be drawn for a one-dimensional periodic potential in x direction by assuming that the potential energy is 0 near the nucleus and equals V_0 halfway between the adjacent nuclei, which are separated by a lattice distance. It is also assumed that the product of $V_0^* \theta$ is constant such that when the potential V_0

increases the width of the sinusoidal's wave, θ (where θ is the full width at its half maxima), decreases accordingly. Under these conditions, a solution to Eq. (18.10) for three-dimensional conditions is given as:

$$\cos ka = P\frac{\sin \alpha a}{\alpha a} + \cos \alpha a \qquad (18.19)$$

where

$$P = \left(\frac{4\pi^2 ma}{h^2}\right)V_0 \theta \qquad (18.20)$$

and

$$\alpha = \left(\frac{2\pi}{h}\right)\sqrt{mE} \qquad (18.21)$$

If the right hand side (RHS) of the equation is plotted for all values of (αa):

$$P\left\{\frac{\sin \alpha a}{\alpha a} + \cos \alpha a\right\} \text{ vs. } \alpha a \qquad (18.22)$$

one gets a continuous sinusoidal wave, as shown in Fig. 18-2. It is observed from Fig. 18-2 that when the value of E is nearly 0, the width of the wave is largest. When the value of E is small, the width of the wave is small, and as the value of E [i.e., Eq. (18.21)] increases, the width of the sinusoidal wave decreases.

However, the solution to Eq. (18.19) must also satisfy the condition given on the left hand side (LHS) of the equation, that is, its value must also be equal to cos (ka). In general, cos x for different values of x oscillates between 0 and ±1. Hence, this puts a restriction to the solution to Eq. (18.19) for all values of αa. All the values of Eq. (18.19) that are greater than ±1 are not permitted because of a restriction of permissible values of cos (ka). Hence, all values shown by continuous lines in Fig. 18-2 are not permitted by this restriction. This suggests

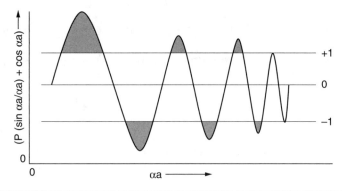

Figure 18-2 Schematic graph showing the variation of $P[\{(\sin \alpha a)/\alpha a\} + \cos \alpha a]$ versus αa.

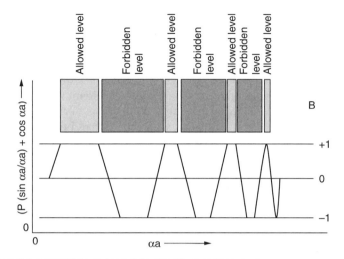

Figure 18-3 Schematic graph showing the variation of $P[\{(\sin \alpha a)/\alpha a\} + \cos \alpha a]$ versus αa after applying the condition that $\cos (ka)$ can have value = ± 1. B shows the effect of such condition in three dimensions. The light gray rectangle shows the position when it is filled with electrons. In between the two rectangles shown in dark gray is the level without electrons.

that there are some values of energy E for which there can be no allowed values of ka (i.e., the colored portion of the graph is not allowed). Hence, electrons cannot possess these energy values.

Alternatively, no electron can be found in these energy regions. It will also be clear that the magnitude of the wave [i.e., $\cos (ka)$] has the same value for either positive or negative values of ka or for any multiple values of 2π. Moreover, the discontinuity in the wave due to the restriction of permissible value of ka to lie within ± 1 occurs at every $\cos (ka)$ where ka equals $n\pi$. In other words, if the density of momentum of the electron is plotted versus the distance in the space of a given system, there would be distinct places where one could not find any density of momentum of the electron.

Thus, there is a forbidden energy region where electrons cannot exist. This period is observed after each $n\pi$ in space. Thus, we arrive at a conclusion that is similar to the one arrived at earlier—when a material is formed due to overlapping of electrons of the constituent atoms, energy bands are generated periodically in the space, electrons can only be found in these energy bands, and no permissible energy levels can be found in between these bands (Fig. 18-3).

18.4   The Band Model to Explain Conductivity in Solids

Can this band model be used to explain the electrical property of materials? To make a material conduct electrically, it should easily allow the flow of electrons within the material, preferably at room temperature. To achieve this objective, the structure of the material must provide a vacant place for electrons to migrate through the lattice. Because the conduction band is situated at an energy level higher than the valence band, naturally it would be easier for material to conduct if there were vacant space available within the valence band for electrons to migrate. One may conclude that if the valence band has empty levels at room temperature, the material could show a conducting property. But it was seen earlier that the valence band is filled. So how can a vacancy be created in the valence band?

Let us examine the formation of a germanium crystal. This is formed by sharing its four valence electrons with four germanium atoms in the space. As a result of this sharing, no vacancy is left in any of the levels present in the valence band, so the material is classified as an insulator. It is assumed that all of the germanium atoms are occupying the expected lattice point. In reality, this is not possible. No matter how pure a compound is, there is always the possibility of having some defects, such as Schottky defects created by the absence of an atom from the system or Frenkel defects caused by the placement of an ion or atom at a site that is not a lattice site (i.e., it occupies an interstitial position). These defects create some vacant sites at the lattice points. To maintain the charge neutrality, the number of cation vacancies must be equal to the number of anion vacancies (in Schottky-type defects). With a Frenkel defect, the number of vacant lattice sites must be equal to the number of ions or atoms present in the interstitial positions. The displacement of ions or atoms from their lattice would make the lattice site positively charged, and the number of such charges would be equal to the number of ions or atoms displaced from their lattice sites. These extra electrons (if cations are displaced) or holes (if anions or neutral atoms are displaced) can help in the migration of electrons through the material. Naturally, the conductivity would be proportional to the number of such defects present in the crystal at a given temperature. Such additional electrons would be created in the conduction band (because the valence band is filled). Likewise, the holes would be created in the valence band because the conduction band is normally empty. Thus, even in highly pure materials, the number of electrons created in the conduction band would be equal to the number of holes in the valence band. Such materials are called *intrinsic-type materials*.

18.4.1   Extrinsic Semiconductors

There is a need to define the term *Fermi level* to understand the reason for a semiconductor to become conducting by doping. Fermi energy is the energy of carriers (either electrons or holes) present in the material. Mathematically, it can be shown that the Fermi energy of the carrier is directly proportional to the number of carriers present in the material, and it does not depend on the temperature. In the intrinsic material,

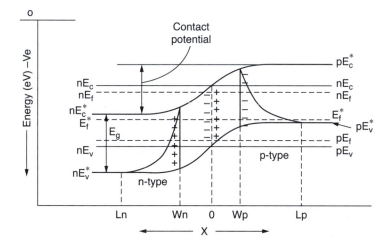

Notes: E_c = conduction band; E_f = Fermi level; E_v = valence band;
L = end of diffusion region width; W = end of space charge width.

FIGURE 18-4 Schematic diagram showing formation of band bending when p- and n-type semiconductors are joined together. The band diagram in black shows the condition before forming the p:n junction. The prefixes n and p are for n- and p-type material, respectively. The plus and minus signs depict the positions of the holes and electrons after forming the p:n junction. The suffixes represent the condition after forming the p:n junction.

because the number of electrons and holes are equal, their Fermi energies are the same for both types of carriers. The Fermi energy in the band model (Fig. 18-4) is represented by a broken line that lies in between the conduction and valence band. It is represented by E_f.

Mathematically, it can also be shown that the Fermi level in intrinsic materials lies at $0.5\ E_g$ in the band model. Because the carrier concentration of electrons and holes in the intrinsic semiconductor is same, the Fermi energy cannot be changed. However, because the Fermi level of a material depends on the carrier concentration, its magnitude can be altered by altering the number of the carrier. Thus, doping the intrinsic material with a suitable dopant can alter the magnitude of Fermi energy. To explain this behavior, let us take the example of pure silicon material.

If a known amount of impurity, such as phosphorous (P^5) or boron (B^3) atoms (known as the *dopant*), is added to the silicon crystal during its purification, then silicon atoms are replaced by an equal number of dopant during the formation of crystalline silicon material. If the dopant has electrons greater than four valence electrons (e.g., the phosphorous atom, which has five valence electrons), then each site of the silicon lattice (i.e., the site that should have been occupied by Si) will occupy a phosphorous atom. Each phosphorous atom has one extra electron than the lattice of the host atom needs (i.e., silicon lattice). This additional electron will occupy the conduction band because the valence band is filled. Because Fermi energy depends on the number of carriers, this additional electron will alter the Fermi level. Thus, doping the silicon with the phosphorous atom will shift the Fermi level from its original position (i.e., from its $1/2E_g$) toward the conduction band. The concentration of electrons can thus be increased to a desired value without affecting the intrinsic concentration of holes. Because this material (i.e., silicon) redundant has excess electrons (i.e., a negative charge due to phosphorus doping), it is termed *n-type silicon*. Such materials are also classified as extrinsic semiconductors.

Likewise, if silicon lattice was doped with atoms having valence electrons less than the silicon atom (e.g., boron), then the addition of each boron atom would create a hole (i.e., a positive charge) at its site, which is as good as creating a hole in the valence band. These materials are classified as p-type material (i.e., positive type). With such materials, the equilibrium concentration of electrons would also be unaffected, but the hole concentration would be in excess of its equilibrium concentration (i.e., intrinsic concentration) by an amount equal to the number of boron atoms per square centimeter added to the intrinsic silicon lattice. For all practical purposes (because the equilibrium concentration of electrons and holes is same), it would be reasonable to assume that in extrinsic materials, the number of extra electrons (in n-type material) or holes (in p-type material) would be equal to the amount of added dopant in the material because the intrinsic concentration of electrons and holes (known as the *equilibrium concentration*) is much less than the additional carriers created by addition of dopant.

The other significant information to note here is that in extrinsic materials, both types of carriers (i.e., electrons and holes) exist simultaneously with one having a larger concentration (equal to the number of dopants added) and the other equal to its equilibrium concentration. When crystal is doped with atoms having valences greater than the host atom, the dopant is called the *donor atom*, and when they are doped with atoms of lower valence, they are called *acceptor atoms*.

Because the concentration of electrons and holes of intrinsic material is not affected by the doping process, the material will continue to exhibit its intrinsic Fermi level at its half band gap position, and an additional Fermi level would be created due to the presence of dopant in the system. Because the Fermi level due to the intrinsic carrier concentration is situated at half its band gap while representing the Fermi level of extrinsic material, the intrinsic Fermi level at the half band gap position is not shown.

18.4.2   Types of Carriers in a Semiconductor

It was shown earlier that doping a semiconductor with aliovalent ions (ions having valence different than the host atom) alters the concentration of electrons (for n-type material) without affecting the equilibrium concentration of the intrinsic hole concentration.

Thus, n-type semiconductors contain both types of carriers, that is, electrons (concentration equal to the dopant concentration + electrons due to intrinsic concetnration) and holes (concentration equal to the intrinsic equilibrium concentration). Likewise, p-type material contains holes (equal to the concentration of the acceptor's concentration + holes due to intrinsic concentration) and electrons (equal to the intrinsic equilibrium concentration). These two types of carriers are classified as *majority carriers* (i.e., electrons for n-type and holes for p-type material) and *minority carriers* (i.e., holes for n-type and electrons for p-type material).

18.5   Theory of Junction Formation

Discussions from previous sections suggest that the Fermi level of an n-type semiconductor would be situated near the conduction band and would be situated near the valence band for p-type materials. This description also suggests that the Fermi energy for electrons in n-type semiconductors is higher (i.e., less negative) than that of p-type semiconductors (i.e., more negative) and vice versa for the holes. What will happen if a p-type semiconductor is brought in contact with an n-type semiconductor with no air gap in between? The n-type material has a large concentration of electrons with its higher Fermi energy (i.e., Fermi energy situated nearer the conduction band). The p-type material, on the other hand, also has large concentrations of vacancies in the valence band (i.e., due to the creation of large concentration of holes), and Fermi energy is situated near the valence band. The energetic electrons of n-type material would therefore tend to flow to the p-type semiconductor. But does this mean that these energetic electrons of n-type material would actually leave n-type material and occupy the vacancies created in p-type material? If this happens, then it would amount to making n-type material positively charged and p-type material negatively charged. Simply by putting two materials in contact, it is not possible to make them charged. On the other hand, it is also not possible to maintain two energy levels of electrons in two materials at two different levels while they are in contact. A strange situation thus has been created (due to creating contact between two materials) in which neither the electron can be permitted to cross over the interface nor the two Fermi levels can maintain their individual levels. A compromise can be visualized. The n-type material can allow its electrons to concentrate at the interface of two materials. Likewise, the p-type material could allow its holes to concentrate at its interface. The concentration of these carriers present at the interface would settle down to a value such that the Fermi levels of two materials would become equalized. It is strange to assume that carriers with a positive charge can stay together at the interface without being annihilated. This situation is similar to placement of army of two countries at a boarder. Soldiers of two different countries are almost ready to annihilate each other, but at the border they stay without fighting because they are under the control of their country's decisions. Annihilation of electrons and holes would amount to ionizing the two semiconductors, which is not permitted by the materials. However, carriers present at the interface will move back to their host lattice as soon as the two materials are disjointed. This situation is like putting money in the bank with the understanding that you can withdraw it from the bank any time you want.

Thus, when two types of materials are joined together, the migration of electrons from n-type material and holes from p-type material occur at the interface. This process continues until the energy levels in the bulk of both materials (i.e., the Fermi levels of both materials) attain an equilibrium level (i.e., when both Fermi levels have equalized). Moreover, both of these carriers coexist at the interface without being annihilated.

Should the change in the displacement of electrons (i.e., majority carrier) in n-type material be abrupt at the interface compared with the bulk, or should it change linearly or exponentially? The most reasonable assumption is that the change in the concentration of majority carrier between the bulk and the interface is exponential in nature. This situation can be shown pictorially by making an exponential bending of conduction band from its original flat, horizontal condition.

As a consequence of the bending of the conduction band, the valence band should also bend accordingly so that the difference between the conduction band and the valence band remains unaltered throughout the material (i.e., from the surface to the bulk). If this does not happen, it would amount to change in the band gap of the material, which is not permissible. In addition, the Fermi level should accordingly shift exponentially in the same fashion as the conduction and the valence band (i.e., downward in n-type material and upward in p-type material). It should bend exponentially by keeping its position from the conduction band (for n-type material) or with the valence band (for p-type material) the same as it had before the forming the contact (because the difference between the these positions depends on the concentration of dopant present in the material, and we have not altered the concentration of aliovalent impurities while joining the two semiconductors together). The nature of bending of the Fermi level, however, would be downward in n-type material and upward in p-type material (see Fig. 18-4).

18.5.1 Impact of Migration of the Carrier Toward the Interface

The lattice site from which carriers have moved will become charged. The lattice site will become negatively charged if a hole has moved and positively charged if electrons have moved to the interface.

This would mean that there would be an exponential decrease in the concentration of the appositively charged lattice sites as we examine the situation inside the material from the interface. In reality, they may not be even exponential in nature. The distribution of the appositively charged lattice site may be random. Any mathematical treatments for such random distribution or even approximated exponential distribution of charged ions become a complicated system to deal with; hence, for our convenience, it is assumed that all appositively charged atoms are situated at a fixed distance in one plane from the surface. This hypothetical visualization of charge distribution near the interface (i.e., its presence in one plane) reveals the possibility of formations of two parallel planes, one at the interface (which is populated with majority carriers) and another plane at a depth of about 100.0 to 400.0 nm (which are populated with appositively charged lattice sites). The region between these two planes is called the *space charge width* designated by w_n (for n-type material) and w_p (for p-type material). If the interface of the two semiconductors is taken on the x-axis as the 0 point, then w_n would be situated at $x = w_n$ and w_p at $x = -w_p$. At the plane at $x = w_n$, the concentration of holes would be much greater than its concentration at $x = \infty$ (known as the equilibrium concentration, h_0).

Within the space charge region, carriers experience a field due to the columbic potential created by the two types of carriers present at the two hypothetical planes separated by space charge region. We also make an assumption that this field becomes 0 beyond the plane at $x = w_n$ (or at $x = -w_p$), whereas within the space charge region due to exponential growth of minority carriers, the columbic field is also exponential in nature. In the diffusion region (i.e., between $x = w_n$ or $x = -w_p$) and at $x = (L_n)$ or $(-L_p)$, there is a noncolumbic field due to creation of the concentration gradient. L_n (or L_p) is the position of the planes (i.e., from the plane of the space charge region) at which the concentration of carrier is the same as that of the bulk concentration. The direction of movement of majority carriers in this region depends on the direction of the concentration gradient established by the minority carriers.

The force present in the diffusion region can be visualized by taking the example of movement of people in a mob. The direction of movement of the people is more or less controlled by the direction of the mob's movement (the mob would always move from its highest population to its lowest population), and any movement opposite to mob's direction is a barrier for an individual's movement. Thus, the majority carrier would find resistance to move toward the space charge region and no resistance in the opposite direction. Alternatively, minority carriers would find no resistance to move toward the space charge region (i.e., toward $x = w_n$ or $x = -w_p$), but high resistance in moving toward the bulk or toward $x > w_n$ or $x > -w_p$. The condition of carriers beyond the diffusion region (i.e., $x > L_n$ or $x > -L_p$) would remain unaltered because there is no concentration gradient.

18.5.2 Graphically Representing These Energy Levels

There is a need to represent the changes in the carriers' concentrations along the x-axis as well as along the cross-sectional area of the surface of the interface (i.e., along the z-axis) and the changes in energy levels of various regions. In other words, representation of two types of information—the concentration of carriers and the energy of the carriers—should be made within the same diagram. In Fig. 18-4, the energy levels of various functions, such as E_c and E_f, are shown by horizontal lines or bend lines.

It has also been suggested that no carriers can be found in between the conduction and the valence band. It has been the understanding that no electrons exist within this forbidden gap. But the Fermi level (or acceptor or donor level) is represented within this forbidden gap (see Fig. 18-4), which could be misunderstood that electrons or holes of corresponding energy are present in this gap. To remove any misconception, there is a need to clarify these points. The energy levels of E_f, E_d, E_a, and so on are drawn within the band gap region only to represent their energy, but the carriers possessing these energies are still situated in the valence band. Unless they are supplied energy equivalent to $E_c - E_f$ (for electrons in n-type material) or $E_f - E_v$ (for holes in p-type material), they would remain in the valence band. In other words, these energy levels shown within the band gap region are only a representation of the energy of the carriers that are lying within the valence band and should never be mistaken for being present in the forbidden gap. Moreover, Fermi energy is written as a level with dotted lines because this is not a definite value of the energy. It depends on the concentration of donor or acceptor. Acceptor or donor levels, on the other hand, are drawn with full lines because their ionization energy depends on the type of dopant present in the crystal and cannot be altered for a given system.

There is a need to show the distributions of concentrations of various carriers within the materials after they have formed the junction. These are presented in x, z directions. These carriers are represented within the forbidden gap region. Hence, it is necessary to understand that these are presentations of the carrier concentrations and have nothing to do with energy level representation.

Thus, two types of representations are made pictorially. One is the representation of energy level of the carriers, and the other is the concentrations of the carriers. The variation in concentration of the carrier is shown with respect to the surface of the interface of the material. Therefore, readers must not be mistaken that the insertion of carriers within the band gap region means that the carrier is present within the forbidden gap. Readers must realize that two types of representations are for different types of axes (one for energy level, and the other for the carrier's concentrations). Representations of both of these behaviors are shown in Fig. 18-4. This type of combination is also referred as the *p:n junction*.

18.5.3   Nature of Potential at the Interface

What would be the magnitude of the potential developed between the charges accumulated at the interface and at plane $x = w_n$ or $x = -w_p$? The driving force for these carriers to accumulate at the interface is the difference in Fermi levels of n- and p-type semiconductors. Therefore, the maximum total potential generated at the interface of the p:n junction should be equal to the difference in the Fermi levels of p- and n-type semiconductors. This potential is known as the *contact potential*.

It is worth remembering that the total contact potential is the sum of two contact potentials, one generated in n-type material and the other generated in p-type material. What would be the maximum contact potential that could be generated in each semiconductor? Can a device be made of such a type that could give a contact potential equal to the band gap's value of the semiconductor? In previous sections, it was observed that the Fermi level of pure semiconductors (i.e., intrinsic semiconductors) cannot be larger than half the band gap of the semiconductor. Therefore, the Fermi level in individual semiconductors constituting the p:n junction cannot be bent more than half of its band gap's value. This condition puts a restriction to the value of contact potential, which is $\frac{1}{2}E_g$.

18.5.4   Does Any Current Flow Through the Interface?

At the equilibrium condition, flow of carriers from n- to p-type material and vice versa occurs in such a fashion that a constant contact potential is established. The majority carrier is holes in p-type material or electrons in n-type material, and their concentrations depend on the concentration of dopant present in the material. On the other hand, the minority carrier in either type of semiconductor is due to thermal excitation of atoms or ions (i.e., due to Frenkel-type defects) and is temperature dependent. This concentration is very low compared with that of the majority carrier, but it may be altered by changing the physical condition (e.g., temperature) of the semiconductor. On the other hand, the concentration of the majority carrier being equal to the concentration of dopant present in the material (of the order of 10^{18} to 10^{20} cm^3) cannot be altered after the material is prepared; therefore, it is a constant quantity.

To avoid confusion, the minority carrier can be designated as a generated carrier (with a suffix of g) due to effects such as thermal excitation, photoexcitations, and so on. Current flowing across the junction because of these carriers can be classified as generation current J_g due to the flow of the minority carriers generated by thermal excitation. The majority current J_m is due to the majority carriers (designated by suffix m). Thus, in a p-type semiconductor, the total current flowing in the system is due to (1) the flux of the minority carrier (i.e., $J_{g\,(n\to p)}$ as well as $J_{g(p\to n)}$) and (2) majority carriers $J_{m(n\to p)}$ and $J_{m(p\to n)}$ flowing from n- to p-type semiconductors as well as flowing from p- to n-type semiconductors. When these two types of semiconductors are joined, an equilibrium condition is set, and the magnitude of the flux of carriers flowing from p- to n-type material is equal to the flux of carrier flowing from n- to p-type material. The magnitude of flux of carriers flowing at the equilibrium condition can be designated by putting a suffix o to each type of carriers.

Therefore, at equilibrium conditions, the p-type semiconductor will contain a flux of J_{pgo} and J_{pmo} due to the minority and majority carriers, respectively. Likewise, in an n-type semiconductor, the flux of carriers consists of J_{ngo} and J_{nmo}. To establish a contact potential of θ_o, the magnitude of current flowing due to the fluxes of carriers flowing from one to other and vice versa should be the same at the equilibrium condition. Therefore, the flux of majority carriers flowing from n- to p-type material would be equal to flux of majority carriers flowing from p- to n-type material:

$$J_{m(n\to p)} = J_{m(p\to n)} \qquad (18.23)$$

Similarly, the flow of the flux of minority carriers flowing from the both side of the semiconductors must also be equal:

$$J_{g(n\to p)} = J_{g(p\to n)} \qquad (18.24)$$

18.5.5 Effect of Application of External Potential to the p:n Junction

The previous calculation suggests that if the p:n junction is connected to an ammeter, there would be no flow of current because there is no net flow of carriers in either direction. What would happen if a small potential ($V_o \sim kT$) is applied to this system such that the p-type semiconductor is connected to a positive terminal of the battery and the n-type semiconductor is connected to a negative terminal of the battery? What would be the impact of the applied potential on the equilibrium condition?

What Would Happen if Positive Potential is Applied to a p-Type Semiconductor of p:n Junction?

By applying a positive potential (V_o) to a p-type semiconductor, the flux of the minority carriers (Fig. 18-5) in the space charge region of either semiconductor (i.e., in the region $x = 0$ and $x = -w_p$ in a p-type semiconductor or $x = w_n$ in an n-type semiconductor) would decrease because its charges would be nullified by the externally applied

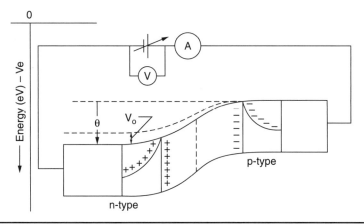

FIGURE 18-5 Schematic diagram showing the position of energy levels when the p:n junction is under a forward bias.

potential (i.e., some of ionized lattice sites of space charge region would become neutral, as was the case before the contact potential was formed). For example, in the n-type material, the ionized lattice site of the space charge region would take electrons from the externally applied potential to get neutral. This is as if the flux of majority carriers at the interface has been reduced accordingly. This decrease in the flux of majority carriers is due to a decrease in the columbic force, which reduces the contact potential by the magnitude of the externally applied potential (i.e., V_o). The new contact potential would thus become $\theta_o - V_o$. What would be the effect of this decrease in the potential on the flow of flux of carriers across the interface?

Flow of Carrier from n- to p-Type Material

In general, the majority carrier experiences the contact potential developed across the interface as a barrier. But a decrease in this barrier height would ease the flow of majority carriers toward the p-type semiconductor. Thus, the majority carrier (i.e., J_{nm}) would find a smaller new potential barrier (i.e., $\theta_o - V_o$) for crossing the interface from the n- to the p-type semiconductor. This decrease in potential would facilitate the flow of majority carriers toward the p-type semiconductor. Thus, the flux of majority carriers flowing from the n- to the p-type semiconductor would increase by a factor of $\exp(eV_o/kT)$. The new flux of majority carrier $J_{m(n \to p)}$ flowing from the n- to the p-type semiconductor can thus be given by:

$$J_{m(n \to p)} = J_{mno} \exp \frac{eV_o}{kT} \tag{18.25}$$

However, minority carriers flowing from the n- to the p-type semiconductor would still find no barrier; hence, its flux would remain unchanged because of the application of external field (i.e., its magnitude) would remain as it was before imposition of the potential (i.e., J_{ngo}):

$$J_{g(n \to p)} = J_{ngo} \tag{18.26}$$

Although the contact potential has decreased, thermodynamically, the minority carrier would still find no barrier in its migration. This condition is similar to a situation for a person falling from either the third or fourth floor to the ground floor. The person would experience no barrier in falling (this is the condition minority carrier experiences). However, climbing from the ground floor to the third floor would certainly be easier than climbing to the fourth floor (this is the condition the majority carrier experiences).

Flow of Carriers from p- to n-Type Material

Likewise, the flow of carriers from p- to n-type material can be calculated. The flux due to the majority carrier (i.e., the hole) would increase by a factor of $\exp(eV_o/kT)$ because the barrier height is decreased by a value V_o. The concentration of its flux $J_{m(p \to n)}$ would be given by:

$$J_{(p \to n)} = J_{pmo} \exp \frac{eV_o}{kT} \tag{18.27}$$

On the other hand, the minority carrier would still find no barrier; therefore, its flux would be the same as it was before the potential V_o was applied. Thus, J_{pg} would be given by:

$$J_{g(p \to n)} = J_{pgo} \tag{18.28}$$

The magnitude of the current flowing due to these carriers would be equal to the product of their concentrations (i.e., their flux) and the electronic charge. Now the equation to represent the flow of current across the interface can be generated.

Flow of Current Due to Holes

The net flow of current due to holes is equal to the difference in the flux of the holes flowing in two directions, that is, the flux of holes (due to the majority carriers) from p- to n-type material minus the flow of flux of holes (due to the minority carriers) from n- to p-type material:

$$I_{hole} = e^o \{J_{m(p \to n)} - J_{g(n \to p)}\} = e^o \left\{ J_{pmo} \exp \frac{eV_o}{kT} - J_{ngo} \right\} \tag{18.29}$$

Flow of Current Due to Electrons

Likewise, the net flow of electrons is equal to the difference of current contributed due to the flow of flux of electrons flowing from n- to p-type material minus the flow of flux of electrons flowing from p- to n-type material:

$$I_{electron} = e^o \{J_{m(n \to p)} \ J_{g(p \to n)}\}$$ (18.30)

i.e.,

$$I_{electron} = e^o J_{nmo} \exp \frac{eV_o}{kT}$$ (18.31)

The total current flowing through the system would be:

$$I_{total} = I_{electron} + I_{hole}$$ (18.32)

$$I_{total} = \left(e^o \left\{ J_{nmo} \exp \frac{eV_o}{kT} - J_{pgo} \right\} \right) + \left(e^o \left\{ J_{pmo} \exp \frac{eV_o}{kT} - J_{ngo} \right\} \right)$$ (18.33)

$$I_{total} = e^o \left\{ \left(J_{nmo} \exp \frac{eV_o}{kT} - J_{nmo} \right) + \left(J_{pmo} \exp \frac{eV_o}{kT} - J_{pmo} \right) \right\}$$ (18.34)

Because there was no net current flowing in the absence of external potential, the net flow of carriers of each type across the interface must be equal:

$$J_{nmo} = J_{pgo}$$ (18.35)

and:

$$J_{pmo} = J_{ngo}$$ (18.36)

Thus, Eq. (18.34) simplifies to:

$$I_{total} = e^o \left\{ J_{nmo} \left(\exp \frac{eV_o}{kT} - 1 \right) + J_{pmo} \left(\exp \frac{eV_o}{kT} - 1 \right) \right\}$$ (18.37)

$$I_{total} = e^o (J_{nmo} + J_{pmo}) \left(\exp \frac{eV_o}{kT} - 1 \right)$$ (18.38)

$$I_{total} = I_o \left(\exp \frac{eV_o}{kT} - 1 \right)$$ (18.39)

where

$$I_o = e^o (J_{nmo} + J_{pmo})$$ (18.40)

where e^o = electronic charge
   I_o = saturation current (or exchange current), which is a constant quantity for a given semiconductor.

If $e^\circ V_o \gg kT$, then Eq. (18.39) reduces to:

$$I_{total} = I_o \exp \frac{eV_o}{kT}$$ (18.41)

Equation (18.41) suggests that if a potential is applied to a p:n junction with the p-semiconductor as positive electrode, then the current flowing through the system increases exponentially with an increase in the applied potential. Such a type of biasing is called the *forward bias*. It should also be realized that the net change in the current is due to an increase in the flux of majority carriers crossing the junction, and there is no change in the flux of minority carriers.

18.5.6 What Would Happen if Negative Potential is Applied to a p-Type Semconductor of p:n Junction?

The next obvious question is that what will happen if a p-type semiconductor of p:n junction is connected to a negative terminal of the battery? Application of such potential would produce an effect that is the reverse of what was observed when positive potential was applied to the p-type semiconductor. The negative potential applied to the p-type semiconductor will try to pull the majority carriers more toward the interface from the bulk of material. This will increase the concentration of electrons at $x = w_n$ and that of holes at $x = -w_p$; thus, the potential difference between the conduction band of p-type material and the conduction band of n-type material would increase by the magnitude of the applied potential V_o. In other words, the equilibrium Fermi level will shift toward the more negative side by increase of negative potential to the p-type semiconductor. In the previous case of forward bias potential, the Fermi level had shifted toward less negative value (i.e., toward E_c of p-type material) (see Fig. 18-4). This increase in applied negative potential to p-type material amounts to an increase in the contact potential of the interface by V_o. That is, the new contact potential would become $V_o - \theta_o$. This is the reverse of the previous condition (i.e., forward bias) in which the contact potential decreased with the increase of potential. The increase in contact potential will have the following effect on the two carriers in each semiconductor.

Flow of Majority Carriers from p- to n-Type Semiconductors

The flux of majority carriers will experience a lesser positive potential at the interface; hence, its flux (J_{pm}) flowing across the junction would be decreased by amount exp $(-eV_o/kT)$. Thus:

$$J_{nm} = J_{nmo} \exp\left(-\frac{eV_o}{KT}\right)$$ (18.42)

Flow of Majority Carriers from n- to p-Type Semiconductors
The flux of majority carriers would likewise experience a greater negative potential because of the increase in the contact potential. Hence, its flux (J_{nm}) crossing the junction would decrease by a factor $\exp(-eV_o/kT)$. Thus:

$$J_{nm} = J_{nmo} \exp\left(-\frac{eV_o}{kT}\right) \tag{18.43}$$

Flow of Minority Carriers from p- to n-Type Semiconductors
The flow of minority carriers from p- to n-type semiconductors would find an increase in contact potential, which is favorable for its flow across the junction. Hence, its flux crossing the junction would be equal to what it had before the potential was applied:

$$J_{pg} = J_{pgo} \tag{18.44}$$

Flow of Minority Carriers from n- to p-Type Semiconductors
The flow of flux of minority carriers likewise would also find the path still easier as before (i.e., before the potential was applied); hence, its flux crossing the interface would be same as its value when no potential was applied:

$$J_{ng} = J_{ngo} \tag{18.45}$$

Thus, the total current passing through the p:n junction due to the flow of flux of electrons would be given by:

$$I_{electron} = e^o\left\{J_{ngo} - J_{pmo}\exp\left(-\frac{eV_o}{kT}\right)\right\} \tag{18.46}$$

Likewise, the total current due to the flow of flux of holes would be:

$$I_{hole} = e^o\left\{J_{ngo} - J_{pmo}\exp\left(-\frac{eV_o}{kT}\right)\right\} \tag{18.47}$$

The net direction of flow of holes and electrons would be in the same direction. Hence, the total current passing through junction would be:

$$I_{total} = I_{electron} + I_{hole} \tag{18.48}$$

$$I_{total} = e^o\left\{J_{pgo} - J_{nmo}\exp\left(-\frac{eV_o}{kT}\right) + J_{ngo} - J_{pmo}\exp\left(-\frac{eV_o}{kT}\right)\right\} \tag{18.49}$$

$$I_{total} = e^o\left\{J_{nmo} - J_{nmo}\exp\left(-\frac{eV_o}{kT}\right) + J_{pmo} - J_{pmo}\exp\left(-\frac{eV_o}{kT}\right)\right\} \tag{18.50}$$

$$I_{total} = e^o \left[J_{nmo} \left\{ 1 - \exp\left(-\frac{eV_o}{kT} \right) \right\} + J_{pmo} \left\{ 1 - \exp\left(-\frac{eV_o}{kT} \right) \right\} \right] \qquad (18.51)$$

$$I_{total} = e^o \left[\left(J_{nmo} + J_{pmo} \right) \left\{ 1 - \exp\left(-\frac{eV_o}{kT} \right) \right\} \right] \qquad (18.52)$$

This can be simplified to:

$$I_{total} = I_o \left\{ 1 - \exp\left(-\frac{eV_o}{kT} \right) \right\} \qquad (18.53)$$

where:

$$I_o = e^o (J_{nmo} + J_{pmo}) \qquad (18.54)$$

where I_o is known as the saturation current (or equilibrium current). If $eV_o > kT$, then $\exp(-eV_o/kT)$ tends to 0. Thus, under this condition:

$$I_{(total)} = I_o \qquad (18.55)$$

Thus, when the p-type semiconductor of a p:n junction is connected with a negative terminal of the battery, there is no change in the current by an increase of potential. This type of biasing is called *reverse bias*. It is also obvious that through the p:n junction, current can flow only if the polarity of the current is such that its positive value is connected to the p-type semiconductor. In the reverse condition, no current can pass through the p:n junction. Extending this conclusion further, if an alternating current is allowed to pass through a p:n junction, the output current would be a direct current only. A p:n junction thus acts like a rectifier.

It would be desirable to spend some time to further examine Eq. (18.41). If a p:n junction is connected to a battery and its p-type semiconductor is connected to the positive terminal of the battery (see Fig. 18-5) and then the current obtained for the various applied potential is plotted, a graph (Fig. 18-6) is obtained.

The log of Eq. (18.41) can be written as:

$$\ln I = \ln(I_o) + \frac{eV_o}{kT} \qquad (18.56)$$

A plot of ln I vs potential V should give a straight line. The slope should give a value $e/(kT)$, and the intercept should give the magnitude of I_o, which is known as an exchange current, a leakage current, or a saturation current. A good p:n junction should give a very low value of the exchange current. Another way to find out the exchange current of a p:n junction is to find out the magnitude of the current under reverse bias (see Fig. 18-6).

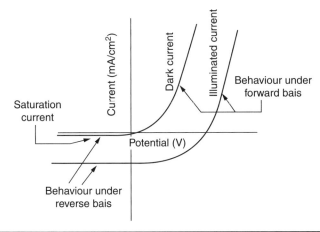

FIGURE 18-6 Current–voltage characteristics of a p:n junction (under dark and light), showing its behavior under reverse and forward bias conditions.

18.6 Effect of Light on the p:n Junction

How can a p:n junction be used for converting solar energy into electrical power? To examine this question, it is necessary to study the effect of illumination of a p:n junction with a light source. It has been seen earlier that a p:n junction has three types of regions on each side of the junction (i.e., on n- and p-types of the semiconductor): the bulk region, diffusion region, and space charge region.

It is appropriate to separately examine the effect of light on each of these three regions. Let us see the effect of illumination of each of these three regions with photons having enough energy to excite electrons from their valence band to the conduction band.

When photons of energy greater than the band gap fall in the space charge region of the n-type semiconductor of a p:n junction, electrons from the valence band are excited to their conduction band, creating a similar number of holes in the valence band. As soon as such electron–hole pairs are formed, they experience a field due to the contact potential θ_o present in the space charge region. The magnitude of columbic field present in the space charge is of the order of few mega volts per centimeter. For example, a p:n junction with a contact potential of 1.0 V and a space charge width of 100.0 nm would produce a columbic field of approximately 0.1 MV/cm. This field is strong enough to force instantaneously photo-generated electrons to move toward the diffusion region before they have a chance to fall back to the valence band or to recombine with holes in this region. Holes naturally follow the opposite path to the electrons. Holes move toward the interface (i.e., toward the front surface of an n-type semiconductor, which is in contact with a p-type semiconductor). The direction of movement of these carriers naturally follows the direction of the columbic field.

When photons of the same energy (i.e., energy $> E_g$) fall in the diffusion region, electron–hole pairs are formed in the same fashion, but instead of experiencing a columbic energy, they experience a concentration potential of minority carriers. Being majority carriers in n-type material, electrons find no resistance in flowing toward the bulk (i.e., toward $x > L_n$) because their concentration is greater than the equilibrium concentration at $x = w_n$ (see Fig. 18-4). However, the direction of flow of photo-generated holes would be in the opposite direction to that of electrons (i.e., toward the space charge region). In other words, their direction would align itself with a concentration gradient set in the space charge region of n-type material. It is important to realize that the force responsible for allowing the photo-generated carrier in the diffusion region is not as strong as it is in the space charge region; it is only proportional to the force created due to the concentration gradient of carriers established in this region. Hence, a large portion of photo-generated carriers is lost because of recombination while they are generated in the diffusion region. Approximately 10 to 20% of the photo-generated carriers of the diffusion region contribute to the total current of the p:n junction; the rest are lost because of their annihilation.

When photons interact with the valence band in the bulk region, because of absence of any kind of field, the photo-generated carrier recombines almost instantaneously and hence contributes nothing toward the photocurrent. Thus, the total effect of these illuminations in the different regions is that a large concentration of electrons gets accumulated at the back side of the n-type semiconductor and holes get accumulated at the interface of the p:n junction.

Similarly, the effect of photon illumination of the three types of regions in the p-type material of the p:n junction can be examined. And it will be seen that the back side of the p-type material gets concentrated with holes, and the interface of the p-type material gets populated with electrons. The concentration of photo-generated electrons and holes accumulated at the interface of the n- and p-type material get annihilated because they are in excess. However, the back surfaces of n- and p-type material get accumulated with photo-generated electrons and photo-generated holes, respectively. In other words, p-type material gets positively charged, and n-type material gets negatively charged. This situation is similar to a p:n junction under a forward bias. If two ends of the p:n junctions are short circuited through a load, the excess carrier would flow, resulting in a current. Thus, light energy has been converted into electrical energy without destroying the material. This type of system is called a *photovoltaic cell*, and if solar energy is used for the illumination of the p:n junction, the system is called a *photovoltaic solar cell*. However, it is necessary to remember that for solar photovoltaic cells, the band gap of the semiconductor should be such that their electrons can be excited by solar radiation. It is observed from the solar spectrum that the maximum intensity of light has a wavelength of about 400 to

800 nm. Theoretically, the efficiency has been calculated for a p:n junction formed with semiconductors of different band gaps operating with solar radiation. This calculation suggests that a semiconductor of band gap 1.4 eV would give the highest efficiency of 30% with solar radiation. Accordingly, scientists have been trying to develop materials with a band gap nearer to this value. Up until now, silicon with a band gap of 1.1 eV has been used to make photovoltaic solar cells, and this type of silicon solar cell is also available commercially.

18.7 Types of p:n Junctions

18.7.1 Silicon Solar Cells

A p:n junction photovoltaic cell can be made with various materials. When a p:n junction is formed with the same material but doped differently to get p- and n-type character (e.g., n and p-type Si), it is called a *homojunction cell.* Homojunction cells are easy to fabricate. For example, in the preparation of a silicon photovoltaic cell, a suitable metal is deposited over about 6-μm-thick p-Si. The Fermi energy of metal must match the Fermi energy of the silicon so that the contact is ohmic in nature. The other side of p-Si is exposed to an atmosphere of phosphorous to convert p-Si to an n-type material. The temperature and time of diffusion are selected to get a depth of about 400.0 nm of n-type material over the p-Si. An aluminum grid is deposited over the surface of n-Si to form a configuration metal (ohmic)/n-Si/p-Si/Al (ohmic-grid). The entire system is then encapsulated in a frame by inserting a silicon polymer that is transparent to visible light.

Silicon solar cells have been made with polycrystalline silicon and amorphous silicon. Commercially available polycrystalline silicon solar cells are reported to possess efficiency of 12 to 15%; the efficiency of amorphous silicon solar cells is about 10%, and after use of 1 year, it decreases to about 8%. The efficiency of polycrystalline silicon solar cells does not change with time. The life of silicon solar cells is expected to be about 20 years. The cost of such solar cells is about $3 to $5 per peak watt (peak watt refers to operating the solar cell at noon).

18.7.2 Cu_xS/CdS_x Solar Cells

When two different materials (e.g., p-CuS/n-CdS, n-CuInSe$_2$/p-Si) are used to make a p:n junction, then such cells are classified as *heterojunction photovoltaic cells.* These cells are relatively difficult to make. p-CuS/n-CdS is also available commercially. One side of a thin film of copper is exposed to H_2S vapor to form a layer of CuS over a copper plate. By controlling the time of exposure with H_2S, the thickness of the CuS formed over the copper plate is maintained. Over CuS, a known amount of Cd metal (~5 to 6 μm) is deposited. This Cu/CuS/ Cd is heated in an inert atmosphere so that all Cd is converted into

CdS by extracting sulfur from underneath (i.e., from the CuS layer). With this process, CuS due to loss of sulfide ion becomes p -type material, and due to diffusion of excess sulphide ion to Cd, sulphide-enriched CdS_x is formed. This CdS_x behaves like n-type CdS. Over CdS, a metal is deposited. The Fermi energy of metal must match the Fermi level of CdS to ensure ohmic contact of the metal with CdS (i.e., not forming a depletion region at the interface of CdS and metal). This entire assembly is then encapsulated in a transparent metal box with the help of a transparent silicon polymer to get a Cu/p-CuS/n-CdS/metal grid heterojunction photovoltaic cell. This type of cell has also been fabricated with an efficiency of 6 to 8%, but the length of its lifetime is still not well established. Because both Cu_xS and CdS_x are relatively prone to oxidation, the lifetime of such solar cells is expected to be around 4 to 6 years. Moreover, because Cd is toxic, this type of cell is not environmentally friendly.

18.7.3 Carbon Solar Cells

Although silicon solar cells are commercially available, their cost has prohibited their popularity for terrestrial applications. The two major components of the cost of a silicon solar cell are the preparation of semiconducting-grade silicon material and fabrication of the p:n junction cell. The technology to fabricate a p:n junction is more or less perfected, so there is not much room for improvement that could reduce the cost of making a p:n junction. Preparation of semiconducting-grade silicon requires almost 50% of the total cost of a silicon solar cell. The major component of this cost is from the high temperature needed (~1700°C continuously for several hours). Therefore, if one could develop a material that possesses all the properties of silicon with a lower temperature needed for its preparation, then the cost of these solar cells could be reduced.

Carbon is in group IV of the periodic table and has properties that are similar to those of silicon. This has attracted scientists to make carbon solar cells. Carbon provides the possibility of getting various band gaps in the range of 0.25 to 5.5 eV. This variation can be controlled by manipulating the concentration of sp^3/sp^2 carbon ratio into a carbon material. This freedom does not exist with silicon because its configuration is fixed (i.e., sp^3 configuration). Moreover, its band gap is 1.1 eV (indirect). For ideal solar cells, the band gap should be 1.4 eV (direct).

Fortunately, semiconducting-grade carbon has been made at a much lower temperature (i.e., 600° to 800°C compared with 1700°C needed for silicon) (Sharon et al., 1995, 1998, 1999). The main hurdle with carbon materials is in finding the exact conditions to get carbon with a desired band gap of 1.4 eV. The other problem associated with carbon is in its low mobility of carriers. In addition, most carbon materials show p-type character even without any external doping. Hence, preparation of an extrinsic semiconducting carbon is a challenge.

Sharon et al. (1996, 1997a, 1997b), Rusop et al. (2005, 2006), and Umeno and Adhikary (2005) have developed a homojunction carbon solar cell of configuration Au/n-C/p-C/Au with an efficiency of 2.1%. It is postulated that this cell will cost 50 cents per peak watt as compared with silicon solar cells, which cost $3 to $5 per peak watt. (Peak watt means the operation of the cell at noon, when the intensity of the sun's radiation is highest.)

A heterostructure carbon solar cell (Islam et al., 2006) is fabricated by depositing phosphorus-doped camphoric carbon thin film (n-type material) on boron-doped crystalline silicon (p-type material) substrate with a pulsed laser deposition technique. A promising thin-film photovoltaic device using organic solar cells offers many advantages, including low cost, lightweight, large area processability, and versatility for applications (Jin et al., 2005).

Recently, a heterojunction device structure has been shown to support a highly efficient photo-induced charge transfer at the junction between electron-donating and hole-accepting materials, leading to a high photocurrent quantum efficiency (≤70%). The device performance, however, is currently limited by the low light absorption, poor carrier transport, and degrading of the materials.

A dispersed heterojunction device containing vertically aligned carbon nanotubes (CNTs) has shown improved carrier transport properties because each constituent-aligned CNT is connected directly to an electrode to maximize the electron mobility. The hole transportation can also be improved by electrochemical or chemical vapor deposition of appropriate conjugated polymers onto the individual vertically aligned CNTs to provide a well-controlled conjugation path and phase morphology.

Single-walled CNTs (SWCNTs) are a promising candidate to act as an electron acceptor in bulk heterojunction organic photovoltaic cells (Kymakis and Amaratunga, 2005). CNTs are functionalized by attaching alkyl chains to the carboxylic acid groups on the open ends of pristine SWCNTs. Diodes (Al/polymer-nanotube composite/ITO) with a low nanotube concentration (<1%) show photovoltaic behavior.

The efficiencies for bulk heterojunction cells are higher than cells without SWCNTs. It is proposed that the main reason for this increase is the photo-induced electron transfer at the polymer–nanotube interface. A dispersed heterojunction solar cell consisting of polymer and dye-coated CNT blend sandwiched between metal electrodes has been developed by Gandhi et al. (2005). The naphthalocyanine (NaPc) dye is used as the main sensitizer, and nanotubes and poly(3-octyl-thiophene) (P3OT) act as the electron acceptor and the electron donor, respectively. The incorporation of the NaPc in the P3OT/SWCNT composite dramatically increases the layer absorption, resulting in a much higher photocurrent.

A flexible transparent conducting electrode has been developed by printing films of SWCNT networks on plastic (Michael et al., 2006).

The printing method produces relatively smooth, homogeneous films with a transmittance of 85% at 550 nm and a sheet resistance of 200 Ω/cm. Cells were fabricated on the SWCNT–plastic anodes on ITO–glass. Efficiencies of 2.5% at air mass 1.5 have been reported.

Golap et al., 2008 have developed a hetrojunction solar cell by using multiwalled carbon nanotubes of lengths of 50–200 nm incorporated in a poly-3-octylthiophene/n-Si heterojunction solar cell. The device with cut-MWNTs shows short circuit current density, open circuit voltage, fill factor, and power conversion efficiency as 7.65 mA/cm^2, 0.23 V, 31%, and 0.54%, respectively

In brief, it can be concluded that there is a need to develop a carbon solar cell that does not use silicon at all. In the development of a homojunction carbon solar cell, there is a need to improve the mobility of carriers to compete with silicon. Because semiconducting carbon can be synthesized with different band gaps (0.25 to 5.5 eV), efforts should be made to make a cascade-type carbon solar cell so that the entire spectrum of solar radiation could be covered. It is worth noting that cascade-type solar cells from other materials create a problem in selecting suitable material to match the energy levels; carbon does not have this problem because contact of carbon to carbon gives ohmic type contact.

If we can achieve in this technology, the efficiency of carbon cascade-type solar cells can be as high as 100%, which can never be achieved by any other material.

18.8   Summary

Conversion of solar energy into electrical energy is today's demand. Before the requirement of energy for terrestrial applications, scientists were trying to develop a portable energy source for space research, with an interest in converting solar energy in space into an electrical energy to provide power to space missions. This led to the discovery and development of silicon solar cells. However, for space research, the reliability of the instrument was more important than the cost. As a result, the first silicon solar cell was produced from a single crystal with a very high cost. Soon after this development, scientists worked hard and were able to reduce the cost of silicon solar cells to a price of 3 to $5 per watt, which is also not very economical for terrestrial applications.

When CNTs were discovered, scientists became interested in trying to make carbon solar cells. Sharon and his group were the first to develop a homojunction carbon solar cell with an efficiency of 2.1%. Efforts are still being made to improve the efficiency to at least 10% so that the cost could be brought down to 50 cents per peak watt.

This chapter has discussed the principles of photovoltaic cells, the physics of semiconductors, and the formation of a p:n junction.

CHAPTER **19**

Absorption of Microwaves by Carbon Nanomaterials

Dattatraya Kshirsagar and Maheshwar Sharon

Nanotechnology Research Centre, Birla College, Kalyan, Maharashtra, India

19.1 Introduction

The electromagnetic spectrum is broadly divided into two regions, namely, radio spectrum from DC to 300 GHz and the optical spectrum extending from 300 GHz to infinity. The term *microwave* is commonly used to designate frequencies ranging from 0.3 to 300 GHz (~1 mm to 100 cm) in the radio spectrum. The microwave region has been further divided into three regions—ultra high frequency (UHF), super high frequency (SHF), and extreme high frequency (EHF)—according to the recommendations of the International Radio Consultative Committee. The positions of microwave bands in the entire radio spectrum are shown in Table 19-1. (The region of microwaves has been pushed into wavelengths lower than 1 mm and are known as *submillimeter waves.*)

Microwaves show almost all the properties of light (i.e., they can be refracted, reflected, and diffracted just as visible light is). Microwaves transmit energy in a very small space. In other words, their energy is confined to a very narrow width. Unless these energies of microwaves are absorbed by some material, there is no loss of energy as they travel from place to place. This is one of the reasons why microwave ovens cook food so quickly. Food grains absorb the entire energy and are cooked. The containers of microwave ovens have to be made of material that does not absorb microwaves.

Microwaves can be focused on another receiver, normally referred as an *antenna*, that might be placed even several miles away.

Band name	Frequency range	Wavelength range	Applications
UHF (ultra-high frequency)	0.3 to 3 GHz	1 to 0.1 m	Television, radar (troposcatter and meteorologic), microwave point-to-point communications, telemetry, medicine, food industry (microwave ovens)
SHF (super high frequency or super short waves)	3 to 30 GHz	10 to 1 cm	Altimeters, air- and ship-borne radar, navigation, and satellite communication
EHF (extreme high frequency or extreme short waves)	30 to 300 GHz	10 to 1 mm	Radio astronomy, radio meteorology, space research, nuclear physics, nucleonics, radio spectroscopy

TABLE 19-1   Various Microwave Frequencies, Corresponding Wavelengths, and Their Applications

This makes it very difficult to intercept signals carried along with it. Another advantage of transferring signals with microwaves is that because of their high frequency, greater amount of information can be put on them (expressed as *increased modulation bandwidth*).

Microwave signals travel in straight lines and are affected very little by the troposphere. They are neither refracted nor reflected by ionized regions in the upper atmosphere. Microwave beams do not readily diffract around barriers such as hills, mountains, and large human-made structures. However, some attenuation occurs when microwave energy passes through trees and houses. The microwave band is well suited for wireless transmission of signals with large band widths.

This chapter discusses in detail the various factors responsible for assisting the absorption of microwaves.

19.2   History

The existence of microwaves was predicted by James Clerk Maxwell from his electromagnetic wave equations but was only experimentally verified by Heinrich Hertz. To honor him, the microwave frequency is represented as Hertz (Hz) per second. (Indian physicist Jagadish Chandra Bose also produced and experimented with waves as short as 5 mm.) The term *microwaves* (standing for the Italian word *microonde*) seems to have appeared in 1932 by Nello Carrara in the first issue of a journal *Mezzanine Alta Frequenza*. The term gained acceptance during the World War II to describe wavelengths of less than about 30 cm (1 GHz). World War II (1939–1945) led to tremendous advancements in microwave technology popularly known as RADAR (radio detection and ranging). Radar was developed to detect objects and determine their ranges (or positions) by transmitting short bursts of microwaves. The strength and origin of "echoes" received from objects that were hit by the microwaves were then recorded. From this information, details about the size and position of the objects can be analyzed.

Microwaves are now popularly used for household cooking in microwave ovens. Also, all communications in space research are being conducted with the use of microwaves. Mobile technology is developed by transmitting sound signals with the help of microwaves, which prevent any noise mixing with the signal. All aviation controls are also governed by microwave technology. To prevent microwaves from detecting the position of objects (especially army planes, rockets, and so on), scientists are developing materials that can absorb the entire range of microwaves so that objects can be coated with such material to prevent their being detected by enemies using radar.

19.3   Microwave Absorption

Microwave absorbers have long been used for civil and military applications because they have the ability to eliminate electromagnetic wave pollution and may reduce radar signatures. The first electromagnetic wave absorber was developed in the mid-1930s. With the development of radar and microwave communication technology, and especially because of the need for anti-electromagnetic interference coatings, the use of self-concealing technology—wave-absorbing materials—has increased in recent years (Li, 2002). For example, if a material can absorb radar waves, it gives the illusion of being invisible on radar screens. Hence, for the stealth defense system and electronic equipment that emits electromagnetic waves, it can be of interest to manipulate the equipment or systems with the absorber coating. Microwaves are used for military applications and wireless area networks, which use the same category of electromagnetic waves. As a result, there has been a growing and widespread interest in microwave-absorbing material. Microwave-absorbing materials and their composites are mostly used in military applications, such as reduction of radar visibility in different types of vehicles. This is also known as *stealth technology*. The most well-known vehicle with this technology is the Lockheed F-117A Nighthawk, which is one of the first stealth aircrafts ever produced (Andersson, 1999).

19.3.1   Classification of Microwave Absorbers

All available materials, such as pure metals, are not good microwave absorbers. Metals absorb nearly no microwave energy because of their high electrical conductivity. On the other hand, water does not reflect microwaves, making it a good absorber. This is because water molecules are highly dipolar, and microwave energy is used up in rotating water molecules. Thus, water molecules do not allow radiation to escape from their environment. This property of water molecules makes it a good microwave-absorbing material. Magnetic materials such as ferrites can also absorb electromagnetic waves because the microwave energy is used up in the magnetic alignment of atoms of ferrites (Fig.19-1). Hence, it is suggested that for good absorption, the material should be able to

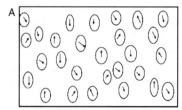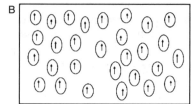

FIGURE 19-1   A schematic representation of effect of microwave absorption by magnetic materials before (A) and after (B) absorbing microwaves.

absorb all the energy of the microwave by transferring energy in rotation of the molecule (e.g., in water), in aligning the atoms of the material (e.g., in ferrite), or by some other method so that no microwave energy is left to escape the environment through which it travels. Magnetic materials such as ferrites are commonly known as *electromagnetic wave absorbers*. Hence, for the design of microwave-absorbing materials, it is essential to use material with suitable properties.

On the basis of the principles of absorption, the microwave absorbers are categorized as dielectric absorbers, magnetic absorbers, resonant absorbers, and graded dielectric absorbers.

Dielectric Absorbers

A dielectric absorber is material in which the electrical and magnetic properties have been altered to allow absorption of microwave energy at discrete or broadband frequencies. Common dielectric materials used as absorbers (called *lossy fillers*) are foams, plastics, and elastomers. They have no magnetic properties, giving them a permeability of 1. High dielectric materials, such as carbon, graphite, and metal flakes, are used to modify the dielectric properties of such materials so they are able to absorb better.

For purposes of analysis, the dielectric properties of a material are categorized as its permittivity, which is a complex number with real and imaginary parts. For dielectric absorbers, loss is primarily generated via the finite conductivity of the material. Incident electromagnetic waves impinging on a conductive surface induce current as the electric field of electromagnetic wave, interacting with mobile electrons within the material. Additional loss may occur via molecular polarization phenomena, such as dipole rotation. The electrical energy is dissipated as heat because of the material's resistance.

Magnetic Absorbers

Magnetic absorbers (e.g., ferrites, iron, and cobalt–nickel alloys) alter the permeability of the material as the absorption of microwave waves occurs. For purposes of the analysis of microwave absorption, magnetic properties such as the permeability are used. The strong magnetism and low resistivity properties of magnetic absorbers make them good microwave absorbers (Makeiff and Huber, 2006).

Resonant Absorbers

The resonant absorber is a metal-backed, single-layered absorber (or consists of a single-layer coating of dielectric layer on a metal plate). In resonant absorbers, a wave incident upon the surface of the layer of dielectric material is partially reflected and partially transmitted (Fig. 19-2).

The transmitted portion undergoes multiple internal reflections to give rise to a series of emergent waves. At the designed frequency, the sum of the emergent waves is equal in amplitude to the wave that

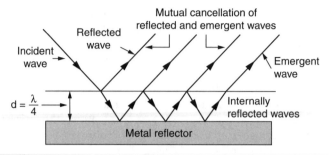

Figure 19-2 Resonant absorber showing an out-of-phase condition existing between reflected and emergent waves.

is 180 degrees out of phase with the initial reflected portion (see Fig. 19-2). This causes absorption by destructive interference at the material's front surface. A dielectric layer with a phase thickness equal to a quarter wavelength ($\lambda/4$) of incident wave (frequency as $\lambda = 1/f$) is required to cause such destructive interference at the material's front surface for absorption of the incident microwave energy. Theoretically, no reflection takes place in these absorbers at the given frequency; but in practice, absorption of greater than 30 dB (99.9%) may be achieved (less reflection and transmission lead to high absorption). These absorbers generally operate over a relatively narrow frequency range.

Graded Dielectric Absorbers

The graded dielectric absorber is the other category of absorber. It consists of different layers of conductive carbon, which are pyramidal in shape in polyurethane foam. The principle of operation of such absorbers is quite different from that of the resonant type. Absorption is achieved by a gradual tapering of impedance from that of free space to a highly absorbing state (lossy). If the transition from the free space to a highly absorbing state is done smoothly, little reflection from the front face will result, making this a good absorber.

Absorption levels of greater than 50 dB can be obtained with pyramids of many materials of graded dielectric absorbers with different thicknesses (i.e., thickness in terms of wavelength). Good levels of reflectivity reduction (>20 dB) can be achieved in the materials less than one third a wavelength thick. In this case, very open-celled foam (10 pores/inch) is used. In this type of absorber, gradual transition (i.e., from free space to a highly absorbing state) is achieved via a conductive carbon coating. Here foam works as a lossy material. These types of absorbing materials are used for "broadband" absorption characterized by a gradual tapering from the impedance of free space to that of the medium over or above the appreciable thickness (in terms of wavelength).

19.3.2   Microwave Absorption Measurement

The electromagnetic field at any point of the transmission line (i.e., a structure used to guide the flow of electromagnetic energy from one point to another) may be considered as the sum of the field of two traveling waves (i.e., the incident wave propagated from the generator and the reflected wave propagated toward the generator). The superimposition of two traveling waves (i.e., incident and reflected waves) gives rise to a standing wave along the path of the traveling wave (known as the *line*). The maximum field strength is found where two waves are in phase, and the minimum field strength is found where the two waves are out of phase. The ratio of electrical field strength of the reflected and incident waves is called the *reflection coefficient*.

The microwave absorption by material in terms of reflection coefficient, or the *transmission coefficient* (i.e., the ratio of electrical field strength of the transmitted and incident wave for the sample), is calculated by measuring the voltage standing wave ratio (VSWR). The VSWR is defined as the ratio of field strength along the line for the reflected wave by the sample and the incident wave on the sample. Generally, a network analyzer or microwave test bench is used for these measurements. Figure 19-3 shows the block diagram of the experimental setup used for measurement of the reflection coefficient or transmission coefficient.

The instrument consists of a microwave source along with an isolator, frequency selector, variable attenuator, matched termination, and detector. The microwave source produces the required frequency. The power of the microwave is selected by the attenuator. (The attenuator is the element that can absorb either all or a portion of the power falling on it without any appreciable reflection.) An isolator allows the microwave energy to be transferred to fall on the sample; at the same time, it differentiates frequencies that are reflected back from the sample so the intensity of the reflected wave can be measured by the detector. The frequency selector (not shown in Fig. 19-3) is used to select the different microwave frequencies that could be transmitted by the microwave source. The selected frequency is made to incident

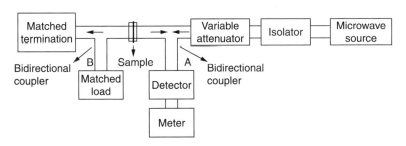

Figure 19-3   Block diagram of a microwave test bench.

on the sample through a wave guide with impedance matching. Here the wave guide provides an alternative to transmission lines for use at microwave frequencies. In fact, wave guides are superior to conventional transmission lines in terms of the attenuation per unit length experienced by the wave propagating through them.

The matched termination is the first type of attenuator that absorbs all of the power falling on them. The detector is used to detect the reflected and transmitted wave from the sample, which is measured using a VSWR meter or a digital multimeter (DMM). Directional couplers are required to measure the incident and reflected or transmitted waves separately, consists of two transmission line, the main arm, and the auxiliary arm, which are electromagnetically coupled to each other.

In microwave absorption measurements, the samples are cut into a rectangular shape a few millimeters in size to fit in a rectangular wave guide of the microwave test bench. By measuring the VSWR for the reflected or transmitted wave and incident wave for different microwave frequencies, the absorption by the sample in terms of reflection or transmission coefficient is measured. The refection coefficient is given by:

$$VSWR = \frac{E_{max}}{E_{min}} \qquad (19.1)$$

or

$$VSWR = \frac{|E_i| + |E_r|}{|E_i| - |E_r|} \qquad (19.2)$$

If the incident and reflected voltage are in phase (i.e., $|E_i| + |E_r|$), the VSWR is maximum (i.e., E_{max}). If the incident and reflected voltage are out of phase (i.e., $|E_i| - |E_r|$), the VSWR is minimum (i.e., E_{min}).
where, E_i = incident voltage
E_r = reflected voltage.
The reflection coefficient ρ is given by:

$$\rho = \frac{E_r}{E_i} = \frac{VSWR - 1}{VSWR + 1} \qquad (19.3)$$

or in decibels:

$$\rho(dB) = -20 \log \frac{I VSWR - 1I}{I VSWR + 1I} \qquad (19.4)$$

To develop inexpensive microwave absorbers, efforts have been made to develop carbon nanomaterials (CNMs) for absorbing microwaves in the range of 8 to 24 GHz (Sharon et al., 2005 and 2006). To explain the application of these equations, the results of one set of

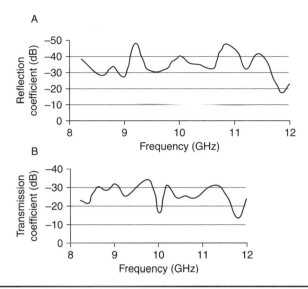

Figure 19-4   (*A*) Variation in reflection coefficient versus frequency.
(*B*) Transmission coefficient versus frequency for a thin film sample of
carbon nanomaterial.

measurements are discussed. Microwave absorption measurements
were done for a thin film sample of CNM. For this sample, the reflection
coefficients ρ were calculated for a frequency range from 8 to 12 GHz.
The results are shown in Fig. 19-4*A*. Similarly, transmission coefficients were measured for the same frequency range, and the results
are shown in Fig. 19-4*B*.

Figure 19-4 shows the plot of the reflection coefficient versus
frequency and transmission coefficient versus frequency. From the
graph, it is clear that the sample has a reflection coefficient in the
range of –30 to –40 dB for a frequency range of 8 to 12 GHz (a –30-dB
reflection coefficient corresponds to ~99% absorption). This indicates a good microwave-absorbing property of the sample. In
Fig. 19-3, if the matched load is interchanged to the directional coupler (A) and the detector and meter to the directional coupler (B),
the transmission coefficient (instead of the reflection coefficient)
for the sample can be measured. The plot of transmission coefficient versus frequency (Fig. 19-4*B*) shows that the sample has a
transmission coefficient in the range of –20 to –30dB. This means
that the loss of microwave energy incident on the sample as a reflection and transmission is in much less, indicating that the energy is
fully absorbed by the sample.

The absorption can be calculated as:

$$\text{Absorption (A)} = 1- [\text{Reflection coefficient} + \text{Transmission coefficient}] \quad (19.5)$$

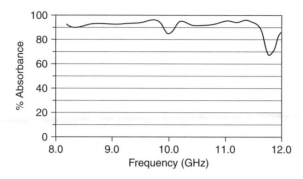

FIGURE 19-5   Percentage absorption of microwaves in the range of 8 to 12 GHz calculated from the data in Fig. 19-4.

The plot of absorption of samples calculated from the reflection and transmission coefficient (see Fig. 19-4) can be plotted versus frequency (Fig. 19-5).

The results shown in Fig. 19-5 suggest that the material showing this behavior is a good microwave absorber for the range of 8 to 12 GHz. A scanning electron micrograph (SEM) of the sample is shown in Fig. 19-6. The micrograph shows the material to be carbon nanofiber (CNF) with a diameter of 250 nm. Dattatray et al. (2006) have shown the absorption of microwave to the tune of 97% by using carbon nanotubes, carbon nanofibers and carbon nanobeads synthesized using oil derived from plant precursors for example seeds of Karanja, Neem, and Mustard.

FIGURE 19-6   Scanning electron micrograph of carbon nanofiber whose microwave properties are shown in Figs. 19-4 and 19-5.

19.4   Microwave Absorption by Carbon Nanomaterial and Its Composite

The morphology of carbon nanotubes (CNTs) has attracted scientists to use it as a microwave absorber (see Fig. 19-6). Nanocomposites made from CNTs with conducting polymers are found to have a higher absorption capacity compared with the individual CNTs. This result has attracted considerable interest in developing CNM as an absorber (Yang et al., 2004). These materials constitute a new form of carbon polymers, which can be produced either as single-walled CNTs (SWCNTs) or multiwalled CNTs (MWCNTs). Individual CNTs are known to exhibit metallic or semiconductor behavior (depending on the number of defects present in the conjugated graphitic surface) and chirality. The high aspect ratio of SWCNTs and MWCNTs allows small amounts to be dispersed in polymer matrices to yield composite materials with superior properties required for absorption. On the other hand, metal fillers, such as iron and nickel (Bi, 2005; Lutsev, 2005) and carbon blacks or fiber filler materials (Yang, 2004; Zhao, 2003), may also be used in microwave-absorbing composites. However, these fillers may improve the effective absorption over a narrow bandwidth. Hence, CNTs are considered a prospective filler material for microwave-absorbing composite. The effectiveness of the absorber depends on the conductivity of the filler (i.e., the larger the conductivity of the filler, the higher the absorption) and the degree of its dispersion in the composite (Makeiff and Huber, 2006).

However, very few attempts have been made to use CNM for studying its microwave-absorption properties (Bi, 2005; Lutsev, 2005; Xing and Yan, 2003). These authors have studied microwave absorption in the range of 5 to 100 GHz. But the absorption pattern has shown that none of the carbon material could absorb microwave with a wide band (Lopes, 2003; Motojima et al., 2004). Hence, there is a need to try to develop materials that could have absorption of microwaves over a larger width of bands so that their application becomes promising.

For obtaining the maximum absorption of microwaves of larger bandwidth, the morphology of carbon needs to be improved so that it has a channel-type structure in which microwaves could get easily trapped and transfer all of their energy to carbon materials. It may be worth trying to impregnate channel-type porous carbon structures with ferrites such as iron or cobalt so that the morphology of carbon material will be able to trap the microwaves. Then its energy could be transferred to magnetic materials with the process shown in Fig. 19-1.

Thus, nanomaterial, especially CNMs (e.g., CNTs, CNF, carbon nanobeads (CNB)) and nanocomposite could be suitable materials for trying the microwave-absorption properties (Sharon et al., 2005; Sharon et al., 2006). Other advantages of developing suitable CNM include its low mass and low density, which means that an even thicker film with insignificant weight can be used for the purpose of absorption, especially for airplanes.

19.5 Some Applications of Microwave Absorbers

Recently, the demand for various kinds of microwave absorbers has increased in the frequency range of 1 to 25 GHz because of their two-fold use: electromagnetic interference (EMI) shielding and as a countermeasure to radar detection. Suppression of EMI and meeting the electromagnetic compatibility (EMC) have become essential requirements in industries dealing with high-speed wireless data communication systems, wireless local area networks (LANs), mobile and satellite communication systems, and automatic teller machines (ATMs) operating in the gigahertz range. Application of appropriate microwave-absorbing material in electronic equipment helps to control the excessive self-emission of electromagnetic waves and ensures undisturbed functioning of the equipment in the presence of external electromagnetic interference. Stealth technology for defense is another area in which these microwave absorbers are used; in this case, they are used for effective countermeasures against radar surveillance. Application of microwave-absorbing coatings on the exterior surfaces of military aircraft and vehicles helps to avoid detection by radar. Therefore, as the science of microwaves develops, its application in various fields will increase day by day, so it would be difficult to discuss all of the applications here. However, some commonly known applications of microwave absorbers are discussed in the next sections.

19.5.1 Applications in Electromagnetic Shielding

EMI shielding refers to the reflection or absorption of electromagnetic radiation by a material, which thereby acts as a shield against the penetration of the radiation through the shield. It is characterized by EMI shielding effectiveness (SE), which quantifies the amount of attenuation typical of a particular material capable of attenuating upon transmission. This is given by:

$$SE = (R + T + A)\ dB \qquad (19.6)$$

where SE is expressed in decibels and is the loss representing the reduction of the level of an electromagnetic field at a point in space after a conductive barrier is inserted between point of interest and the source. R is the sum of initial reflection losses in decibels from both surfaces of the shield exclusive of additional reflection losses, A is the absorption or penetration loss in decibels within the barrier itself, and T is the sum of initial transmission losses in decibels. Carbon nanocomposite materials are the most promising material used as electromagnetic interference shielding (EMIS) in various applications (Wang et al., 2005).

19.5.2   Applications in Commercial Electronics

Absorbers are increasingly being used for reduction of interference in commercial electronics. High-frequency wireless devices often use powerful transmitters and sensitive receivers in close proximity inside a cavity or housing. Spurious signals may cause leakage or system interference, which degrades performance. In such cases, carbon nanocomposite absorbers inside the cavity may be used to absorb unwanted reflections. Applications for absorbers may also be found in wireless LAN devices, network servers, very small aperture terminal (VSAT) transceivers, radios, and other high-frequency devices.

19.5.3   Applications in Radar-Absorbing Materials

As the name implies, radar-absorbing materials (RAMs) are coatings whose electric and magnetic properties have been selected to allow the absorption of microwave energy at discrete or broadband frequencies, which is very useful for the reduction of radar cross-section (RCS). RCS is the measure of a target's ability to reflect radar signals in the direction of the radar receiver. That is, RCS is a measure of the ratio of backscatter power per sterdian (unit solid angle) in the direction of the radar (from the target) to the power density that is intercepted by the target. In other words, RCS is a description of how an object reflects an incident electromagnetic wave. In military applications, a vehicle's low RCS may be required to escape detection during covert missions. These requirements have led to the very low-observable or stealth technology that reduces the probability of detection of aircrafts. But the design of RAM for such applications is limited by constraints on the allowable volume and weight of the surface coating. Now it may be possible to design a broadband radar-absorbing structure in limited volume with CNM or CNM composite.

19.6   Summary

In this chapter, microwave technology has been introduced in a very simple fashion, including its properties and the range of frequencies that are normally used. The chapter has emphasized the properties needed by the material to absorb microwaves. The principle of absorption and the techniques used for its measurement have been discussed. Finally, the various applications of microwaves have been highlighted.

CHAPTER 20

Carbon Nanosensors

Bhushan Patil

Nanotechnology Research Centre, Birla College, Kalyan, Maharashtra, India

Maheshwar Sharon

Nanotechnology Research Centre, Birla College, Kalyan, Maharashtra, India

Madhuri Sharon

MONAD Nanotech Pvt Ltd, Mumbai, India

20.1 Introduction

With industrial development, we keep on adding materials to the environment that are not eco friendly, such as pesticides and herbicides, which are causing an alarm to public health. Hence, the need to develop technology to assess the presence of these chemicals in our food chain has become essential. Similarly, there is an increasing demand to discover a faster method for detecting chemicals such as sugar in our blood and urine. Thus, there is an increasing demand to develop quick and preferably portable sensors that can detect the presence of inorganic materials, organic chemicals, and so on that are present in very small quantities in the sample. Such sensors are being developed by using two aspects of properties of materials: (1) the adsorbing properties of chemicals to be analyzed and (2) promoting or changing the physical or chemical properties of materials to which testing chemicals have been adsorbed or directly showing their electrochemical or physical properties (Fig. 20-1).

The sensors (see Fig. 20-1) are made up of three major parts:

1. Molecular recognition and chemical reactions occur on a **sensitive membrane**.

2. The **transducer** is an interface between the sensor and environment that transforms molecular adsorption and chemical reactions into measurable physical parameters. Carbon nanomaterial (CNM) may act separately as the sensitive membranes and the transducer, or it can even function as both a sensitive membrane and a transducer.

3. An **electronic unit for signal processing** is also necessary.

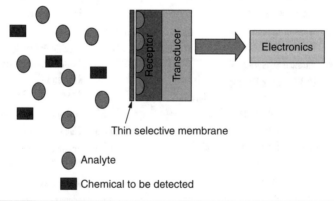

Analyte

Chemical to be detected

Figure 20-1 Schematic diagram of a sensor in which the receptor is deposited over a thin film of a membrane, which assists in the electron transfer reaction with the chemical to be detected. This sends the signal to the transducer, which in turns sends it to electronic instruments for measurement.

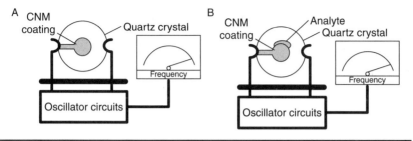

Figure 20-2   Schematic diagram of a quartz crystal microbalance (QCM) sensor before (*A*) and after (*B*) addition of the analyte.

A good sensor should be sensitive to the measured property but insensitive to any other property in such a way that it does not influence the measured property. The base of the sensor could be a micro-sized metal or a conducting polymer strip coated with materials that could initiate adsorption of materials to be analyzed. CNMs are the preferred materials for this purpose because they provide a large surface area, good conductivity, and various other physical properties. This strip (i.e., base plate coated with sensitive material) is connected with a suitable instrument (Fig. 20-2) that measures the appropriate changes in properties of the sensitive coated material (i.e., changes in the properties due to adsorption of the chemicals that are to be analyzed). Table 20-1 lists various properties that are used in sensors for assay. Various instruments have been developed that may effectively measure the physical, chemical, electrochemical, or photochemical properties of the strip over which the chemical in question is adsorbed (see Table 20-1).

Analyte	Method of assay
Glucose	Amperometric
Urea	Potentiometric
Hepatitis B	Chemiluminiscent immunoassay
Cholesterol	Amperometric
Penicillin	Potentiometric
Sodium	Glass ion-selective electrode
Potassium	Ion-exchange selective electrode
Calcium	Ionophore ion-selective electrode
Oxygen	Fluorescent quenching sensor
Ph	Glass ion-selective electrode

Table 20-1   Different Types of Analytes and Their Specific Detection Techniques

20.2   Fundamentals of the Workings of Nanosensors

The working of nanosensors depends on two parameters, sensitivity and selectivity. Sensitivity S can be defined as the changes of some measurable values I caused by changes in the concentration C of registered analyte molecules:

$$S = \frac{dI}{dC} \tag{20.1}$$

Moreover, the sensitivity and selectivity depend on adsorption and desorption kinetics and the noise in signals. Adsorption occurs when a gas or liquid solute accumulates on the surface of a solid or a liquid (adsorbent), forming a film of molecules or atoms (the adsorbate).

Based on the thickness of the adsorbent layer on the coating of sensitive material, adsorption can be calculated with the Langmuir adsorption isotherm (developed in 1916) for monolayer adsorption on CNM:

$$\theta = \frac{KaC}{1 + KaC} \tag{20.2}$$

where $Ka = [S-M]/[S][M]$
         $S - M$ = site occupied by the molecules,
           S = the empty sites
          M = free molecules in the environment

The constant Ka is the Langmuir adsorption constant and increases with an increase in the strength of adsorption and with a decrease in temperature. In the case of nano-biosensors made up of antibody functionalized CNM as an electrode or field effect transistor (FET), it adsorbs a single layer of adsorbate on the CNM surface. Hence, for this kind of sensors, the sensitivity of the sensors can be calculated with the Langmuir adsorption equation.

The Brunauer Emmett Teller (BET) equation is used for multi-layer molecular adsorption on CNM. This theory is an extension of the Langmuir theory, which is a theory for monolayer molecular adsorption, to multilayer adsorption with the following hypotheses:

1. Gas molecules physically adsorb on a solid in layers infinitely.

2. There is no interaction between each adsorbed layer.

3. The Langmuir theory can be applied to each layer.

The resulting equation is expressed by BET equation:

$$\theta = \frac{\theta_{max} K_b C}{(C_{max} - C)[(1 + (K_b - 1)C/C_{max}]} \tag{20.3}$$

where θ_{max} = maximum coverage,
       C_{max} = concentration of analyte molecule,
        K_b = equilibrium constant,
        C = concentration of analyte molrcule.

The sensitivity is directly proportional to the isotherm of adsorption. As the adsorption decreases, the sensitivity varies accordingly; hence, the sensitivity is not a constant. This adsorption kinetics can be given by:

$$N = N_o[1 - \exp(-t/\tau_a)] \qquad (20.4)$$

where N = number of occupied binding sites
N_o = total number of binding sites,
τ_a = characteristic time of adsorption.
t = time at which N was measured

The response time ($t_{response}$) is 2.3 τ_a, or the time when the signal reaches 90% of its maximum adsorption.

Small particles, in addition to producing carriers at the surface, facilitate the adsorption of reacting species over the surface. Hence, not only does N_o increase, but N also increases by using CNM in a sensor. In the case of CNM, the major benefit is its high surface-to-volume ratio. This peculiar characteristic of CNM increases the surface area for adsorption. Single-walled carbon nanotubes (SWCNTs) have very good adsorption capacity. Different carbon structures have different surface areas. CNTs can have a surface area of about 1580 m^2/g, carbon nanofiber (CNF) with a 80-nm diameter has a surface area of about 134 m^2/g, and carbon nanobeads (CNBs) have a surface area of about 16.49 m^2/g. Activated carbon also has a large surface area of around 3000 m^2/g. However, these values are approximate and need not be true for any CNT, CNF or CNB.

A good quality sensor must have good adsorption capacity within a short time frame to measure signals. Furthermore, it is equally important for the sensitive membrane to return to the initial state, which depends on the desorption rate of the analyte molecules. Hence, for a good sensitive membrane (i.e., coating layer) or sensor must achieve an initial state of physically measurable parameters after removing the source of analyte from the environment. Based on the number of usages, these sensors can be divided into single-shot (one-time use) and multiple-use sensors. The output of these sensors might be in analog or digital signals. However, desorption of the adsorbate does not affect single-shot sensors (single-use sensors).

Selectivity can be defined as the extent to which the method can be used to determine particular analytes in mixtures or matrices without interferences from other components of similar behavior. In other words, it is a ratio of an amount of reagent and of interferent associated with the receptor.

The recovery time of the sensor or desorption of the adsorbate from the sensor membrane is in a reciprocal relationship with the binding constant (or selectivity) That is, the higher the selectivity of adsorption, the longer the desorption (recovery) time. Quick desorption can be initiated in some cases by illuminating the sensor with ultraviolet (UV) or with the help of thermal heating.

The selectivity of a sensor depends on the noise in the signals. Noise is the unwanted signals disturbing the sensitivity of the sensor. Noise can be calculated by using variation of relative conductance change in the baseline using root mean square deviation (RSMD):

$$RMSD = \sum (Y_i - Y)^2 \qquad (20.5)$$

where Y_i is the measured data and Y is the corresponding value calculated from the curve-fitting equation.

$$RMS\ noise = \sqrt{\frac{RSMD}{N}} \qquad (20.6)$$

where N is the number of data points used in fitting the curve.

As per the International Union of Pure and Applied Chemistry (IUPAC) definition, when the signal-to-noise ratio equals 3, the signal is considered to be a true signal. Therefore, the detection limit can be extrapolated from the linear calibration curve when the signal equals three times the noise.

$$Detection\ limit = 3 * \frac{RMS}{Slope} \qquad (20.7)$$

where the slope is calculated from the linear plot of concentration of analyte versus change in the signal obtained from the measuring set up.

The overall variation of sensitivity for fabricated devices is about 6%, which is comparable to and even better than that of metal oxide- or polymer-based sensors, which show excellent reproducibility of SWCNT sensors. (Stark 2002)

20.3 Types of Carbon Used for Making Nanosensors

Various types of CNMs, such as SWCNTs, multiwalled CNTs (MWCNTs), vertically aligned carbon, CNF, glassy carbon, and diamondlike carbon, have been used for making nanosensors. These materials help in adsorbing analyte more effectively to give a strong signal by the instruments used for measuring the concentration of analyte present in the solution.

20.4 Physical Sensors

Physical sensors work on analysis of change in physical properties such as frequency, force, and resonance.

20.4.1 Specific Properties of Carbon Nanomaterials Related to Physical Sensors

Various types of physical sensors have been developed by using some of the specific properties of CNMs:

- CNMs have a high Young's modulus (~1 TPa), high tensile strength (SWCNTs, ~13.53 GPa; MWCNTs, 150 GPa), and high mechanical resilience, which enables them to be used as probes in scanning tunneling microscopy and atomic force microscopy. The higher strength and flexibility of CNMs makes them suitable material for physical sensors. It also increases the life of the sensors.

- CNTs have a high aspect ratio (length vs. diameter) of approximately 1000. CNF also has a high aspect ratio, which facilitates imaging in narrow and deep crevices of samples.

- It is predicted that CNTs will be able to transmit up to 6000 W/m/K at room temperature. If this is compared with a good thermal conductivity metal such as copper that only transmits 385 W/m/K, CNT is surely a better material because this feature allows the sensor to work with a low operating power.

The properties of analyte are measured either physically or chemically. Physical sensors using physical properties include probe sensors, gravimetric sensors, surface acoustic wave (SAW) sensors, quartz crystal microbalance (QCM) sensors, and actuators.

20.4.2   Probe Sensors

The mechanical and chemical properties of CNTs enable them to be used in probes for measuring the chemical force of single molecules and even for nanoelectrodes for biosensing of specific molecular recognition.

For targeted molecule detection, functionalization of open ends of CNTs is carried out. MWCNT and SWCNT tips are modified to trap biologic molecules. The benefits of using CNTs as probes are their cylindrical shape and small tube diameter, which enable imaging in narrow, deep crevices and improve resolution compared with conventional probes. CNTs have a low Euler buckling force and high flexibility, and their hydrophobic nature facilitates their use in water medium. Moloni et al. (2000), Nguyen et al. (2004) and Baughman et al. (2002) have used MWCNTs as tips in scanning probe microscopy. Because of the high elasticity of CNTs, the tips do not experience crashes on contact with the substrate, which increases the life of the sensor probe. Any impact causes a buckling of CNTs, which is reversible after the reaction of the tip with the substrate. Implanting single CNTs on the probe is somewhat difficult; however, successful attempts have been made by Hafner et al. (1999b) to grow individual CNTs onto Si tips using chemical vapor deposition. Not only CNTs, but vertically aligned CNFs also have the potential to be used as electrochemical probes.

20.4.3   Gravimetric Sensors

The working principles of gravimetric analysis are based on interaction of acoustic waves with material and transformation of added mass or chemical or biosensor reactions on the surface. The two types of gravimetric sensors are QCM and SAW sensors.

Quartz Crystal Microbalance Sensors

QCM sensors are made up of a thin, vibrating AT-cut quartz wafer sandwiched between two metal excitation electrodes. In most of the QCM sensor, quartz is covered with gold because gold's resonant frequency is 1.729 MHz, and the concentration of free electrons in gold metal is 5.90×10^{22} cm^{-3}. Gold also has a high electrical conductivity, low reactivity with chemicals, good resistance to oxidative corrosion, and high malleability. QCM sensors work in gaseous media or a vacuum, and they are highly sensitive to mass changes ($\sim$1 ng/cm^2).

QCM sensors have been widely used to monitor the changes in mass adsorption by measuring the shift of its resonant frequency. This mass adsorption depends on the chemical nature and physical properties of the coating material. Hence, rigid and homogeneous coating is a major requirement to avoid dumping oscillations. (This is one of the reasons for selecting gold as a coating material.) Zhang et al. (2004). have made a humidity sensor with MWCNT-coated QCM. This QCM sensor is prepared by spraying MWCNTs on silver-coated quartz wafers (Fig. 20-3).

The working principles of QCM sensors are based on the Saurbrey equation.

$$\Delta f = -2 f_x^2 \frac{\Delta M}{A \sqrt{Pq \cdot \mu q}} \tag{20.8}$$

where ΔM is the added mass of analyte, Δf is the changes in resonance frequency, f_x is the initial resonance frequency of the quartz crystal, Pq is the density of quartz (2.648 g$\cdot$cm^{-3}), μq is the shear modulus of quartz ($2.947 \cdot 10^{11}$ g$\cdot$cm$^{-1}\cdot$s^{-2}), and A is the area of overlapping carbon electrode.

To fulfill the linearity of the Saurbrey equation, the frequency shift should not exceed 1% from main resonance frequency. If the change in frequency is linear, then Eq. (20.8) becomes:

$$\Delta f = -2.3 \times 10^{-6} f_x^2 \frac{\Delta M}{A} \tag{20.9}$$

Penza et al. (2006) studied the properties of SWCNTs as QCM sensors and reported a 10-MHz QCM resonator using CNT.

Because CNTs have an increased surface area, they increase the adsorption, which increases the difference in mass before and after

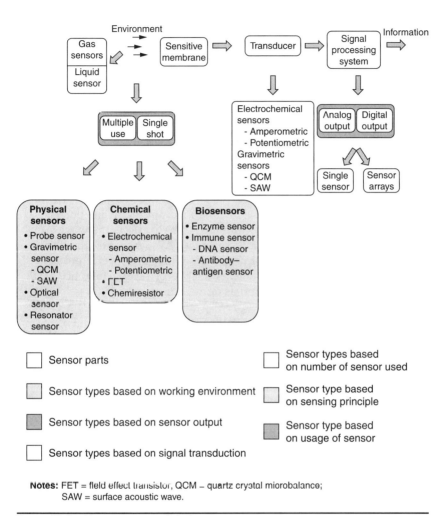

Notes: FET = field effect transistor, QCM = quartz crystal microbalance; SAW = surface acoustic wave.

FIGURE 20-3 Nanosensors and their types.

making contact with analyte, resulting in an increase in ΔM. Hence, in Eqs. (20.8) and (20.9), M increases, resulting in an increase in the change in resonance frequency. Jia et al. (2007) studied the effects of different coating materials (e.g., bare gold, MWCNTs, modified Au) on the change in resonance frequency and found that the change in frequencies with bare gold was –267; it was –364 for MWCNT and Au.

As per these results, if the coating area is kept constant, then CNM-coated QCM will be more sensitive than gold-coated QCM. One major reason for using CNM is its higher surface-to-volume ratio, which increases adsorption sites. Furthermore, CNM has a high affinity of adsorption for analyte at a very low concentration.

Surface Acoustic Wave Sensors

In 1887, Lord Raleigh discovered the surface acoustic wave mode of propagation, which accelerates the use of SAW in different sensors. SAW sensors are based on a piezoelectric substrate on which two interdigited transducers (IDTs) are fixed; one induces an electric field to excite surface acoustic waves in CNM, and the other receives changes in the electric field. The surface in between the IDT of the sensor is modified by coating CNM on a specific substrate. Penza et al. (2004) demonstrated that CNT acoustic sensors are highly sensitive to a wide range of polar and nonpolar organic solvents up to a sub-ppm detection limit at room temperature.

SAW sensors produced on CNM have a longitudinal and a vertical shear component that couples with a medium that is in contact with the device's surface. Such coupling strongly affects the wave's amplitude and velocity. This feature enables SAW sensors to directly sense the mass and mechanical properties. There are several possibilities of SAW-sensing measurements, such as being able to notice a delay in the time of receiving signals or measuring changes in velocity. Velocity changes can be monitored by measuring the frequency or phase characteristics of the sensor, which can be correlated to the corresponding physical quantity being measured. The sensitivity of SAW sensors is up to two orders of magnitude higher than that of QCM sensors (Fig. 20-4).

20.4.4   Resonator Sensors

Resonator sensors give frequency signals to reflect chemical information around the device. These sensors are made up of three layers: the conductor, insulator, and sensitive membrane. Chopra et al. (2002). developed an ammonia sensor in which microwave-resonance CNTs were coated on a circular disk. By adsorption of ammonia onto CNTs,

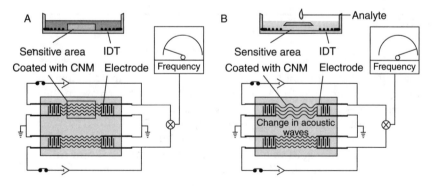

Notes: CNM = carbon nanomaterial; IDT = interdigited transducer.

Figure 20-4   Schematic diagram of a surface acoustic wave sensor showing its functioning before (*A*) and after (*B*) addition of the analyte.

the resonant frequency changes. The downshift in frequency is used to find out the quantity of ammonia present. Experiments were carried out using SWCNTs and MWCNTs. SWCNT-coated disks had a higher sensitivity than MWCNT-coated disks. Ong et al. (2002) have made a CNT-based gas sensor for CO_2 and O_2 by using an MWCNT–SiO_2 mixture as a sensing material. This sensor is the wireless sensor.
The benefits of such resonator sensors include:

1. Because the change in frequency is the measure of analysis, sensors can become wireless by using a radiofrequency receiver.

2. Radiofrequency identity (RFID) can be used in the sensor for tracking of items, and automation in data storage is possible.

20.5 Chemical Sensors

Chemical sensors sense organic as well as inorganic molecules with a relatively small molecular weight. Various types of chemical sensors have been developed by using some unique properties of CNM, including:

- CNTs are unique one-dimensional quantum wires with an extremely high surface-to-volume ratio. Furthermore, CNM is very sensitive to molecular adsorption of various particles from the environment.

- CNTs and graphite have also been used because of similarities between them. CNTs have good electrical conductance along with tube axis, and the sidewalls of CNTs show inert chemical properties as basal planes of graphite. Additionally, end caps of CNTs are similar to graphitic edge planes and are much more reactive because of the dangling sp^2 bonds.

- Due to the curvature of carbon graphene sheets in nanotubes (Fig 20-5), the electron cloud changes from a uniform distribution around the carbon, resulting in an asymmetric distribution inside and outside the cylindrical sheet of the CNT. Because the electron clouds are distorted, a rich π electron conjugation forms outside the tube, making CNTs electrochemically active. Similar curves are found in CNBs that change the electrical properties of CNBs.

- In electrochemical analysis, it is necessary that materials used as electrodes do not themselves get oxidized or reduced in the potential region of interest. In addition, water (which is normally used as a solvent) should not get oxidized on the electrode in the potential range where the analyte is going to be used for the analysis. The range of potential in which neither

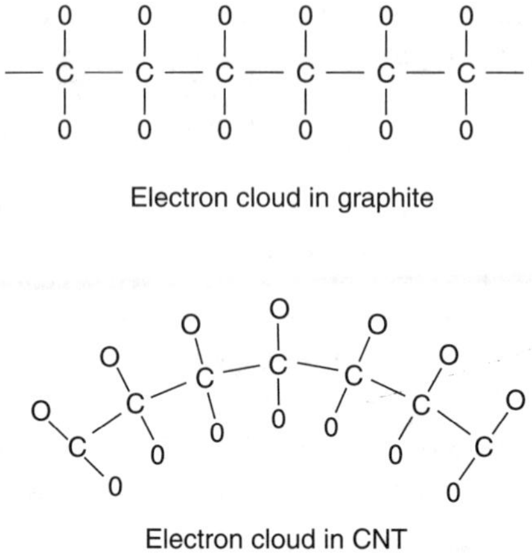

FIGURE 20-5
Diagram showing
the shift in an
electron cloud
because of the
curvature of carbon
nanotubes (CNTs).

Electron cloud in graphite

Electron cloud in CNT

water gets oxidized or reduced nor the electrode materials get decomposed is known as the *potential window*. The advantage of CNM is that it has wide electropotential window. Sharon el al. (1998) reported that carbon synthesized from kerosene in contact with H_2SO_4 and NaOH shows a potential window range of 2.5 to –1.5 V versus standard calomelelectrode (SCE). Furthermore, they have compared the potential window of carbon (synthesized from kerosene) with platinum and glassy carbon by considering these as working electrodes, platinum as a counter electrode, and 100-mM H_2SO_4 as an analyte. In this experiment, the potential window of kerosene carbon was higher than that of the platinum and glassy carbon (reported to be 2.91, 2.02, and 2.79 vs. SCE, respectively).

- The metallic, semiconductor, and insulator properties of carbon can be set as required by controlling the parameters during the synthesis of carbon. Sharon et al. (1998, 1999) showed that CNM prepared from camphor shows a metallic or graphitic nature when pyrolyzed above 750°C by converting sp^3 to sp^2 (transition energy needed for σ bond to π bond). Conductance increased, and the band gap decreased with an increase in temperature. As per the requirements, CNM can be made as p- or n-type material for electrochemical sensors.

- The properties of CNTs depend on the diameter and chirality. The current carrying capacity of CNTs is about 1 TA/cm³; Cu has a current carrying capacity 1 GA/cm³. The electric current–carrying capacity of CNT is about 1000 times higher than that of copper wires. The flow of electrons through CNT is without collisions, thus showing ballistic conduction, which enables the use of the quantum mechanical properties of electron wave functions. These properties of CNTs enhance their utilization in electrochemical sensors.

- The detection limit of CNT-based nanosensors exceeds that of conventional solid-state sensors.

- CNM provides a high adsorption surface area and a high surface-to-volume ratio, which enables sensing of analyte at a minute level.

- CNM can change electrical properties at room temperature, which reduces the power consumptions of operation.

- In contrast, conventional macro and micro devices that deal with the assembly of relatively large amount of samples of carbon nanosensors are almost reaching the size of biomolecules.

- The inert nature of CNM opens up new prospects for multiple analyte sensing (see Fig. 20-5).

20.5.1 Electrochemical Sensors

SWCNT-based chemical sensors were first reported by Collins et al. (2000), who observed a high oxygen sensitivity of the electrical resistance of SWCNTs. There are different types of chemical sensors, including electrochemical sensors.

Electrochemical oxidation/reduction reactions take place at a specific potential. The electrode potential at which a reaction takes place can be used to identify the species. The quantity of the species undergoing the electrochemical reaction can be calculated from the magnitude of current produced during the reaction. Thus, in an electrochemical reaction, it is possible to identify the species as well as measure its quantity. Based on this basic principle, various instrumentation techniques have been developed for measuring the concentration of the species present. It would be beyond the scope of this book to deal with each type of instrument used for such purposes. However, the magnitude of the current that one gets with a specific electrochemical reaction in addition to its quantity present in the solution also depend on the specific surface area of the electrode at which the reaction is being conducted, as well as the ease with which the species in question adheres with the electrode. To increase the surface area of the electrode and its adhering properties, metal or metal oxide, ceramic, and alumina plate electrodes are coated with conducting carbon materials (e.g., CNTs, CNFs, CNBs, glassy carbon) or conducting polymers. Thus, the electrochemical sensors rely on the efficiency of the working electrode capacity to adsorb the species on its surface so that a large quantity of the species can be obtained, which in turn gives larger electrical signals such as current.

The major benefits of using electrochemical sensors are their low cost, the simplicity of their construction, and the reliability of their performance. These qualities have made them very common sensors. Based on the design of fabrication, electrochemical sensors can be classified as electrochemical cells with three electrodes (amperometric or potentiometric), FETs, and chemoresistors.

Electrochemical Cells with Three Electrodes

Electrochemical cells with three-electrode systems can be classified as amperometric or potentiometric. Amperometric technique measures the current changes in between the electrodes. The current is proportional to the concentration of analyte being considered. This measurement is done at a constant current. In potentiometric technique each species gets reduced or oxidized at a specific potential. This potential is considered the signature of the species. Hence, if the current is measured at the potential where the electrochemical reaction is occurring, then the magnitude of the current can be used to calculate the concentration of species undergoing the electrochemical reaction.

Field Effect Transistor Sensors

FET sensors were constructed with SWCNTs as channels to conduct the source drain current. FET sensors control the flow of electrons (or electron holes) from the source to the drain by affecting the size and shape of the SWCNT, which is influenced by voltage (or lack of voltage) applied across the gate and source terminals. This SWCNT is the "stream" through which electrons flow from the source to the drain. Sometimes functionalization of the CNT sidewalls is required. Modified FET sensors that use enzymes are known as Enzyme FET (EnFET). Modified FET sensors show a better solubility for further application in chemistry or biochemistry.

Shiraishi et al. (2001) have found work functions of MWCNTs and SWCNTs to be 4.95 and 5.05 eV, respectively. They have also reported that the work function of CNTs is 0.1 to 0.2 eV larger than that of highly oriented pyrolytic graphite (HOPG) (Fig. 20-6).

These FET sensors are used to measure polar molecules such as NH_3, HCl, CO, NO, or NO_2. Kong et al. (2000) were the first to use single SWCNT FET sensors to detect NO_2 and NH_3 and reported that their electronic properties depend on the chemical environment. These sensors are referred to as chemFET sensors. Different chemFET sensors are used to measure concentrations of light metal ions, K^+, Na^+, Ca^{++}, and ammonium ion (NH_4^+). Because NO_2 is a strong oxidizer, hole carriers in SWCNTs (10, 0) increase the current. But because of lone pairs of electrons in NH_3, SWCNTs (10, 0) do not have affinity for NH_3.

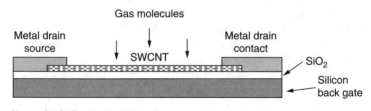

Notes: SWCNT = single-walled carbon nanotube.

FIGURE **20-6** Schematic diagram of a field effect transistor sensor.

Furthermore, the sensitivity of the sensor can be increased by the functionalization of CNTs with polyethyleneimine (PEI). This modification increases sensitivity for NO_2 (~100 ppt) and minimizes the sensitivity for NH_3 (~11 ppt). The fundamental behind the increase in sensitivity is that PEI-coated SWCNTs become n-type material because of electron transfer or donation by high-density amine groups on PEI. Hence, the conductivity of the sensor decreases with adsorption of NO_2.

Li et al. (2006) were the first to report the use of double-walled CNTs (DWCNTs) in FET sensors. For use in FET sensors, p-type DWCNTs were converted into n-type DWCNTs through functionalization of the DWCNTs with ferrocene. During functionalization, the DWCNTs were modified by encapsulating ferrocene inside the tube. Thus, n-type semiconducting DWCNTs were formed. These ferrocene molecules get decomposed inside the DWCNTs, and unipolar, n-type, magnetic Fe-filled DWCNTs are obtained.

Chemiresistors

Chemiresistors measure the current change while a constant voltage is applied across the two ends of sensing material, which reflects the change in resistance or conductance before and during the exposure of chemical species. CNTs as chemiresistors were first used for organic vapor sensing, in which resistance changes in CNTs were continuously monitored.

To prepare an SWCNT-incorporated chemiresistor (Li et al., 2003) a cast of SWCNTs on an IDE is made. These sensors are sensitive up to parts per billion concentrations of NO_2 and nitrotolune. The IDE is fabricated with the photolithographic method. Interdigited electrode (IDE) of 10 µm width are made by evaporating nanoparticles of Ti and gold of 20 and 40 nm size, respectively, on SiO_2. SWCNTs are synthesized by high-pressure carbon monoxide disproportionation (HiPCo). After purification, suspension of these SWCNTs is prepared with dimethylformide (DMF). This suspension is dropped on IDE. Dimethyformamide (DMF) is evaporated, and the IDE with the SWCNT network is used as a chemiresistor.

20.6   Biosensors

Biosensors sense large biological molecules. Biosensors can be defined as devices that transform different biochemical reactions (including adsorption of molecules) into measurable physical parameters.

20.6.1   Specific Properties of Carbon Nanomaterials Related to Biosensors

Various types of biosensors have been developed by using some of the specific properties of CNM:

- CNM has the specialty to connect biomolecules to macro and micro solid-state devices so the information of bio-events can be transformed into measurable signals. To integrate nanoscale CNTs with functional devices, lithography, photo-lithography, and functionalization techniques are followed to pass the information to the macroscopic world. MWCNT and SWCNT tips are modified to trap biological molecules such as amyloid-β-protofibrils. Sometimes it is necessary to grow CNTs in a specific location. This is possible by using catalyst spots defined by UV or e-beam lithography.

- A large surface-to-volume ratio and good electrical conductance make CNT an attractive option for making enzyme-based electrodes in biosensors. The change in the chemical state of the surface because of the adsorption of chemicals or biological agents may result in a decline of electrons (holes) not only near the surface but also in the entire volume of the nanostructure with an associated change in its Fermi level position within the band gap. By modifying CNM, self-assembly in sensor chips is achieved, so the sensors may be used numerous times.

- Hydrophobic forces on the sidewalls of CNT have been found to be important for the insertion process. Along with van der Waals forces, hydrophobic forces play a dominant role in the DNA–CNT interaction.

Major conditions required to design correct combination of biological molecules with transducer surface are as follows:

- **Maximum preservation of the activity of immobilized biomolecules:** Many times, protein changes its structure (it either decomposes or untangles) or may lose functioning. Hence, immobilized biomolecules must keep their structure and function as long as possible in both working and storage conditions. Similarly, the life of the sensor should be at least few months. CNM is uniform, well-structured, and chemically inert; hence, functionalization of CNM with appropriate proteins such as antibodies helps maintain the functioning conditions of the sensor as long as possible in working as well as storage conditions. CNTs serve both the functions (i.e., as large immobilization matrices and as mediators) to improve electron transfer between the enzyme site and the electrochemical transducer.

- **High density of immobilized molecules on the transducer surface:** Thick layers of sensitive membranes are better than a monolayer in absorbing maximum numbers of analytes. Still, the film thickness should be based on the permeability of the analyte; hence, it must be optimum. An alternative is to use highly porous materials, such as porous CNM, for adsorption. CNM synthesized from bagasse, cotton, and tea leaves has shown a porous structure and has very good adsorption capacities (Sharon et al. 2007). In the same manner, a highly porous substrate along with a highly accessible monolayer-sensitive coating offers maximum sensitivity. Furthermore, CNTs can be functionalized from inside and outside the CNT walls, which increases the area of sensitive coating.

- **Exposition of the majority of active sites to the analyte:** An ordered molecular structure with a uniform orientation of the sensitive host biomolecules is favorable. CNM fits in very well here because it has a uniform structure, and its curve nature increases the sites for bonding of the analyte.

- **Good adhesion of biomolecules to the transducer surface:** Carbon is known for its good adsorption capacity. Furthermore, functionalization of CNM solves the adhesion problem and provides ample space for adsorption of analytes.

- **Simplicity and low cost of manufacturing sensitive membranes:** CNM can be deposited on various substrates such as silicon wafers, ceramic plates, and quartz plates. These deposited films can act as sensitive membranes in a sensor.

As the name suggests, the key element in a biosensor is a sensitive membrane consisting of bioactive molecules that provide selective binding of analyte molecules and their reactions. Nature has lot of nanosensors in diverse forms, such as the sense of smell, especially in animals. A common example is the highly sensitive smell perception in dogs, which is performed using nanoreceptors. Certain plants also use nanosensors to detect sunlight, and microbes and algae use nanosensors to detect toxic molecules in aquatic conditions. Various fish use nanosensors to detect minuscule vibrations in the surrounding water, and many insects detect sex pheromones using nanosensors. Ants are very tiny creatures, but through their nano sensing device they are able to detect sugar placed anywhere in the house, and also are able to differentiate between sugar and other food grains. Mimicking the sensors used by ants can be extremely useful in detecting the presence of things like heroin, etc.

Researchers are also interested in mimicking such biological sensors as a sensing material by functionalization of natural receptors with CNM because CNT sidewalls are hydrophobic and chemically inert in nature. Moreover, biofunctionalization is needed to reduce the problems of specific recognition and possibly biocompatibility. Hence, for biological sensing, immobilization of biomolecules with specific functionalities on the sensing devices is necessary and can be targeted through following mechanisms:

- Antibody functionalization
- Nucleic acid hybridization
- Enzymatic functionalization
- Cellular interaction by biological barcodes such as DNA and RNA

Most of the time, functionalization needs opening up of end caps of CNT. This can be done by annealing the CNT sensors in open air for a suitable duration and temperature. This not only opens up the end caps but also removes the residual solvent and amorphous impurities from the CNTs formed during their synthesis. Such purification results in an improvement of the sensor sensitivity. Because CNT end caps have higher numbers of pentagonal rings than the sidewalls of CNTs, their functionalization forms covalent bonds at the end of caps or noncovalent bonds at the CNT sidewalls. Similarly, functionalization of CNBs forms covalent bonds with compounds because of the spherical shape and large number of pentagonal rings.

The interest in the use of CNTs is because of their confined space in the inner channel, particularly for use in a liquid environment. SWCNTs have been demonstrated as molecular channels for water transport. Geo et al. (2005), have shown that a DNA molecule could be spontaneously inserted into SWCNTs through a water medium. To get appropriate results, sometimes certain compounds are added to amplify the sensitivity for a specific material. For example, Ru $(bpy)_3^{2+}$ has been added for guanine oxidation, which requires a high potential. It amplifies signals, excluding the noise produced by carbon oxidation and water electrolysis.

Along with the sensitivity, the selectivity of CNT sensors can be enhanced by using a layered construction made up of an outer layer interacting specifically with target molecules and an inner layer transmitting the charges induced by this interaction to the CNTs in the core of the biosensors. Most biosensors contain expensive natural bioactive molecules (e.g., antibodies, enzymes, DNA, RNA fragments); therefore, efforts are ongoing to synthesize artificial chemicals with similar properties.

There are several kinds of biosensors, including enzyme, immune, DNA, and antibody–antigen sensors.

20.6.2 Enzyme Sensors

Various enzyme-based biosensors are based on electrochemical and amperometric principles. In most enzyme electrodes, Nafion is used along with CNT. Nafion is a perfluorinated sulfonic acid ionomer with a good biocompatibility and ion exchange properties. It has proven very effective as a protective coating for enzyme sensors. Based on molecules, sensors are known as glucose, pesticide and heavy metal, urea, triglyceride, butylcholine, and creatinine sensors.

20.6.3 Glucose Sensors

Glucose sensors are used to detect glucose levels in blood. To make these sensors, CNM is functionalized with glucose oxidase (GOx). Sidewall functionalization of the MWCNT array increases the enzyme loading in electrochemical glucose sensors. Various processes are used to prepare electrodes for glucose sensor, including sonication of Nafion along with CNTs to coat CNT with Nafion. Nafion coating on CNT electrodes reduces the leaching of GOx and extends the life of the sensor. GOx catalyses the oxidation of β-D-glucose to D-glucono-1,5-lactone using oxygen (O_2) as an electron acceptor. During this reaction, hydrogen peroxide (H_2O_2) is evolved, which is electrochemically detected at an appropriate electrode. Because oxidation of 2-deoxy-D-glucose, D-mannose, and D-fructose is catalyzed at a slow rate compared with oxidation of GOx, a very high specificity is shown for β-D-glucose.

Furthermore, Wang et al. (2003) reported another modified electrode created with deposition of platinum nanoparticles onto Nafion-containing CNT/GOx. When comparing CNT electrodes and platinum electrodes, this modified electrode is highly sensitive, with a detection limit of 0.5 mM glucose and a response time of 3 seconds. Wang et al. have reported that this glucose sensor is highly sensitive and works with a low potential requirement (0.05V vs. Ag/AgCl). The reaction at the electrode is as follows:

$$\text{Glucose} + O_2 + H_2O \xrightarrow{\text{GOx}} \text{Gluconate} + H^+ + H_2O_2 \quad (20.10)$$

Alternatively, a platinum plate may be coated with CNTs that can adsorb suitable enzyme, which converts glucose to gluconic acid. In this process, the release of electron is measured to find out the presence of sugar in the blood (Fig. 20-7).

Rivas et al. (2007) have reported the development of an electrochemical glucose biosensor constructed on a glassy carbon electrode modified with MWCNTs dispersed in perfluorosulfonated polymer, Nafion. The sensitivity of this sensor is 0.035 mM.

Another method of coating Nafion to CNTs for use as a glucose sensor is by dispersing Nafion with CNTs in concentrated sulfuric acid followed by casting them onto a glassy carbon electrode (GCE). This results to well-connected CNT networks.

An electrode made by combining CNF with MWCNTs prepared by dipping the fiber into a sonicated phosphate buffer solution containing CNTs and Nafion has also been reported.

20.6.4 Pesticide and Heavy Metal Sensors

Pesticide compounds, which are commonly used in agriculture and for food preservation, are among the most toxic substances. Early detection of pesticide residue has become imperative.

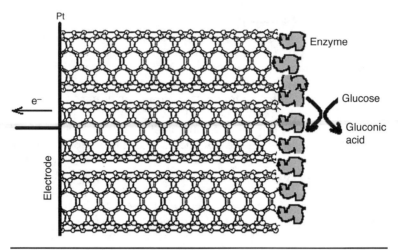

Figure 20-7 Schematic representation of a sensor for detecting sugar in blood. A platinum plate is coated with carbon nanotubes over which a suitable enzyme is adsorbed. This helps to extract electrons from glucose by converting it into gluconic acid.

Lin et al. (2005) have developed a CNT-modified, screen-printed electrode for amperometric detection of organophosphorous pesticides. The activity of acetylcholinesterase is inhibited, and the hydrogen peroxide of the co-immobilized choline oxidase is detected by CNTs. The reaction at an electrode is:

$$CH_3 COO (CH_2)_2 N^+ (CH_3)_3 + H_2O \xrightarrow{\text{Acetylcholinesterase}} HO (CH_2)_2 N^+(CH_3)_3$$
$$+ CH_3COO^- + H^+ \tag{20.11}$$

Wang et al. (2003) have produced a pesticide sensor by using organophosphorous hydrolase (OPH). Amperometrically OPH converts into p-nitriphenol, which is oxidized at a CNT-modified electrode. Oxidation of phenolic products improves the stability of sensors and increases their sensitivity. These sensors give a stable anodic current for up to 60 minutes. Furthermore, under optimum conditions, these sensors can detect as low as 0.15 µmol/L paraoxon and 0.8 µmol/L methyl parathion with respective sensitivities of 25 and 6 nA/mmol/L.

Deo et al. (2005) have also developed a pesticide sensor by using OPH. They prepared the electrode by mixing 100 µL of a 5% Nafion solution with 900 µL of a phosphate buffer (pH, 7.4). They then added MWCNTs and sonicated the solution for 30 minutes. A 20-µL aliquot of this CNT solution was cast on a cleaned glassy carbon electrode. The electrode was air dried at room temperature for 1 hour. The OPH enzyme was immobilized by casting a 10-µL solution of OPH (50 IU/mL) in Nafion (0.5% in ethanol) onto the modified (10-μL solution of CNT) glassy carbon electrodes, and the solvent was evaporated. The enzyme-modified electrode was dried at room temperature.

Other sensors, including triglyceride, urea, butyrylcholine (BCH), and creatinine sensors, can be made by functionalization of CNM with lipase, urease, butyrlcholinesterase, and creatinine deiminase, respectively.

$$\text{Triglycerides} + 3H_2O \xrightarrow{\text{Lipase}} \text{Glycerol} + 3 \text{ Fatty acids} \quad (20.12)$$

$$CO(NH_2)_2 + 3H_2 \xrightarrow{\text{Urease}} CO_2 + 2NH_4^+ + 2OH^- \quad (20.13)$$

$$CH_3(CH_2)_2N^+(CH_3)_3 + H_2O \xrightarrow{\text{BCH esterase}} HO(CH_2)2N^+(CH_3)_3$$
$$+CH_3(CH_2)_2COO + H^+ \quad (20.14)$$

$$\text{Creatinine} + H_2O \xrightarrow{\text{Creatinine deiminase}} NH_4^+ + \text{N-Methylaydantonine} \quad (20.15)$$

These signals are measured by recording the change in current or in pH and then converting them into measurable physical parameters.

20.6.5 DNA Sensors

DNA is a nanodiameter coding molecule of living organisms. It is the most important part of living systems and controls the functioning of the whole body. Therefore, detecting functioning or malfunctioning body parts can be done by sensing DNA. Moreover, DNA can also be used in nanosensors. CNM provides an excellent platform to attach DNA. Surface-confined MWCNTs have been shown useful to facilitate the adsorptive accumulation of the guanine a nucleo-base and greatly enhances its oxidation signal. A similar enhancement of the guanine DNA response was reported at MWCNT paste electrodes and at SWCNT-coated glassy carbon electrodes. Within a short period, trace nucleic acids can be detected.

DNA probes are also used for DNA hybridization. An indicator-free AC impedance measurement of DNA hybridization based on DNA probe–doped polypyrrole film over an MWCNT layer has been developed.

DNA attached to FET sensors is used for electronic tongue and electronic nose. Staii et al. (2005) created new classes of nanoscale biosensors that are based on ssDNA as the chemical recognition site and SWCNT-FETs as the electronic readout component. SWCNT-FETs with a nanoscale coating of ssDNA respond to gas odors that do not cause a detectable conductivity change in bare devices. As per the requirements of sensing, the base sequence of ssDNA can be changed. ssDNA/SWCNT-FET sensors are very efficient and fast and can detect a variety of odors within a fraction of seconds. The arrays of nanosensors can be prepared to detect various gas molecules up to 1 ppm.

A self-assembly technique is being developed to be applied to DNA- CNM nanosensors.

20.7   Future Challenges

The future challenges for the development and use of carbon nano-sensors include:

- Controlled growth of the desired CNM structure
- Reproducibility of CNM for consistent and accurate signals
- Control of the chirality of CNTs, which affects the electrical signals of CNT-based sensors
- Simplification of CNM modification techniques to make CNM sensors commercially viable

PART V

Applications of Carbon Nanomaterials in Biosystems

CHAPTER 21

Biosystems and Nanotechnology

Madhuri Sharon

MONAD Nanotech Pvt. Ltd., Powai, Mumbai, India

21.1   Introduction

The rapidly developing nanosciences and nanotechnology reveal many exciting features of materials at the nanometer scale (10^{-9} m) and the possibility of manipulation of such features to create novel materials or products that were previously unthinkable. The belief is getting stronger with every passing day that nanosciences and nanotechnology development will become a new industrial revolution.

Three landmark achievements of the twentieth century have contributed greatly to the understanding of nano-biotechnology. These developments include:

- Transmission electron microscopes and atomic force microscopes
- Discovery of DNA by James Watson and Francis Crick
- Arc, plasma, and chemical flame furnaces, which are used to produce nanoparticles

Nanomaterials can be described as "novel materials whose size of elemental structure has been engineered at the nanometer scale." Materials in the nanometer size range commonly exhibit fundamentally new physicochemical behavior. Moreover, involvement in the properties of materials at the nanoscale enables the creation of materials and devices with enhanced or completely new characteristics and functionalities.

Nanoforms of material and the functioning of living beings at a nano level existed in nature long before humans were able to identify them. Advances in synthetic chemistry and physical equipment, such as scanning electron micrography and transmission electron micrography, have been driving forces in the development of an integration of biology and manoscience.

The unique physicochemical properties of nanomaterials (as mentioned in Chap. 12) are not found in their bulk materials. In general, they have a much higher reactivity, and because of their ultra small size and increased surface area, they can easily penetrate skin or cells, rapidly distribute in the human body, and even directly interact with organelles within cells. Their huge surface area-to-volume ratio increases the chemical activities and therefore, allows nanomaterials to become efficient catalysts.

These increased chemical and biologic activities have resulted in many engineered nanoparticles that are being designed for specific purposes, including diagnostic and therapeutic medical uses and environmental remediation. Nanostructured materials coupled with liquid crystals and chemical receptors offer the possibility of cheap biodetectors that might be worn as a badge. Such badges could change color in the presence of a variety of chemicals and have applications in hazardous environments.

21.2 Bio-nanotechnology

However, rather than believing that the nanosciences have given input into the development of modern-day biotechnology, it would be more appropriate to say that biosystems functioned at a nano level right from the beginning of evolution.

If we take an example of a cell, which is the unit of a biosystem or an organism, and consider the cell as a micro factory, organelles as nanomachines, and molecules as custom-made nanoproducts with specified tasks, we realize that biosystems function so that:

- Smaller is faster and better.
- Nano-level existence enables technology to do new things in almost every conceivable discipline.
- Biosystems use various bonding forces and thermal motion, and gravity and inertia have a negligible effect on assembly and synthesis, which can be copied or mimicked.

Nucleotides, sugar, and phosphates (Fig. 21-1) are three basic molecules that, when joined together by H bonding, van der Walls forces, and covalent bonds, create a living entity functioning as a system that can duplicate itself.

21.3   Functioning of Cells at the Nano Level

The entire functioning at a molecular level takes place in the cellular factory's various organelles (Fig. 21-2):

- The **nucleus**, the biggest cell organelle (800 nm), consists of a nuclear envelope chromatin, and a nucleolus. The nuclear envelope plays a role in selective permeability to control movement in and out of the nucleus. Chromatin contains instructions that control cell metabolism and heredity.

- **Mitochondria**, which supply energy for chemical activities, are also known as the powerhouses of the cell. The mitochondria are composed of a modified double-unit membrane (protein and lipids); the inner membrane is folded to form cristae, which function as a site of cellular respiration (i.e., the release of chemical energy from food).

- **Ribosomes** are nonmembranous (20 to 30 nm) spherical bodies composed of RNA and protein enzymes. They are the sites of protein synthesis.

- The **endoplasmic reticulum** is a sheet of unit membrane with ribosomes on the outside, forming tubular network 1 to 2 nm wide throughout the cell. It transports protein and other chemicals within and outside the cell and provides a large surface area for organization of chemical reactions and synthesis.

- The **cell membrane** is composed of protein and lipid molecules. It acts as a boundary layer to contain cytoplasm in the cell. Its interlocking surfaces bind cells together, and it is selectively permeable to select chemicals that pass in and out of the cell and the organelles.

- **Chloroplasts** are the basic producers for living beings on this planet. They are composed of a double layer of modified membrane (protein, chlorophyll, and lipids). The inner membrane invaginates to a form layer called *grana*, where chlorophyll is concentrated. The inner membrane of chloroplasts are the sites of photosynthesis.

Apart from above-mentioned functions, biosystems have been using nanomaterials for millions of years. For example, nanostructured elements that play a vital role are magnetotactic bacteria, ferritin, and

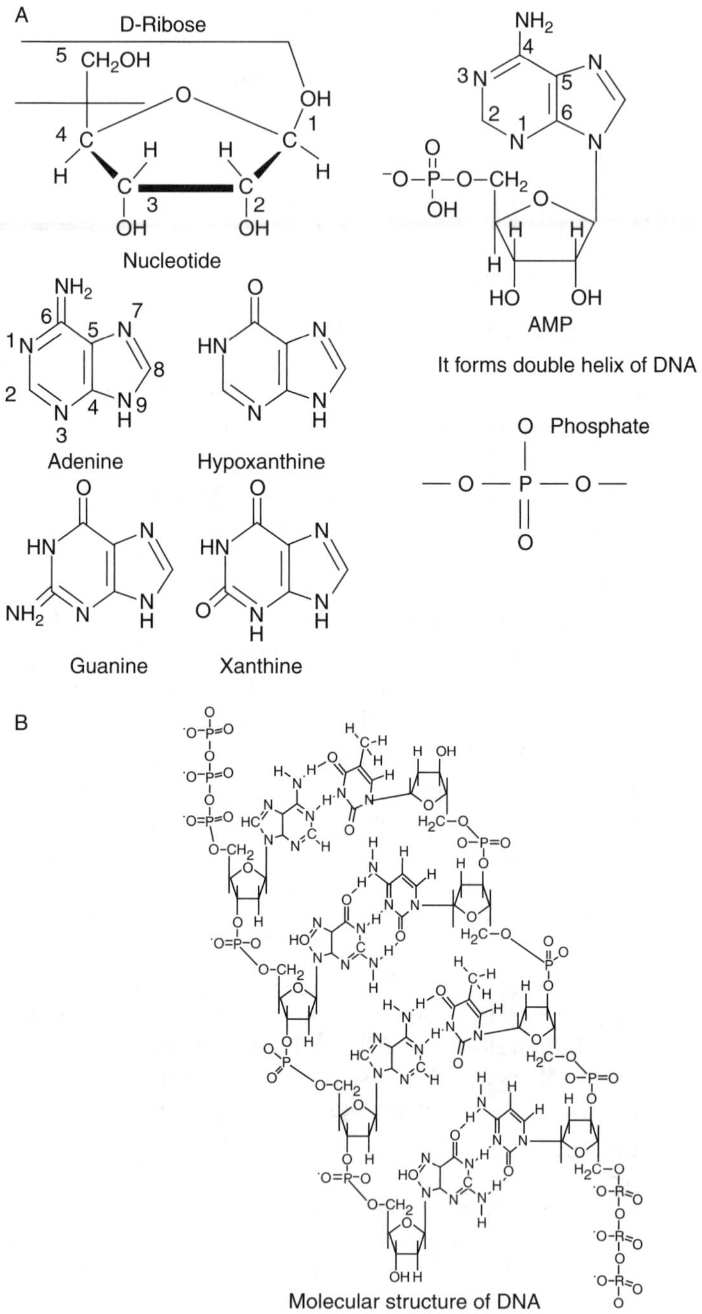

Notes: AMP = Adenosine monophosphate.

FIGURE 21-1 Formation of a living DNA molecule from chemicals (sugar, nucleotides, and phosphates). (*A*) It forms a DNA double helix. (*B*) Molecular structure of DNA.

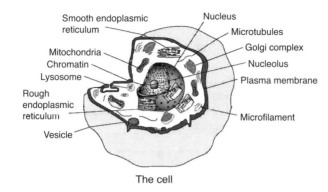

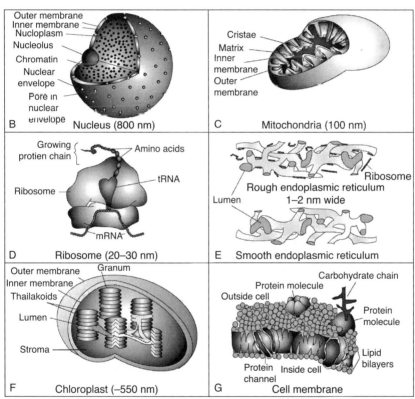

FIGURE 21-2 A biosystem at the nano level, including (*A*) the cell, (*B*) nucleus, (*C*) mitochondria, (*D*) ribosome, (*E*) smooth endoplasmic reticulum, (*F*) chloroplast, and (*G*) cell membrane.

molluscan teeth. Several species of aquatic bacteria use the Earth's magnetic field to orient themselves. They are able to do this because they contain chains of nanosized single-domain magnetic (Fe_3O_4) particles. Because they have established their orientations, they are able to swim down to nutrients and away from what is lethal to them (e.g., oxygen). Another example of nanomaterials in nature is the storage of iron in a bioavailable form within the 8-nm protein cavity of ferritin. Hence, it was very appropriate of Dickson to state: "Life itself could be regarded as nanophase."

By looking at the varied activities that go on in biosystems generating their own power, one can engineer and manufacture nanoscale devices with atomic precision to:

- Create nanomachines with kits and reagents
- Design a nanoassembler for synthesis of molecules
- Use biologic templates for making nanomaterials
- Construct a nanomachine that can recognize and kill diseased or cancerous cells
- Build a molecular size sensor to detect acidity, poisonous metal, and so on

Hence, there are two ways of manifesting nanotechnology: (1) by mimicking nature or using biomaterial for development of nanoscience and (2) application of nanosciences for biotechnologic purposes. Some successful applications include:

- **Diagnostic equipment** including biosensors, nanoprobes (DNA-based biochips) for cancer detection, nano-robots, and X-ray devices
- **Surgical supplements** including nanomedicinal devices, bioactive nanomaterial in bone grafting, nanotweezers
- **Nanocomposites** (polymers) for dental restoration and tissue engineering
- **Fluorescent nanocrystals** as markers and fluorescence assays for drug discovery
- **Gene delivery** for creating transgenic plants or cloning instead of using a vector
- **Pollution control** by killing pathogenic microbes and through filtration and physicochemical methods
- **Antimicrobial activity** of nanomaterials
- **Anticarcinogenic activity** of nanomaterials
- **Drug delivery** using carbon nanomaterial (CNM), synthesized organic hydrophilic or hydrophobic polymers, and glycolipid-based bio-nanomaterials
- **Biosynthesis** of nanomaterials using microbial cells

It is beyond the scope of this book to discuss the applications of all of the known nanomaterials in biosystems. Hence, the discussion is restricted to the application of CNM in biosystems, which is presented in the next few chapters (see Fig. 21-1).

21.4   Nanotechnology and Future Fantasies

Nanotechnology is at the junction of the past and the future. The first nanotechnologists were perhaps the medieval artisans who added minute quantities of gold and silver in molten glass to produce the luminous colors that are still preserved on the stained glass windows of many cathedrals.

Today, intentional efforts are geared toward maximizing the benefits available by harnessing nanosciences and nanotechnology to what may sound difficult even at a fantasy level.

When experts predicted that nanotechnology would revolutionize the world, scientists from all disciplines joined the race to become a part of this new craze. Many journals, Web sites, special reports, companies, and products with the nano- prefix mushroomed, and billions of dollars were invested worldwide. Even fiction books based on nanosciences and nanotechnology are flooding the market. Science fiction portrays nanotechnology as both our savior and our destroyer. Are these ideas pure fantasy, or could they become a future reality?

Marlow's book entitled *Bacteria Building Nanobots* (based on extremely general physical principles, such as the conservation of energy and the limits of available energy vs. the characteristic energies of chemical bonds) fantasizes that rogue nanobots can replicate at a faster rate than already-existing self-replicating organic life forms such as bacteria, so they can destroy the planet in a matter of days.

Michael Crichton has fantasized that adaptable nanoparticles, programmed as predators, self-reproduce, learn, and evolve into an inexplicable menace that baffles the brilliant scientists who created them.

The idea of rampaging and malevolent robots that are thousands of time smaller than the point of a pin is still the stuff of futuristic fiction. But as an emerging science, nanotechnology has already given rise to new approaches to physics, new promise for breakthroughs in medicine, and new hopes in communications and consumer goods.

When we talk about nanotechnology, we are talking small. As we get more prolific in working with this tiny technology, we will actually produce nanomachines that will produce even smaller machines. Actually, these machines will be so small that molecules will produce them automatically.

This is not science fiction; this is real stuff that is expected to be developed in about five years that include:

- Car tires that will need air only once a year
- Self-assembly of small electronic parts (based on artificial DNA or guest host systems)

- Artificial semiconductors based on protein
- Complete medical diagnostic laboratories based on a single computer chip less than one inch square

Scientists have predicted that within 10 years, we can expect to see the following developments:

- Erasable and rewritable paper for programmable books, magazines, and newspapers
- Light, efficient ceramic car engines
- "Smart" buildings that self-stabilize after earthquakes and bombings
- Inexpensive solar power that heats and lights cities by using roads and building windows as sun collectors

And some have fantasized that in 10 to 15 years, we will see:

- Paint-on computer and video displays
- Cosmetic nanotechnology, including permanent hair and teeth restoration
- Handheld super computers

Moreover, foresight nanotechnology has a vision that nanotechnology will influence following six major areas that will affect our lives:

1. **By meeting global energy needs with clean solutions while protecting the environment:** Nanotechnology will help to solve the dilemma of energy needs and limited planetary resources through more efficient generation, storage, and distribution.

2. **By providing abundant clean water globally:** The demand for fresh water is increasing. Considering the current rate of consumption and projected population growth, about two thirds of the world will be affected by drought by the year 2050. Nanotechnology can help solve this problem through improved water purification and filtration.

3. **By increasing health and the longevity of human life:** Humans are living longer lives, yet infectious diseases and cancer continue to kill millions annually. Because of an aging population, there could be a 50% increase in new cancer cases by the year 2020. Nanotechnology will enhance the quality of life for human beings through medical diagnostics, drug delivery, and customized therapy.

4. **By healing and preserving the environment:** As a set of fundamental technologies that cuts across all industries, nanotechnology can benefit the environment in a wide variety of ways. Stronger, lighter-weight materials in transportation can reduce fuel use, nanostructured fibers reduce staining and therefore laundering, and low-cost nanosensors will make pollution monitoring affordable. In the longer term, manufacturing processes using productive nanosystems should be able to build our products with little, if any, waste.

5. **By making powerful information technology available everywhere:** Humans will need to cooperate as we respond to disasters and critical threats to our survival. A "planetary nervous system" fostering rapid communication and cross-cultural relationships is needed. Nanotechnology applications in electronics will increase access through reduced cost and higher performance of memory, networks, processors, and components.

6. **By enabling the development of space:** Heavy demands on resources and raw materials are creating challenges on Earth; however, these items are plentiful in space. Current obstacles to developing space are cost, reliability, safety, and performance. Nanotechnology will solve these problems through improved fuels, smart materials, uniform products, and environments.

We can conclude that nanotechnology is not a delusion or a fantasy; rather, it is a reality that is about to be manifested.

CHAPTER 22

Applications of Carbon Nanomaterials in Biosystems

Madhuri Sharon

MONAD Nanotech Pvt. Ltd., Powai, Mumbai, India

Amol Kakade

Sir H.N. Hospital and Research Center, Mumbai, India

22.1   Introduction

Nanotechnology is the application of knowledge of nanosciences, which have inputs from all the branches of science. Nanotechnology is the ability to build nanosized complicated shapes and machines

357

with every atom in its specified place. We have had several "revolutions" in technology—industrial, agricultural, medical, and computer—within the past two centuries. But each of these revolutions has only given us a small fraction of the capabilities we could have. Nanotechnology will let us finish the job by being much more precise in our design and fabrication of machines and by using better materials. For example, chemists and biologists will be able to create molecules with every atom precisely placed, and engineers will be able to build incredibly complicated and useful nanomachines without any wasted space.

It is beyond the scope of this book to discuss the applications of all the known nanomaterials in biosystems. Hence, this chapter discusses only the application of the carbon nanomaterials (CNMs) in biosystems.

22.2   Entry of Carbon Nanomaterials in Bioapplication

The discovery of CNMs has the potential to revolution biosystems research because CNMs can show superior performance because of their impressive structural, mechanical, and electronic properties such as small size and mass, high strength, and high electrical and thermal conductivity.

These distinct properties are being exploited such that they can be used for applications ranging from sensors and actuators to composites. As a result, in a very short duration, CNMs appear to have drawn the attention of both industry and academia. However, certain challenges need proper attention before CNM-based devices can be realized on a large scale in the commercial market. The challenges associated with CNMs that remain to be fully addressed for their maximum utilization for biosystems applications are discussed here.

The known forms of nanosized carbon—carbon nanotubes (CNTs), carbon nanofiber (CNF), carbon nanobeads (CNBs), carbon nanoshell, and fullerenes—are discussed in detail in Chap. 1.

One property of CNMs that has influenced their application in biosystems is their surface morphology, which makes them a suitable material for adsorption, absorption of gases and liquids, data storage, template for the various biologic reactions, and so on. Moreover, functionalization alters many of the characters of CNMs. Functionalized CNMs have developed solubility in both polar and nonpolar solvents, expanding their application in various biosystems (e.g., drug delivery, biosensors, biochips).

Apart from the surface morphology, their electrical, electronic properties, conducting, and semiconducting properties make CNMs different from other materials. The band gap of CNMs and their photochemical properties have found wide application in purification of water and degradation of organic and inorganic materials.

In the past decade, enormous progress has been made in developing new fabrication techniques and materials for developing small devices. One of the most promising applications of miniaturization technology is in the biosystems industry. Today the biosystems industry is characterized by irreconcilable demands such as environmental cleanliness, better health care, and cost-effective diagnostics and treatments.

22.2.1 Diagnostic Tools and Devices

Radiation Oncology

The traditional method of generating X-rays includes a metallic filament (cathode) that acts as a source of electron when it is heated (resistively) to a very high temperature. The accelerated electrons that are emitted are bombarded on a metal target (anode) to generate X-rays, as shown in Fig. 22-1. The advantage associated with this method is that it works even in non–ultra-high vacuum ambiences, which contain various gaseous molecules.

However, this method has several limitations, including a slow response time, high consumption of energy, and a limited lifetime. Recent research has reported that field emission is better for extracting electrons than thermionic emission (Romero et al., 2000). This is because electrons are emitted at room temperature, and the output current is voltage controllable. In addition, giving the cathode the form of tips increases the local field at the tips, and as a result, the voltage necessary for electron emission is lowered (Yue et al., 2002). An optimal cathode material should have a high melting point, low work function, and high thermal conductivity. CNTs meet all of these requirements and hence are used as a cathode material for generating free-flowing electrons.

Electrons are readily emitted from their tips either due to oxidized tips or because of curvature when a potential is applied between a CNT surface and an anode (de Vita et al., 1999; Saito et al., 1997)

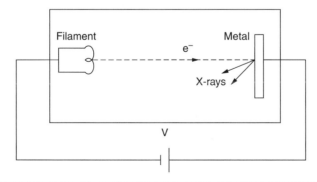

FIGURE 22-1 The traditional method of generating X-rays.

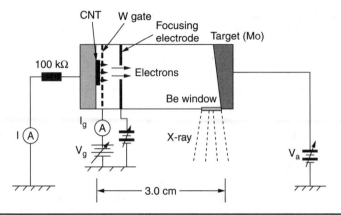

FIGURE 22-2 Schematic drawing of a carbon nanotube (CNT)–based microfocus X-ray tube.

generated continuous and pulsed X-rays using a CNT-based field emission cathode. A schematic diagram of a CNT-based microfocus X-ray tube is shown in Fig. 22-2. X-ray radiation with continuous variation of temporal resolution as short as nanoseconds can be readily generated by these systems.

The advantages of CNT-based X-ray devices are that they have a fast response time, fine focal spot, low power consumption, possible miniaturization (and hence portability), longer life, and low cost. The devices can readily produce both continuous and pulsed X-rays (>100 KHz) with a programmable wave form and repetition rate. A field emission X-ray tube can produce sufficient X-ray intensity to image the human anatomy (Fig. 24-3). The technology can be used for portable and miniaturized X-ray sources for industrial and medical applications.

Moreover, CNT-based X-ray devices minimize the need for cooling required by the conventional method (Senda et al., 2004; Sugie

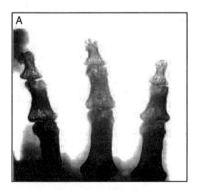
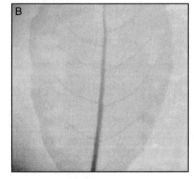

FIGURE 22-3 X-ray image of (A) human finger bone and (B) plant leaf.

et al. 2001; Yue et al., 2002). Miniaturized X-ray devices can be inserted into the body by endoscopy to deliver precise X-ray doses directly at a target area without damaging the surrounding healthy tissues. Malignant tumors that are highly localized during the early stage of their development can be easily mapped by miniaturized X-ray devices.

Sensors

Sensors are devices that detect a change in physical quantity or event. Many studies have reported the use of CNTs as pressure, flow, thermal, gas, chemical, and biologic sensors. Liu and Dai (2002) demonstrated that piezo-resistive pressure sensors can be made with the help of CNTs. They grew single-walled CNTs (SWCNTs) on suspended square polysilicon membranes. When uniform pressure was applied on the membranes, a change in resistance in the SWCNTs was observed. Because CNTs have a temperature coefficient almost two orders of magnitude lower than that of silicon and have an increased sensitivity, highly efficient pressure sensors incorporating CNTs can be fabricated.

Fabrication of piezo-resistive pressure sensors that incorporate CNTs can bring dramatic changes in the biosystems industry because many piezo-resistance–based diagnostic and therapeutic devices are currently in use. Pressure sensors can be used in eye surgery, hospital beds, respiratory devices, patient monitors, inhalers, and kidney dialysis machines (Brodie and Schwoebel, 1994; Joseph et al., 1997). During eye surgery, fluid is removed from the eye, and if required, it is cleaned and replaced. Pressure sensors measure and control the vacuum that is used to remove the fluid and provide input to the pump's electronics by measuring barometric pressure. Hospital bed mattresses for burn victims contain pressure sensors that regulate a series of inflatable chambers. To reduce pain and promote healing, sections can be deflated under burn areas. Pressure sensors can also be used for detection of sleep apnea (a cessation of breathing during sleep). The pressure sensor monitors the changes in pressure in inflated mattresses. If no movement is found for a certain period, the sleeper is awakened by an alarm (Joseph et al., 1997).

Pressure-sensing technology is used in both invasive and noninvasive blood pressure monitors. Many patients who use inhalers activate their inhalers at an inappropriate time, resulting in an insufficient dose of medication. Pressure sensors in the inhalers identify the breathing cycle and release the medication accordingly (Brodie and Schwoebel, 1994).

During kidney dialysis, blood flows from the artery to the dialysis machine and flows back into the vein after cleaning. Waste products are removed from the blood through osmosis and move across a thin membrane into a solution that has blood's mineral makeup (Sakai, 1997). Using pressure sensors, the operation of the dialysis system can be regulated by measuring the inlet and outlet pressures of both the blood and the solution.

Intelligent pressure sensing systems play an important role in portable respiratory devices that consist of both diagnostic (spirometers, ergometers, and plethysmographs) and therapeutic equipment (ventilators, humidifiers, nebulizers, and oxygen therapy equipment). They serve patients with asthma, sleep apnea, and chronic obstructive pulmonary disease. They measure pressure by known fluid dynamic principles (Romero et al., 2000).

New classes of nanoscale chemical sensors are being developed, which are based on ssDNA as the chemical recognition site and SWCNT field effect transistors (FETs) as the electronic readout component (Staii et al., 2005). SWCNT FETs with a nanoscale coating of ssDNA respond to gas odors that do not cause a detectable conductivity change in bare devices. Responses of ssDNA/SWCNT FETs differ in magnitude for different gases and can be tuned by choosing the base sequence of the ssDNA. SsDNA/SWCNT FET sensors detect a variety of odors, with rapid response and recovery times on the scale of seconds. The arrays of nanosensors could detect molecules on the order of 1 ppm. The sensor surface is self-regenerating; samples maintain a constant response with no need for sensor refreshing through at least 50 gas exposure cycles. The nanosensors can sniff molecules in the air or taste them in a liquid. This remarkable set of attributes makes sensors based on ssDNA-decorated nanotubes promising for "electronic nose" and "electronic tongue" applications ranging from homeland security to disease diagnosis.

Biosensors

Sotiropoulou and Chaniotakis (2003) have used CNTs as an immobilization matrix for the development of an amperometric biosensor. The biosensor was developed by growing aligned multiwalled CNTs (MWCNTs) on platinum (Pt) substrates. The platinum substrate served as the transduction platform for signal monitoring, and the opening and functionalization of large CNT arrays allowed for the efficient immobilization of the model enzyme (glycine decarboxylase in this case).

A schematic diagram of a CNT array biosensor is shown in Fig. 22-4. The CNTs to be used as arrays were purified by treatment with acid or air. The acid treatment resulted in the removal of impurities, including amorphous carbon, that occurred during the production procedure. The lengths of the CNTs were also reduced by approximately 50%. Air oxidation resulted in production of thinner CNTs because of the peeling of the outer graphitic layers from the CNT. The use of nanotube-based sensors will avoid problems associated with the larger implantable sensors, which can cause inflammation, and may eliminate the need to draw and test blood samples. The

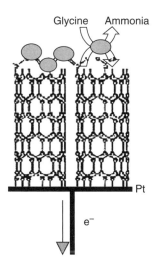

Glycine   Ammonia

Pt

e^-

devices can be administered transdermally (i.e., through the skin), avoiding the need for injections during space missions.

Biosensors can also be used for glucose sensing. CNT chemical sensors for liquids can be used for blood analysis, such as for detecting sodium or finding the pH value (Adrian, 2003). CNTs can also be used in flow sensors (Ghosh et al., 2003; Liao et al., 2003). Ghosh et al. (2003) found that the flow of a liquid on bundles of SWCNTs induces a voltage in the direction of flow.

In the future, this finding can be used in micromachines that work in a fluid environment such as heart pacemakers that need neither heavy battery packs nor recharging. Flow sensors can also be used for precise measurements of gases used by respiratory apparatuses during surgery and automatic calculation of medical treatment fees based on output data, leading to reduced hospital costs and more accurate calculation.

22.2.2   Water Purification

The provision of sufficient clean water for human consumption, agriculture, and industrial processes is an ongoing and increasing challenge as a result of population growth, extended droughts, and numerous competing demands. Nanotechnology offers several possible novel, improved, and efficient methods for purifying water. CNTs have shown promise in water purification by three different methods: photocatalytic degradation of pollutants, as an adsorbent for various chemicals, and by providing an ultrafiltration unit for purifying water.

Photocatalytic membranes have been produced and tested in a pilot plant for their efficacy in degrading triazine herbicides. Use of these compounds, particularly in areas with sandy soils, has contaminated underground aquifers with compounds such as atrazine.

Composite membranes produced by nanotechnology and containing titanium dioxide and tributyl- and triisopropyl vanadate have been exposed to sunlight, resulting in the oxidation and destruction of atrazine in water at a concentration of 1 ppm (Bellobono et al., 2005). Other systems containing titanium dioxide nanoparticles have been reported to degrade polychlorinated biphenyls (PCBs) and other organic pollutants in water (Savage and Diallo, 2006).

Nanomaterials can also be used to adsorb or sequester pollutants and remove them from water. Sorbents are already widely used in water purification. Inclusion of nanosorbents can be much more effective because they have a much larger surface area than conventional bulk particles. Various chemical groups can also be added to nanoparticles to improve their specificity in removing certain pollutants. MWCNTs have been found to adsorb three to four times the amount of heavy metals (copper, cadmium, and lead) than powdered or granular activated carbon. Chitosan nanoparticles containing tripolyphosphate adsorb even greater amounts of lead. Other nanosorbents have been devised to remove arsenic and chromium from water. CNTs and nanoporous activated carbon fibers can effectively adsorb organic pollutants, such as benzene and fullerenes, and can also adsorb polycyclic aromatic compounds, such as naphthalene (Savage and Diallo, 2006).

Ultrafiltration and reverse osmosis are now used to remove impurities from water. Nanotechnology can enhance the effectiveness of these processes, and nanofiltration processes are being developed for desalination. CNT filters can effectively remove bacteria and viruses from water, and other nanostructured membranes have been reported to remove organic pollutants, uranium, arsenic, and nitrates.

22.2.3 Probes

Probes are devices that are designed to investigate and obtain information on a remote or unknown region or cavity. Many studies have reported on the use of CNTs for making probes (Moloni et al., 2000; Nguyen et al., 2004). One such example is their use in atomic force microscopy (AFM). CNTs are highly suitable materials for AFM probes because the AFM-generated image is dependent on the shape of the tip and surface structure of the sample of interest.

An optimal probe should have vertical sides and a tip radius of atomic proportions. AFM tips made of silicon or silicon nitride are pyramidal in shape and have a radius of curvature around 5 nm. Compared with them, nanoprobes made of CNTs have a high resolution because their cylindrical shape and small tube diameter enable imaging in narrow and deep cavities. In addition, probe tips made of CNTs have mechanical robustness and a low buckling force. A low buckling force lessens the imaging force exerted on the sample and therefore can be applied for imaging soft materials such as biologic samples. Use of CNTs in AFM enhances the life of the probes by minimizing

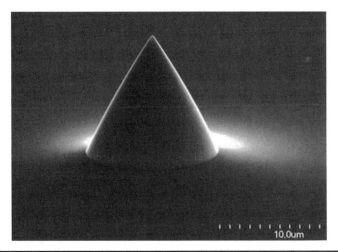

FIGURE 22-5   Scanning electron micrograph of a nanoprobe.

sample damage during repeated hard crashes onto substrates (Baughman et al., 2002) (Fig. 22-5)

22.2.4   Quantum Dots

Quantum dots are tiny light-emitting particles 2 to 10 nm in size. They are a new class of biologic labels. Semiconductor quantum dots can be used for quantitative imaging and spectroscopy of single cancer cells. They could possibly be applied for disease diagnostics, cancer imaging, molecular profiling, drug and biochemical discovery, disease staging, targeted therapeutics, and high-throughput drug screening. This is because of their dimensional similarity with biologic molecules such as nucleic acids and proteins and their size-tunable properties. In addition, they allow longer periods of observation and do not fade when exposed to ultraviolet light. They can also be used for tracking many biologic molecules simultaneously because their color can be tailored by changing the size of the dot.

Researchers have shown that CNTs can form quantum dots. Tans et al. (1997) and Bockrath et al. (1997) simultaneously observed that SWCNTs can form quantum dots. Coulomb blockade and a quantization of electron states were shown by their transport experiment that implied that a CNT quantum dot had been formed. Later in 2002, Buitelaar et al. showed that MWCNTs can also form clean quantum dots where the level of separation exceeds the charging energy.

Recently, a structure has been proposed to fabricate a quantum well of a few nanometers by using the electromechanical properties of SWCNTs (Nojeh et al., 2003). When embedded in biologic fluids and tissues, quantum dot excitation wavelengths are often quite constrained. Therefore, excitation and emission wavelengths should be selected carefully based on the particular application (Lim et al., 2003).

22.2.5   Biopharmaceuticals

Molecular Carriers

The diversity of available chemistry and cell-penetrating structures makes CNTs viable candidates as carriers for the delivery of drugs, DNA, proteins, and other molecular probes into cells. One of the prerequisites for such a task is the ability of the carrier to bind to biologically relevant molecules. Early experimental studies regarding interactions between MWCNTs and protein revealed the self-organization of streptavidin molecules and the growth of its helical crystals on the nanotube surface. Similarly, DNA molecules may be adsorbed on MWCNTs, and small protein molecules, such as cytochrome c and β-lactose I, can be inserted within the interior cavity of open CNTs.

As an alternative to the binding of molecules to the outside of the CNTs, it would be convenient to fill the interior cavity of tubes, whose open ends might be capped to generate a nanopill containing a drug for delivery to the cell. A template method has been used to synthesize nano test tubes, which are nanotubes with one end closed and the other open. Such an approach might be construed as a first step toward the development of a nanopill, in which the substance to be delivered is introduced and then bottled by releasing the open end.

An important issue in intracellular drug delivery is the poor permeability of the plasma membrane to many drugs. Thus, various carriers, including polyethylene glycol, peptides, and lipids, have been developed to facilitate the cellular entry of the intracellular drug. Water-soluble SWCNTs were functionalized with a fluorescent probe, fluorescein isothiocyanate (FITC), and tracking of SWCNTs was done. When murine and human fibroblast cell lines were exposed to SWCNT FITC, the nanotubes were shown to accumulate within the cells. Similarly, SWCNTs that were covalently functionalized with biotin and reacted with streptavidin were internalized within human promyelocytic leukemia (H60) cells, human T cells, Chinese hamster ovary, and 3T3 fibroblast cell lines. Although the mechanism of the CNT cell entry remains undefined, these experiments suggest the viability of CNTs as carriers for delivering relatively large molecules to mammalian cells. (For details, see Chapters 19 and 20.)

Antimicrobial Activity

An antimicrobial agent is a substance that kills or slows the growth of microbes such as bacteria (antibacterial activity), fungi (antifungal activity), viruses (antiviral activity), or parasites (antiparasitic activity).

Certain nanoparticles are disruptive to bacteria and viruses simply by virtue of their physical nature. Silver, carbon, copper, platinum, and zinc, the old antibacterial agents, have had their effectiveness enhanced by being made nanoparticulate, an approach already available in sprays that can be inhaled to protect against various pathogens. The versatile properties of CNMs are used for killing microorganisms that are harmful to the environment. CNMs functionalized with different organic and inorganic groups are used to prevent the growth of microorganisms.

Adsorption of photons with energy equal to or greater than the band gap energy (e.g., of a semiconductor) results in transient formation of electron–hole pairs. In the presence of appropriate electrons and holes, when scavengers are adsorbed at the surface of a semiconductor carbon nanoparticle, the valence band holes function as powerful oxidants, and conduction band electrons function as moderately powerful reductants. Thus, an illuminated semiconductor in aqueous medium produces oxidizing free radicals in addition to photo-generated electron hole pairs at the interface of the semiconductor–electrolyte solution. It has been reported that He-La cells and bacteria can be killed by these oxidants or reductants produced at the illuminated surface of a powdered semiconductor CNMs. Various possible reactions occurring at the interface of CNMs (floated) with water can be enumerated as in Table 22-1.

When O_2 is present in aerated solutions, it could in principle also accept a conduction band electron to form the superoxide radical (O_2^-). It is accepted that oxygen plays an important role in semiconductor-mediated reactions by trapping the conduction band electron as superoxide ion (O_2^-) and thus delaying the electron hole recombination process. Superoxide is probably formed transiently in small

Reaction Number	Reaction
1	$CNMs + h\upsilon \rightarrow e^-_{photo} + h^+_{photo}$
2	$O_2 + e^-_{photo} \rightarrow O_2^-$
3	$2O_2^- + 2H^+ \rightarrow H_2O_2 + O_2$
4	$O_2^- + H_2O_2 \rightarrow O_2 + OH^- + OH$
5	$H_2O_2 + e^-_{photo} \rightarrow 2OH^-$
6	$OH^- + h^+_{photo} \rightarrow OH$
7	$H_2O + h^+_{photo} \rightarrow OH + H^+$
8	$Bacteria_{cellwall} + h^+_{photo} \rightarrow CO_2$
9	$Bacteria_{cellwall} + OH \rightarrow CO_2$

TABLE 22-1 Various Possible Reactions Occurring at the Interface of Carbon Nanomaterials Floated with Water

amounts during normal respiratory processes, and it is also produced by light through mediation of a sensitizer (CNMs) and by one electron transfer to oxygen. Reactions 4 to 7 can produce a highly oxidizing free radical OH via reaction 3, which is supported by the works of Kellogg and Fridowich (1975) and Kormann et al. (1988). Formation of hydroxy radical is further supported by use of the electron spin resonance spin trapping technique in the semiconductor electrolyte interface. It is also suggested that these radicals form not only via valence bond holes but also via H_2O_2 from O_2^-. Hydroxy radicals posses strong oxidation potential and are capable of attacking any of the organic substances present in the cell. These free radicals (OH) or superoxide radicals (O_2^-) oxidize proteins, lipids, and polysaccharides into CO_2. The exact mechanism of the process that kills microorganism is unknown. These reactions probably proceed through interaction of free radicals with the microbial cell wall because they can be inhibited by free radical scavengers.

The inhibition of the bactericidal activity by superoxide dimutase may indicate that O_2 and H_2O_2 react with each other to form hydroxy radicals and single oxygen as the lethal compounds. Free radicals can react with an inactive enzymes and damage intracellular components and sensitive macromolecules in the cell. Therefore, they can also be assumed that exopolypeptide or polysaccharides may undergo photo degradation reactions by either reaction 8 or 9 because the slime layers of bacteria are known to get destroyed by the oxidation initiated by the superoxide radicals and hydroxy radicals.

Formation of CO_2 from bacterial cell degradation, reported by Saito et al. (1997) supports the possibility of reaction 8 because for a given intensity of illumination, the yield of reaction 8 would depend on the probability of bacteria coming in direct contact with illuminated semiconductor, and this probability would increase with increase in number of particles present in the solution. On the other hand, if reaction 9 was the rate-determining step, then killing of bacteria would be almost independent of the number of CNMs present in the solution (but this does not occur). Whatever the actual mechanism might be, it is clear from these observations that CNMs are able to kill bacteria with the photocatalytic process.

Discovery of Drug Molecules

Traditional trial-and-error methods have a very high lead time, so it takes several years for a new drug to reach the market. The use of CNT for discovery of drug molecules sounds like a very far-fetched approach at present. However, indirectly, input of CNTs may be valuable in drug discovery. The critical bottlenecks in drug discovery may be overcome by using arrays of CNT sensors and current information technology solutions (e.g., data mining and computer-aided drug design) for identification of genes and genetic materials for drug discovery and development (Jorgensen, 2004).

22.2.6 Implantable Materials and Devices

Implantable Nanosensors and Nanorobots

There are certain cases, such as diabetes, in which regular tests by patients themselves are required to measure and control the sugar level in the body. Pediatric and elderly patients may not be able to perform this test properly. Another similar example is regular tests of persons exposed to hazardous radiation or chemicals. The objective is to detect the disease in its early stage so appropriate action for higher chances of success can be taken. Implantable sensors and nanorobots can be useful in health assessment (Bhargava, 1999). CNT-based nano-sensors have the advantages that they are thousand times smaller than even micro-electromechanical system sensors and consume less power. Therefore, they are highly suitable as implantable sensors.

Implanted sensors can be used for monitoring pulse, tempera-ture, and blood glucose, and for diagnosing diseases (Shandas and Lanning, 2003; Shults et al., 1994). In addition, CNTs can be used for repairing damaged cells or killing them by targeting tumors by chem-ical reactions. Implantable nanosensors can also monitor the heart's activity level and regulate heartbeats by working with an implantable defibrillator.

CNTs can also be used for investigation of retinal diseases caused by loss of photoreceptors. One way of compensating for the loss of photoreceptors is by bypassing the destroyed photoreceptors and artificially stimulating the intact cells in the area. Another possible area related to the application of CNTs that can be investigated is cochlear implants related to hearing problems. According to Bhar-gava (1999), implanted nanorobots can have following possible applications:

- To cure skin diseases. A cream containing nanorobots could remove the right amount of dead skin, remove excess oils, add missing oils, and apply the right amount of moisturizing compounds.

- To protect the immune system by identifying unwanted bac-teria and viruses and puncturing them to end their effective-ness

- To ensure that the right cells and supporting structures are at the right places

- As a mouthwash to destroy pathogenic bacteria and lift food, plaque, and tartar from the teeth to rinse them away

Actuators

Actuators are devices that put something (e.g., a robot arm) into action. They do so by converting electrical energy to mechanical energy. The direct conversion of electrical energy to mechanical energy through a material response is crucial for many biomedical applications, including microsurgical devices, artificial limbs, artificial ocular muscles, and pulsating hearts in addition to robotics, optical fiber devices, and optical displays. The main technical requirements of these actuators are low weight, low maintenance voltage, large displacements, high forces, fast response, and long life cycle (Gao, 2000). Different materials that have been previously investigated for use in electrochemical actuators are ceramics (piezoelectrics), shape memory alloys, and polymers. However, these materials have certain limitations. For example, piezoelectrics have low stiffness and electromechanical coupling coefficients.

Baughman et al. (1999), Gao (2000), and Vohrer et al. (2004) have suggested that CNTs can act as actuators. CNT actuators can work under physiologic environment, low voltages, and temperatures as high as 350°C. Nanotube-based polymer composites have promise as possible artificial muscle devices because of their incredible strength, stiffness, and flexibility in addition to their relatively low operating voltage. The mats of nanotubes expand and contract when operated as assembled electrodes in an electrochemical cell. When a potential is applied, charging of the electrodes takes place, and there is a linear change in the CNT length because of introduction of an electronic charge on the tube and a restructuring of the double layer of charge in the double layer outside the tube (Inganas and Lundstrum, 1999).

The biocompatibility, crystallinity, and morphology of the composites have been evaluated using scanning electron micrography, transmission electron micrography, hot-stage microscopy, and polarized light. Thermal analysis has also been performed. Methods of characterization have included thermal analysis using thermal gravimetric analysis and differential scanning calorimetry. The results of all these analyses were promising.

Baughman et al. (1999) were the first to evidence the actuator property of CNTs. They used actuators based on sheets of SWCNTs. The CNT electromechanical actuators (also known as *artificial muscles*) generated higher stresses than natural muscles and higher strains than high-modulus ferroelectrics. MWCNTs are excellent candidates for electromechanical devices because of their large surface area as well as their high electrical conductivity.

Gao (2000) was the first to show an electromechanical actuator based on MWCNTs. The actuator sheets were prepared by aligned arrays of parallel, nonbundled MWCNTs perpendicular to the sheet planes. The MWCNTs had a length of 5 to 40 μm and a diameter of 10 to 60 nm. The actuators used the electrostatic repulsion between electrical double layers associated with parallel MWCNTs.

Vohrer et al. (2004) developed an experimental setup for the measurement of the actuation forces and the displacement of CNT sheets. In their setup, vertical elongation or forces of bucky papers could be observed, which was a prerequisite for the optimization of artificial muscles for industrial applications. The fastest actuation time observed by them was approximately 3 seconds. The parameters that affected the electromechanical properties of CNT were the raw material (SWCNTs and MWCNTs; MWCNTs have a contorted structure compared with SWCNTs), different production techniques (e.g., arc-discharge material shows very low actuation), the purification grade (not only the amount of carbon particles but also the amount of remaining catalysts decrease the actuation time and actuation amplitude), the chirality and diameter of CNTs, the homogeneity of the nanotubes' distribution, the alignment of the nanotubes, the size and thickness of the produced bucky paper (it was observed that thinner bucky papers react faster than thicker ones at comparable thickness values), the type of electrolyte (concentration and viscosity), the electrode material, the surface area of the electrode, the arrangement of electrodes, the surface electrode resistance, the applied voltage, and the polarity (Vohrer et al., 2004).

Nanofluidic Systems

If the implantable fluid injection systems are large, the functions of surrounding tissues are adversely affected. However, tiny nano-dispensing systems can dispense drugs on demand using nanofluidic systems, miniaturized pumps, and reservoirs. Currently, limited attention has been given to understand the fluid mechanics at the nanoscale because fluid mechanics at the nanoscale is in infant stage. The research so far reveals that MWCNTs show great potential for use in nanofluidic devices. This can be attributed to their extremely high mechanical strength coupled with their ability to provide a conduit for fluid transport at near–molecular length scales. Furthermore, there is a lack of defects on their inner surface (Megaridis et al., 2002). The nano-dispensing systems using CNTs can be applied for chemotherapy in which precise amounts of drugs are targeted directly at the tumors when the patient falls asleep. Other potential areas where fluid dispensing systems could be applied are lupus, AIDS, and diabetes.

Surgical Aids

Surgery using macro instruments can be cumbersome for both the surgeon and the patient. On one hand, the patient experiences severe pain, scarring, and a long healing time because of large cuts; on the other hand, the surgeon requires high concentration for a long period to perform the surgery accurately. Sometimes it may lead to surgical error because of the surgeon's fatigue. In many cases, surgical error may result because of the limited view of the organs by the surgeon. In addition, macrosurgical instruments are not suitable for certain delicate cases such as surgeries related to the heart, brain, eyes, and ears. One of the solutions is laparoscopic surgery, which uses a small entry port, long and narrow surgical instruments, and a rod-shaped telescope attached to a camera. However, laparoscopic surgery requires highly skilled surgeons for efficient surgery. Research needs to be carried out to investigate if smart instruments (e.g., forceps, scalpels, and grippers with embedded sensors to provide improved functionality and real-time information) using CNTs can be developed that can aid surgeons by providing specific properties of tissue to be cut and provide information about the performance of their instruments during surgery. The usefulness of CNTs for optically guiding surgery should also be investigated. This can lead to easy removal of tumors and other diseased sites. Another option is the use of molecular nanotechnology (MNT) or nanorobotics in surgery. In nanorobotics, surgeons move joystick handles to manipulate robot arms containing miniature surgical instruments at the ports. Another robot arm contains a miniature camera for a broad view of the surgical site. This results in less stress for surgeons and less pain for patients; at the same time, high precision and safety are achieved. MNT allows in vivo surgery on individual human cells. Nanorobotics-based surgery can be used for gallbladder, cardiac, prostate, cardiac bypass, colorectal, esophageal, and gynecologic surgery. However, nanorobotic systems for performing surgery require the ability to build precise structures, actuators, and motors that operate at a molecular level to enable manipulation and locomotion.

Nanotweezers, which that can be used for manipulation and modification of biologic systems such as structures within a cell, have already been created using CNTs, and they have the potential to be used in medical nanorobotics. CNT nanotweezers can be used for manipulation and modification of biologic systems such as DNA and structures within a cell. They can also be used as nanoprobes for assembling structures. It will be helpful in increasing the value of measurement systems for characterization and manipulation at a nanometer scale. Application of CNTs as nanoprobes for crossing tumors, but not crossing into healthy brain tissue should also be investigated because the presence of cancer in a brain tumor may result in weakening of the blood–brain barrier.

In addition, Cumings and Zettle (2000) have demonstrated that nested CNTs can make exceptionally low-friction nanobearings. These nanobearings can be used in many surgical tools. Therefore, there is a need for research to be extended to investigate the application of CNTs in other surgical tools.

22.2.7 Therapeutic Uses

In Vitro Cell Tracking

Using perfluorocarbon (PFC) nanoparticles 200 nm in size to label endothelial progenitor cells taken from human umbilical cord blood enables their detection by magnetic resonance imaging (MRI) in vivo after administration (Partlow et al., 2007). The MRI scanner can be tuned to the specific frequency of the fluorine compound in the nanoparticles, and only the nanoparticle-containing cells are visible in the scan. This eliminates any background signal, which often interferes with medical imaging. Moreover, the lack of interference means one can measure very low amounts of the labeled cells and closely estimate their number by the brightness of the image. Because several PFC compounds are available, different types of cells could potentially be labeled with different compounds, injected, and then detected separately by tuning the MRI scanner to each one's individual frequency. This technology offers significant advantages over other cell-labeling technologies in development. Laboratory tests have showed that the cells retain their usual surface markers and are still functional after the labeling process. The labeled cells have been shown to migrate to and incorporate into blood vessels, forming around tumors in mice. These nanoparticles could soon enable researchers and physicians to directly track cells used in medical treatments using unique signatures from the ingested nanoparticle beacons. They could prove useful for monitoring tumors and diagnosing and treating cardiovascular problems.

Cardiovascular Diseases

PFC nanoparticles provide an opportunity for combining molecular imaging and local drug delivery in cardiovascular disorders. Ligands such as monoclonal antibodies and peptides can be cross-linked to the outer surface of PFCs to enable active targeting to biomarkers expressed within the vasculature. PFC nanoparticles are naturally constrained by size to the circulation, which minimizes unintended binding to extravascular, nontarget tissues expressing similar epitopes. Moreover, their prolonged circulatory half-life of approximately five hours allows saturation of receptors without the addition of polyethylene glycol or lipid surfactant polymerization. The utility of targeted PFC nanoparticles has been demonstrated for a variety of applications in animal models and phantoms, including the diagnosis of ruptured plaque, the quantification and antiangiogenic treatment of atherosclerotic plaque, and the localization and delivery of antirestenotic therapy after angioplasty (Lanza et al., 2006).

Allergic Disorders

Carbon nanoparticles can aggravate antigen-related airway inflammation. The effect may be mediated, at least partly, through the increased local expression of IL-5 and eotaxin and by the modulated expression of IL-13, macrophage chemoattractant protein (MCP)-1, and IL-6 and regulated on activation and normal T cells expressed and secreted (RANTES). Carbon nanoparticles 14 nm in size enhance total IgE and antigen-specific production of IgG1 and IgE. These results suggest that carbon nanoparticles can be a risk for exacerbation of allergic asthma. The aggravating effect may be larger with smaller particles (Inoue et al., 2005).

Brain Repair

Saido (2003) has grown nerve cells from the hippocampus region of the brain on substrates containing networks of CNTs and found that the CNTs improved neural signal transfer between the cells. The idea of putting together CNTs and neurons came first of all because of their structural similarities. Neurite elongations are reminiscent of the cylindrical shape of CNTs. And because CNTs can be either conducting or semiconducting, in principle they could be used as assistive devices to functionally and structurally reconnect neurons that no longer communicate with each other.

To deposit MWCNTs onto a glass substrate, the researcher functionalizes the tubes with pyrrolidine groups, boosting their solubility in the organic solvent dimethylformamide. The team then places small drops of a solution of the CNTs onto glass cover slips. After the solvent has evaporated, the application of a heat treatment defunctionalizes the CNTs, leaving a coating of nonfunctionalized CNTs on the glass.

The researchers attach hippocampal neurons both to CNT-coated glass cover slips and to uncoated cover slips. Then the researchers monitor the growth of the neurons for 8 to 10 days. The amount of growth on both substrates appears similar.

The neurons developed on CNTs and directly on glass also have similar electrophysiologic characteristics (e.g., resting membrane potential, input resistance, and capacitance) and similar intrinsic excitability. But neurons grown on CNTs display a sixfold increase in the frequency of postsynaptic currents.

Saido (2003) for the first time demonstrated that a large improvement in neural signal efficacy due to the presence of the CNT substrate. This result will prompt the development of new tissue engineering strategies such as the development of materials suited to functionally reconnecting injured neurons or to directly improving neural signal transfer. An immediate impact of these findings could be in the design of chronic neural implants.

Monitoring Viruses

Scientists have taken another step toward using CNTs as fast-acting biosensors to detect the presence of viruses. An evidence of straightforward, ambient covalent immobilization of a viral ligand–receptor–protein system onto individual and bundled SWCNTs has been developed. The result allows researchers to use biofunctionalized SWCNTs so that specific viruses will bind to the structure. Furthermore, the covalent nature of the functionalization means that the receptor proteins are able to resist extended washing and remain immobilized on the surface of the CNTs. A series of FETs has been fabricated based on the functionalized structures and one of many devices to a complementary adenovirus known as the Ad 12 Knob virus. The target protein was absorbed by FET, modifying the current–voltage (I-V) characteristics of the device and providing a simple means of detecting the virus. When exposed to a nonspecific protein, the I-V characteristics of the FET remained unchanged. This methodology can be further developed to reveal the presence of serotype 12 and all other possible CAR (Coxsackie B virus–adenovirus receptor)–binding adenoviruses, as well as subgroup B Coxsackie viruses. In other words, the researchers have found a model system.

22.3 Challenges

CNMs appear to be the frontrunner that has the potential to dominate biosystems research. However, challenges remain that need to be addressed before the full potential of CNMs for biosystems applications can be realized. These challenges include:

- There is a lack of detailed understanding of the growth mechanism of CNMs. As a result, an efficient growth approach to structurally perfect nanotubes at large scales is currently not available.

- It is difficult to grow defect-free nanotubes continuously to macroscopic lengths (Gao, 2000).

- Control over nanotubes' growth on surfaces is required to obtain large-scale ordered nanowire structures (Inganas and Lundstrum, 1999).

- Controlling the chirality of SWCNTs by any existing growth method is very difficult.

- The above-mentioned limitations result in a high cost of production for pure and uncontaminated nanotubes with uniform characteristics.

In short, the optimization of production parameters and the control of the growth of nanotubes have yet to be mastered. In addition to the challenges at the fabrication level, the low-dimensional geometry results in structural instability, resulting in buckling, kink formation, and collapse. In addition, the toxicology of CNMs is not well understood. CNMs may be associated with health risks. CNMs are lightweight and therefore become airborne, so they may become agglomerated in the lungs, filling the air passages, which may lead to suffocation or lung cancer. This warrants an in-depth study about the toxicology of CNMs to come up with a final conclusion with respect to their acceptance by the human immune system. Finally, the time from proof of concept in the laboratory of CNM-based devices to the commercial marketplace should be reduced as the competition from other novel materials and technologies continue to emerge.

An important concern regarding the use of CNTs in biology and medicine is their toxicity. Now CNTs are mainly under investigation in laboratories, but if there is widespread commercialization, the exposure of the general populace to this material must not occur without adequate testing. Thus, toxicologic studies will be needed, and several studies have indicated that CNTs could have undesirable effects on human health. The exposure of cultured human skin cells to SWCNTs has been shown to cause oxidative stress and loss of cell viability, indicating that dermal exposure may lead to skin conditions. This is perhaps to be expected because graphite and carbon materials have been associated with increased dermatitis_and keratosis.

The toxicity of MWCNTs, CNFs, and carbon nanoparticles was tested in vitro on lung tumor cells and clearly showed that these materials are toxic, while the hazardous effect was size dependent (Magrez et al., 2006). Because these studies used very high concentrations of SWCNTs that were directly instilled into the lungs of the animals, further testing is required to establish their inhalation toxicity. Moreover, cytotoxicity is enhanced when the surface of the particles is functionalized after an acid treatment. Water-soluble SWCNTs have been functionalized with the chelating molecule diethylentri-aminepentaacetic acid and labeled with indium (111-In) for imaging purposes (Singh et al., 2006). Intravenous administration of these functionalized SWCNTs (f-SWCNTs) followed by radioactivity tracing using gamma scintigraphy indicated that f-SWCNTs are not retained in any of the reticuloendothelial system organs (liver or spleen) and are rapidly cleared from the systemic blood circulation through the renal excretion route. The observed rapid blood clearance and half-life (3 hours) of f-SWCNTs have major implications for all potential clinical uses.

Moreover, urine excretion studies using both f-SWCNT and functionalized multiwalled CNTs followed by electron microscopy analysis of urine samples revealed that both types of nanotubes were excreted as intact nanotubes. The next steps for this research are to prolong the blood circulation of CNTs to give them enough time before excretion to get to a target tissue. The researchers will also consider pharmaceutical development of functionalized CNTs for drug delivery. In one study, carbon nanoparticles, including CNTs (but not C60 fullerenes) stimulated platelet aggregation and accelerated the rate of vascular thrombosis in rat carotid arteries (Radomski et al., 2005). All particles resulted in upregulation of glycoprotein IIb/IIIa in platelets. In contrast, particles differentially affected the release of platelet granules, as well as the activity of thromboxane–, adenosine diphosphate–, matrix metalloproteinase–, and protein kinase C– dependent pathways of aggregation. Furthermore, particle-induced aggregation was inhibited by prostacyclin and S-nitrosoglutathione, but not by aspirin.

Thus, some carbon nanoparticles have the ability to activate platelets and enhance vascular thrombosis. These observations are of importance for the pharmacologic use of carbon nanoparticles and support the safety of C60 fullerenes. Manufactured SWCNTs usually contain significant amounts of iron as impurities that may act as a catalyst for oxidative stress. Because macrophages are the primary responders to different particles that initiate and propagate inflammatory reactions and oxidative stress, the interaction of SWCNT (0.23 wt% of iron) with macrophages has been studied (Kagan et al., 2006). Nonpurified SWCNTs more effectively converted superoxide radicals generated by xanthine oxidase or xanthine into hydroxyl radicals compared with purified SWCNTs. Iron-rich SWCNTs caused significant loss of intracellular low-molecular-weight thiols (glutathione [GSH]) and accumulation of lipid hydroperoxides in macrophages. Catalase was able to partially protect macrophages against SWCNT-induced elevation of biomarkers of oxidative stress (enhancement of lipid peroxidation and GSH depletion). Thus, the presence of iron in SWCNTs may be important in determining redox-dependent responses of macrophages.

Estimates of the airborne contaminations and glove deposits of SWCNTs during production have indicated a very low workplace exposure (perhaps one million times lower than the concentrations used in the toxicity studies). To future toxicologists, studies of CNMs will be challenging and should be conducted to determine safe exposure limits. Because similar measurements were performed on carbon soot– and carbon black–based toners, it seems likely that the appropriate measurements can be performed in a timely manner.

22.4 Summary

The uses and possible future uses of CNM in various domains of biosystems have been discussed in this chapter. CNM has found its way in diagnostic devices such as X-rays, sensors, and probes (AFM). Uses of CNM in other health care areas, including water purification, surgical aids, and antimicrobial activity, have been envisaged. Pharmaceutical and therapeutic uses of CNM are also being sought. There are many futuristic visions of using CNM in the field of implantable nanosensors and nanorobots, actuators, and nanofluidic systems. Scientists all over the world are engaged in achieving such targets.

CHAPTER 23

Contribution of Carbon Nanomaterial in Solving Malignancy

Sejal Shah

Nanotechnology Research Centre, Birla College, Kalyan, Maharashtra, India

Madhuri Sharon

MONAD Nanotech Pvt. Ltd., Powai, Mumbai, India

Carbon nano tubes are versatile structures that over the next decade are going to make a huge impact in medicine.

—*Kirk Ziegler (2005)*

23.1   Introduction

Nanotechnology is now on its way to popularization into the stream of medicine with an assurance that every disorder will be solved with the blink of an eye. This idea was in the flight of imagination a decade ago, but the incredible persistence of scientists to gain knowledge and the desire to explore the technology have led to far better progress than expected.

Cancer treatment is a subtle issue when it comes to diagnosis, and the treatment of HIV is still an incomplete success. After so much study, we have not yet succeeded in curing patients suffering from dreadful diseases such as AIDS. But a ray of hope has been shown by the drug-delivery application of nanotechnology, in which nanomaterials such as quantum dots, a one-dimensional nanoparticle, and carbon nanomaterials (CNMs) structured as tubes, fibers, beads, and so on are used. In this chapter, the role of CNMs and their significance in cancer treatment and prevention are discussed.

23.1.1 Carbon Nanomaterials: A New Substitute for Metallic Macrosyringes to Treat Cancerous Cells

A needle stuck into your arm is no fun, but that thin, hollow tube of metal is an essential tool of medicine, delivering drugs and vaccines. Now imagine a needle 1 million times smaller, one capable of piercing not through your skin, but through the wall of a cancer cell. Although seemingly impossible, such nanoscale needles exist; they are called carbon nanotubes (CNTs), and they are well on their way to becoming as an essential tool for twenty-first century oncologists as metal needles are for today's physicians.

The ultimate utility of CNTs should extend beyond their use as nanoscale needles that can deliver drugs directly into tumor cells (after being taken orally or injected through needle). Researchers are already using this novel form of carbon to create miniature X-ray machines that can snake through the bloodstream and deliver cancer-killing radiation directly to tumors. CNT heaters that bake tumors to death and CNT-based sensors capable of spotting even the smallest tumors are also in the offing.

Nanotechnology is a multidisciplinary field that covers a vast and diverse array of devices derived from the knowledge of engineering, biology, physics, and chemistry. These devices include nanovectors for the targeted delivery of anticancer drugs and imaging contrast agents. Nanowires and nanocantilever arrays are among the leading approaches under development for the early detection of precancerous and malignant lesions from biologic fluids. These nanodevices may provide essential breakthroughs in the fight against cancer.

In 1991 Iijima discovered the existence of CNTs, which were found to be related to buckyballs, the soccer ball–shaped form of carbon discovered six years earlier by the Nobel Laureate Sir Harold Kroto, the late Richard Smalley, and Robert Curl, Jr.

CNT come in two basic forms—single-walled CNTs (SWCNTs) and multiwalled CNTs (MWCNTs). SWCNTs, which have the greatest potential for use in cancer research and clinical oncology, can be thought of as sheet of graphites rolled into a cylinder of about 1.2 to 2.0 nm in diameter. As such, they resemble a piece of chicken wire rolled into a tube, with carbon atoms arranged as a series of connected hexagons. MWCNTs are similar except that they are made up of multiple concentric tubes that look like tree rings when viewed on end.

SWCNTs are simple in structure, consisting of repeating hexagonal arrangements of carbon atoms attached to each other. The interior of a CNT is hollow, but it can be loaded with a wide variety of molecules.

In the materials science world, CNTs are famous for their unique physical properties. SWCNTs can behave like metals, and in fact, few metals conduct electricity or heat as readily as do CNTs. However, with a slight twist in their structure, CNTs become semiconductors. CNTs are the strongest fibers ever created. Their physical properties have shown surprising performance, but their chemical modification can bring about more changes in the properties, therein giving insight into the physical properties that were formerly beyond reach.

23.1.2   Confronting Different Tricks in Targeting Cancerous Cells

Although outstanding advances have been made in fundamental cancer biology, these advances have not translated into comparable advances in the clinic. Inadequacies in the ability to administer therapeutic agents with high selectivity and with minimum collateral damage have largely accounted for the discrepancies in cancer therapeutics. Furthermore, it is most striking to recognize that only one to 10 parts per 100,000 of intravenously injected monoclonal antibodies reach their targets in vivo. Therefore, new therapeutic technologies that can potentially reach their targets with maximum selectivity and minimum collateral damage need to be explored. Nanoparticles, nanowires, and SWCNTs have generated interest among oncologists because of their small size and unique physical properties, and because of the potential to develop minimally invasive therapeutics without affecting patients' quality of life.

Two general synergistic goals should be striven for to increase the efficacy per dose of any therapeutic or imaging contrast formulation:

- To increase its targeting selectivity
- To endow the agent(s) comprising the therapeutic formulation with the means to overcome the biologic barriers that prevent it from reaching its target

An ideal therapeutic system would be selectively directed against cell clusters that are in the early stages of the transformation toward the malignant phenotype. The realization of such a system faces formidable challenges, including:

- Identification of suitable early markers of neoplastic disease
- Understanding their evolution over time
- Deployment of markers in screening and early detection protocols
- Development of technology for the biomarker-targeted delivery of multiple therapeutic agents
- Simultaneous capability of avoiding biologic and biophysical barriers

If nanotechnology is properly integrated with established cancer research, there will be extraordinary opportunities to meet these challenges.

Cancer development is a very complex process. A wide range of factors may contribute to the development of cancer. Factors that lead to the development of cancer may range from oxygen free radicals to micronutrient deficiencies, mutations of signaling network, increased cell division, and overweight. Today, cancer-related nanotechnology research is progressing on two main fronts; laboratory based diagnostics and in vivo diagnostics and therapeutics.

Nanoscale devices, such as nanowire grids, that are designed in the laboratory are fabricated with sophisticated lithographic techniques. After fabrication, the silicon nanowires are coated with monoclonal antibodies directed against various tumor markers. With minimal sample preparation, substrate binding to even a small number of antibodies produces a 100-fold increase in detection sensitivity over current solution phase diagnostic techniques.

23.1.3   Nanoscale Activities in Living Systems

Nanoscale devices present in living cells as organelles are 100 to 10,000 times smaller than cells. Based on the concept of nanosized organelles, nanosized devices are being fabricated. They are similar in size to large biologic molecules ("biomolecules") such as enzymes and receptors. For example, hemoglobin, the molecule that carries oxygen in red blood cells, is approximately 5 nm in diameter. Nanoscale devices smaller than 50 nm can easily enter most cells, and those smaller than 20 nm can move out of blood vessels as they circulate through the body. Because of their small size, nanoscale devices can readily interact with biomolecules on the surface as well as inside cells. By gaining access to so many areas of the body, they have the potential to detect disease and deliver treatment in ways unimagined until now. Because biologic processes, including events that lead to cancer, occur at the nanoscale inside cells, nanotechnology offers a wealth of tools that are providing cancer researchers with new and innovative ways to diagnose and treat cancer. The research tools provided by nanotechnology include:

- Prevention of precancerous cells by imaging agents and diagnosis from becoming malignant at early stages
- Tools to kill cancerous cells and identify biologic targets
- Multifunctional targeted devices to deliver drugs directly to cancer cells
- CNMs as therapeutic agents
- Systems to provide real-time assessments of therapeutic and surgical efficacy
- Novel methods to manage symptoms that reduce the quality of life

Knowledge of the possible advantages of each of these research tools will make it possible to recognize the trends and opportunities that can be beneficial in curing and fighting against cancer. Thus, the first problem that needs to be overcome is the cytotoxic effects of CNMs on cells.

23.2   Carbon Nanomaterial Causing Cytotoxic Effects on Malevolent Cells

CNTs are new members of carbon allotropes similar to fullerenes and graphite. Because of their unique electrical, mechanical, and thermal properties, CNTs are important for novel applications in many disciplines of science, including biologic systems, electronics, aerospace, and the computer industry. The cytotoxicity of CNMs has been a much-publicized topic. There have been various reports against and for the use of CNMs by different schools of scientists. It is imperative to have a close look at this aspect before studying the possibility and potentiality of use of CNMs for cancer treatment.

23.2.1   Effect of Carbon Nanomaterial on Different Cell Lines

Carbon Nanomaterial and Skin Cells

Victims of CNMs have experienced an increased incidence of skin diseases, including carbon fiber dermatitis and hyperkeratosis, when exposed to graphite and carbon materials. While assessing the cytotoxicity of as-grown unpurified CNT using human epidermal keratinocyte cells, Shvedova and Castranova (2003) have shown adverse effects of SWCNTs within 18 hours of exposure.

Oxidative stress and cellular toxicity forming free radicals, accumulation of peroxidative products, antioxidant depletion, and loss of cell viability were envisaged as the causes for the damaging effect of CNMs. Moreover, exposure to SWCNTs also resulted in morphologic changes in the ultrastructure of the cultured skin cells. Shvedova et al. (2004) concluded that the adverse effects of unrefined SWCNTs might have lead to dermal toxicity because of accelerated oxidative stress in the skin of exposed workers.

Molecular Characterization of Multiwalled Carbon Nanotubes

The increasing use of nanotechnology in consumer products and medical applications underlines the importance of understanding its potential toxic effects on people and the environment. Although both fullerenes and CNTs have been demonstrated to accumulate to cytotoxic levels within the organs of various animal and cell types, the molecular and cellular mechanisms for cytotoxicity of this class of CNM are not yet fully apparent.

To address this question, Ding et al. (2005) did molecular characterization of the cytotoxic mechanism of MWCNTs and nano-onions on human skin fibroblast. Whole-genome expression array analysis and high-content image analysis phenotypic measurements on human skin fibroblast cell populations exposed to multiwalled carbon nano-onions (MWCNOs) and MWCNTs have shown that CNM induces cell cycle arrest and increases apoptosis and necrosis when human fibroblast cells were exposed to MWCNOs and MWCNTs at cytotoxic doses. The gene expression profile of these cells indicated that the exposure to the CNM disturbs a number of cellular pathways.

The expression array analysis indicates that multiple cellular pathways are perturbed after exposure to these nanomaterials. Moreover, there are also distinct qualitative and quantitative differences in gene expression profiles with each material at different dose levels. For MWCNOs, the most effective concentrations recognized were 6 and 0.6 µg/mL; for MWCNTs, 0.6 and 0.06 µg/mL were believed to show effective results. MWCNO and MWCNT exposure activates genes involved in cellular transport, metabolism, cell cycle regulation, and stress response. Whereas MWCNTs induce genes indicative of a strong immune and inflammatory response within skin fibroblasts, MWCNO changes are concentrated in genes induced in response to external stimuli.

Promoter analysis of the micro-array results have demonstrated that interferon and p38/ERK-MAPK (extracellular signal-regulated kinase–mitogen-activated protein kinase) cascades are critical pathway components in the induced signal transduction contributing to the more adverse effects observed upon exposure to MWCNTs compared with MWCNOs (Fig 23-1).

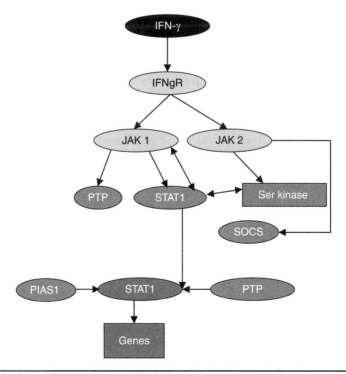

Figure 23-1   The interferon (IFN)-γ pathway.

23.3 Toxicity versus Nontoxicity

23.3.1 Toxicity Due to Structural Manipulations

The structural similarity between asbestos (a highly toxic fibrous form of a normally benign silicate mineral) and CNTs has raised concerns about the potential biotoxicity of widely used SWCNTs and MWCNTs. Therefore, systematic in vitro study with different CNMs (CNTs, carbon nanofibers [CNFs], carbon nanobeads [CNBs], and other various shaped carbon nanoparticles) has been performed using human tumor cell line H596.

The cytotoxicity was evaluated by MTT (di-Methyl Tri-azolediphenyl Tetrazoliumbromide) assay, in which formation of formazan crystals in living cells (but not in dead ones) is used to quantify the number of living cells. The toxic effects of CNM (0.02 µg/mL^{-1}) were noticed within 24 hours of exposure, and the number of viable cells decreased as a function of concentration. Microscopic examination of the cells showed retraction of the cytoplasm and decrease of the nuclei size, which are typical for irreversible cell injury and cell death (Fig. 22-2).

Because chemically functionalized CNTs are often used for DNA and protein binding and drug transport, the results of their effects on cell proliferation are particularly important. A range of CNTs functionalized by carboxyl, carbonyl, and hydroxyl have shown a significant reduction in the number of viable cells. Although the exact mechanism leading to cell death is still unknown, the results clearly show that more emphasis should be put on the cell toxicity study of these materials. Further work in that direction is particularly important

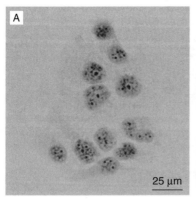
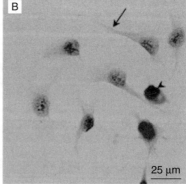

FIGURE 23-2 Effect of carbon nanotubes on cells. (*A*) A typical control image of H596 cells. (*B*) H596 cells after 24 hours of treatment. The cells have lost their mutual attachments and have retracted their cytoplasm (*arrows*), and the nuclei have become smaller.

in cases in which they are used in biomedical applications such as drug delivery and magnetic resonance imaging (MRI) screenings.

Movement of CNTs across the outer membranes of cells have been found to occur so easily that it leads to a concern that if a higher amount of CNTs is administered, they may not only become toxic but may also clog the pathway. In fact, some experiments have shown that under certain circumstances, CNTs can become toxic. The most well-known of these studies was reported by NASA Johnson Space Center, Wyle Laboratories in Houston, and The University of Texas Medical School at Houston, which showed that inhaled CNTs could induce tumor-like growths in the lungs of mice.

The cellular toxicity of CNMs (MWCNTs, CNFs, and other carbon nanoparticles) as a function of their aspect ratio and surface chemistry was tested in vitro on lung tumor cells. The results indicated that a toxic or hazardous effect is size dependent. Moreover, cytotoxicity is enhanced when the surface of the particles is functionalized after an acid treatment.

In similar studies, Magrez et al. (2006) have added that increasing concentrations of MWCNTs, CNFs, or other carbon nanoparticles to three different types of cultured human lung tumor cells caused changes in cell proliferation and overall cellular health. Evidence of toxicity was obvious within 24 hours after dosing each cell line with all the three materials. It was suggested that "dangling bonds" could be responsible for the toxicity of CNMs. Because carbon atoms not bonded to other carbon atoms are available to react with biomolecules, highly reactive dangling bonds are more prevalent in the carbon nanoparticles tested than in either CNFs or CNTs.

23.3.2   Nontoxicity Due to Functionalization

At low levels of injected CNTs used in biomedical applications, toxicity has not been an issue. However, the toxic effects of CNM were observed in unpurified CNMs, which possibly had leftover metal catalysts as well as amorphous carbon. Such toxicity is not encountered with highly purified CNTs. Weismann (2003) has shown that CNTs functionalized to make them dissolve easily in water show very little toxicity, even at very high concentrations. Cells that had taken up as many as 70,000 functionalized CNTs showed no signs of toxicity. Nonetheless, any imaging or therapeutic agents developed using CNTs would have to undergo the same rigorous toxicologic studies that all agents for use in humans must undergo.

Two types of functionalized CNTs (f-CNT) by 1,3-dipolar cycloaddition reaction and the oxidation/amidation treatment have been found to be taken up by B and T lymphocytes as well as macrophages in vitro without affecting cell viability. Subsequently, the functionality of the different cells was analyzed carefully. It was discovered that CNTs, which are highly water soluble, did not influence the functional activity of immunoregulatory cells, but CNTs with oxidation, which instead possess reduced solubility and form mainly stable water suspensions, preserve lymphocytes functionality while provoking secretion of proinflammatory cytokines by macrophages.

Chen et al. (1999a) were among the first to describe the mechanism of CNM toxicity at the molecular level. Although there have been other previous reports on CNM toxicity on various cells, there has been no detailed, comprehensive molecular biology study to give a complete picture genomewide. Their results proved that apoptosis (programmed cell death), cell cycle delay, cellular transport, and inflammation are linked to the treatments. Chen et al. (1999b) found that exposure to the CNTs and nano-onions activated genes involved in cellular transport, metabolism, cell cycle regulation, and stress response. MWCNTs induced genes related to a strong immune and inflammatory response, and the presence of nano-onions caused most of the changes in genes induced in response to external stimuli. CNTs appeared to be 10 times more toxic than nano-onions. An add-on advantage to this research is that it has a combination of high-content imaging analysis and high-throughput genomics. The whole genome profiling of the treated cells gives a very detailed picture of the effects.

23.4   Contribution of Carbon Nanomaterials to Solving Various Aspects of Malignancy

Various types of CNMs, such as buckyballs and CNTs, have shown promise as drug-delivery tools and imaging agents. However, the toxicity associated with some of these materials has raised questions about their ultimate utility in clinical oncology.

CNMs are arguably the most celebrated products of nanotechnology to date, encompassing fullerenes, CNTs, CNFs, and a wide variety of related forms. These nanomaterials can enter the human body through inhalation, skin contact, ingestion, or intentional injection and may affect microorganisms, plants, or animals if released into the environment in significant quantities. Table 23-1 gives a clear description about the routes of drug delivery and their specificities.

This combination of new materials, multiple exposure routes, and environmental and transport issues creates a complex set of research questions that scientists have just begun to tackle. Early studies of nanomaterial toxicity have produced apparently conflicting results and raised more questions than they have answered. Both biologic systems and real nanomaterial formulations are complex, so fundamental progress in this new field requires teaming of toxicologists and materials scientists. Several research groups are also developing CNTs as targeted delivery vehicles for anticancer drugs. CNTs can ferry proteins as well as deliver anticancer drugs into cells and act as miniature thermal scalpels that can bake cancer cells to death.

Route	Absorption pattern	Special utility	Limitations and precautions
Intravenous	Absorption is circumvented Potentially immediate effects	Valuable for emergency use Permits titration of dosage Suitable for large volumes and for irritating substances if diluted	Increased risk of adverse effects Must inject solution slowly as a rule Not suitable for oily and insoluble substances
Subcutaneous	Prompt from aqueous solutions Slow and sustained from repository preparations	Suitable for insoluble substances and implantation of solid pellets	Not suitable for large volumes Possible slough from irritating substances
Intramuscular	Prompt from aqueous solutions Slow and sustained from repository preparations	Suitable for moderate volumes, oily vehicles, and some irritating substances	Precluded during use of anticoagulant medication May interfere with interpretation of certain diagnostic tests (e.g., CPK level)
Oral ingestion	Variable; depends on many factors	Most convenient, safe, and economical	Requires patient cooperation Absorption is potentially erratic and incomplete for drugs that are poorly soluble and absorbed slowly

CPK = creatinine phosphokinase.

TABLE 23-1 Drug Delivery Routes and Their Specificities

23.4.1 Prevention of Precancerous Cells by Imaging Agents and Diagnostics from Becoming Malignant at Early Stages

This is a burning field where scientists are working on developing a method to detect cancer at very early stages so that its cure could be achievable. One landmark achievement has been the development of a highly sensitive immunodetection of cancer biomarkers by amplifying CNTs (Yu et al., 2006). Another landmark was using a protein-coated nanocantilever (Gupta et al., 2006).

In an attempt to increase the sensitivity of cancer biomarker detection and to decrease the need for large samples from which to detect those molecules, a "forest" of SWCNTs is used to detect lower levels of prostate-specific antigen (PSA), which is a protein produced by the cells of the prostate gland. When the prostate gland enlarges, PSA levels in the blood tend to increase. PSA levels may increase because of cancer or benign (not cancerous) conditions. Because PSA is produced by the body and can be used to detect disease, it is sometimes called a *biologic marker* or *tumor marker*. Moreover, this new system requires between 5 and 15 times less sample than does the other commercial detecting systems.

Yu et al. (2006) have used bundles of SWCNTs with chemically reactive groups on the ends of the CNTs to which enzymes or antibodies capable of reacting or binding to specific biomolecules are attached. An antibody that binds to PSA, which is used in detecting prostate and breast cancer, was used for this purpose. When antibody-labeled CNTs are added to a serum sample, then any PSA present in the serum sample will get bound to the antibody. Using electrical impulses, the number of PSA molecules bound to CNTs can be measured.

The advantage of this system is that it can detect PSA at levels as low as 4 pg/mL in a sample size of 10 μL of serum. The existing standard assay has a detection limit of 10 to 100 pg/mL and requires as much as 50 to 150 μl of sample. In addition, this system may provide quantitative detection of PSA from laser microdissected tissue samples of 1000 cancer cells, something that current technology cannot achieve.

Scientists have tried another route of detecting cancer by fabricating and using a cantilever. CNTs have many promising properties, including their cylindrical and hollow tube shape; the number of walls enclosing a CNT, making it an SWCNT or MWCNT; and the arrangement of carbon atoms with respect to tube axis (there are many kind of chiral tubes), making it a suitable material for preparing a cantilever. The mechanical (e.g., bending or bucking and high resilience to mechanical strains) and thermal (thermal expansion under temperature) properties of CNTs show that a cantilever can be fabricated from them. Moreover, MWCNTs can adsorb light energy selectively, depending on the light polarity.

Yang et al. (2006) have coated nanocantilevers with molecules capable of binding specific substrates (i.e., DNA complementary to a specific gene sequence) and have detected (using optical techniques) a single molecule of DNA or protein from the change in the surface stresses that bend the nanocantilever. Recently, DNA-labeled magnetic nanobeads have been used to detect DNA and proteins that may serve as diagnostic or prognostic indicators of cancer. Much interest in using nanoparticles for cancer detection arose with the use of semiconductor nanocrystals as a tool for laboratory diagnosis of cancer. On an unconventional front, researchers have used empty RNA virus capsids from cowpea mosaic virus and flock house virus as potential contrast agents for MRI by attaching homing molecules, such as monoclonal antibodies or cancer cell–specific receptor antagonists, and reporter molecules. Therapeutic agents loaded inside the capsid may also serve as drug-delivery vehicles.

Ushering in entirely new approaches to molecular recognition, Fritz et al. (2000) pioneered the concept that bimolecular binding events yield forces and deformations that might be detected and recognized by appropriately selective sensing nanostructures. Primary examples of such devices are micro- or nanocantilevers, which deflect and change resonant frequencies as a result of affinity binding and as a result of nucleic acid hybridization events occurring on their free surfaces.

Majumdar et al. (2001) have also used microcantilevers to detect single-nucleotide polymorphisms (SNI's) in a 10-mer DNA target oligonucleotide without the use of extrinsic fluorescent or radioactive labeling and simultaneously demonstrated the applicability of microcantilevers for the quantitation of PSA at clinically significant concentrations. The specificities and sensitivities of these assays do not yet offer substantial advantages over conventional detection methods, but the use of nanoparticle probes might allow for individual single-pair mismatch discrimination.

The breakthrough potential afforded by nanocantilevers resides in their extraordinary multiplexing capability. It is realistic to envision arrays of thousands of cantilevers constructed on individual centimeter-sized chips, allowing the simultaneous reading of proteomic profiles or, ultimately, the entire proteome.

23.4.2 Tools to Kill Cancerous Cells and Identify Biologic Targets

A very serious attempt is being made to develop CNT-based therapeutics and imaging agents. At the same time, cancer-related applications are also being explored that will not involve getting CNTs and their cargoes into cells. Because of their excellent electrical conductivity, CNTs are poised to become the key component of ultrafast, miniaturized diagnostic gear that may soon be able to detect the earliest signs of cancer from a pinprick of blood right in a doctor's office.

Wang et al. (2004) at Arizona State University are using electrodes modified with CNTs to create highly sensitive devices capable of detecting specific sequences of DNA, including those associated with the breast cancer gene *BRCA1*. Also, Sirdeshmukh et al. (2004) have attached antibodies to the surfaces of CNTs that can bind molecules shed by tumors into the bloodstream. In both cases, the idea is to incorporate the tumor-sensing CNTs into a small electrical circuit, which would signal the presence of the tumor marker with a change in electrical conductivity.

Use of nanoshells, and more recently, SWCNTs to kill cancer cells by using infrared radiation and imaging have proven to be successful approaches for cancer cell killing in vivo and have increased nano-technologists' interest in killing cancer cells. In both nanoshell and SWCNTs, lasers with very high power (35 W/cm² for nanoshells and 1 to 4 W/cm² for SWCNTs) are used to heat the nanoshells and SWCNTs at a thermal heat of 55° to 70°C. However, the level of temperature increases in SWCNTs to an extraordinary level to create nanobombs that explode at low laser intensities. By exploding SWCNTs that are co-localized with cancer cells, one can selectively destroy the SWCNTs as well as the cancer cells. This approach may be the only way of using SWCNTs (Fig. 23-3).

Nanobombs are being developed to be used in cancer treatment. Nanobombs are made by bundling CNTs; when they are exposed to light, heat is generated. With a single CNT, the heat generated by the light is dissipated by the surrounding air. But in bundles of CNTs, the heat cannot dissipate as quickly, and the result is an explosion on the nanoscale.

Nanobombs are created from thermal energy confinement in SWCNT bundles and subsequent vaporization of liquids between SWCNTs in bundles, creating pressures that cause an eventual explo-sion. This phenomenon works in a host of different liquids such as alcohol, de-ionized water, and phosphate-buffered saline solutions. By hydrating SWCNTs in sheets as well as by co-localizing them to cells, potent nanobombs can be created to completely explode cancer cells. This process is highly localized, with minimal collateral damage to neighboring cells. Nanobombs are created from the optical thermal transitions in SWCNTs. Hydrating SWCNTs and exposing them to light cause the thermal energy to heat the water molecules, which in turn creates pressure inside the SWCNT bundles, causing them to explode. When co-localized with cells, nanobombs completely destroy the cells, causing 100% cell destruction. This method is sim-ple, robust, and highly effective in killing cancer cells. Furthermore, this technique, which uses low laser power compared with compet-ing techniques, can produce larger explosions that may ensure com-plete destruction of cancer cells (Figs. 23-4 to 23-6).

After spending some time with this novel material, Wilson et al. (2002) found that ultra-short CNTs (20 to 100 nm long, which is 1000 times shorter than standard nanotubes) are best suited for creating nanobombs. In this size range, the CNTs are readily taken up by cells, and as importantly, they are also cleared rather rapidly from the body, which is essential for imaging agents to minimize the amount

A

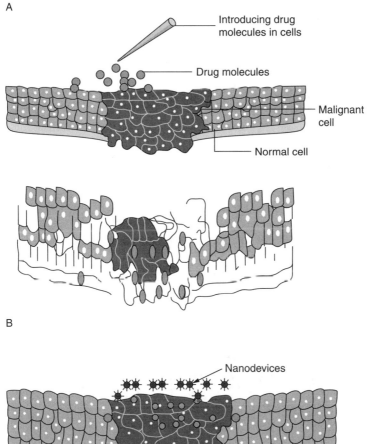

Introducing drug
molecules in cells

Drug molecules

Malignant
cell

Normal cell

B

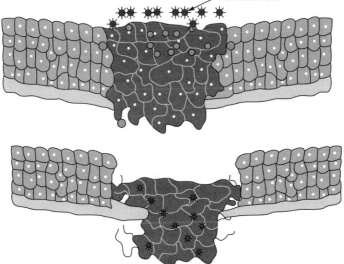

Nanodevices

FIGURE 23-3   Comparison of the traditional cancer treatment method (*A*)
versus nanotherapy (*B*). In the traditional method of treating malignant cells,
tumors as well as normal cells are destroyed.

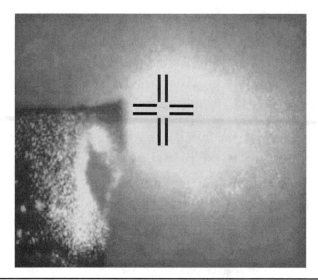

FIGURE 23-4   Nanobomb explosions when SWCNT is irradiated with near-infrared light.

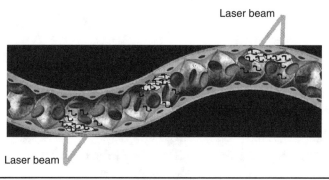

FIGURE 23-5   Targeted killing of cancer cells in a blood vessel using carbon nanotubes.

of nonspecific background signal that is received from any agent that is not binding to a tumor and is just circulating in the bloodstream.

Wilson et al. (2002) have also developed methods for loading a variety of metal ions into these nanotube capsules. Nanocapsules loaded with gadolinium ions show promise as MRI contrast agents, and those with radioactive iodine or astatine (an element in the same family as iodine) are being developed as cancer-killing therapeutics.

23.4.3   Multifunctional Targeted Devices to Deliver Drugs Directly to Cancer Cells

Dai and Flesher (2005) have reported a number of advances using nanotubes as a drug-delivery device. In one experiment, for example,

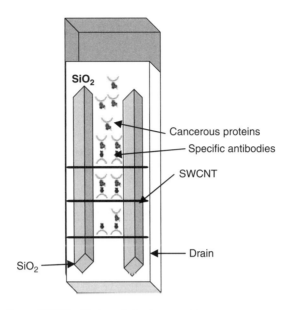

SiO₂

Cancerous proteins

Specific antibodies

SWCNT

Drain

SiO₂

Notes: SWCNT = Single-walled carbon nanotube.

FIGURE 23-6 Antibodies attached to carbon nanotubes can diagnose cancer by detecting specific proteins found only on the surfaces of cancer cells.

the investigators showed that CNTs could ferry a protein known as cytochrome c across the cell membrane. More importantly, the transported protein, which can trigger cell death, retained all of its biologic activity after it was inside the cell. One formulation under development adds a monoclonal antibody that recognizes melanoma cells to the outside of the nanotube. Another nanotube formulation will target leukemia cells.

23.4.4   Targeted Drug Delivery

Nanotubes have been targeted to cancer cells by attaching folic acid to their surface. Folic acid binds to a folic acid receptor protein found in abundance on the surfaces of many types of cancer cells. Folate is just an experimental model that was used, and folic acid is a stable form of folate that is biologically accessible (Dai and Felsher 2005). Apparently, folate plays a role in the cycle that results in the remethylation of homocysteine, as shown in Fig. 23-7.

In reality, there are more interesting ways to target the cancerous cells such as attaching an antibody to a CNT to target a particular kind of cancer cell. One example is lymphoma, or cancer of the lymphatic system. Similar to many cancers, lymphoma cells have well-defined surface receptors that recognize unique antibodies. When attached to a CNT, the antibody would play the role of a Trojan horse

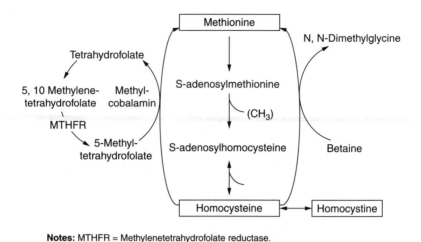

Notes: MTHFR = Methylenetetrahydrofolate reductase.

FIGURE 23-7   The folate pathway.

(Dai and Felsher, 2005). It would be interesting to determine if shining near-infrared (NIR) light on the animal's skin would destroy lymphatic tumors while leaving normal cells intact.

23.4.5   Carbon Nanomaterial as a Therapeutic Agent for Cancer Treatment

An interesting property of CNTs is that they absorb NIR light waves, which are slightly longer than visible rays of light and pass harmlessly through cells. When abeam of NIR light falls on a CNT, electrons in the CNT become excited and begin releasing excess energy in the form of heat. A solution of CNTs under a NIR laser beam would heat up to about 70°C (158°F) in 2 minutes. Dai and Flesher (2005) harnessed this property of CNTs to use them as therapeutic agents. When CNTs were placed inside cells and irradiated by the laser beam, the cells were quickly destroyed by the heat. However, cells without CNTs showed no such effects when placed under NIR light.

Carbon nanobombs hold great promise as a therapeutic agent for killing cancer cells, with particular emphasis on breast cancer cells, because the shockwaves from nanobombs kill cancerous cells as well as the small blood vessels that nourish the diseased cells. After the nanobombs are exploded and kill cancer cells, immune system cells known as macrophages can effectively clear the cell debris and the exploded nanotubes along with them.

23.4.6   Specific Targeted Drug Delivery and Nanovectors

SWCNTs with different chiral vectors have dissimilar properties such as optical activity, mechanical strength, and electrical conductivity.

Intravascular injectable nanovectors are a major class of nano-technologic devices of interest for use in cancer. Their envisioned use is for in vivo, noninvasive visualization of molecular markers of early stages of disease A targeted delivery of therapeutic agents would be associated with a concurrent substantial reduction in deleterious side effects and by a combination of the interception and containment of lesions before they reach the lethal or even the malignant phenotype, with minimal or no concurrent loss of quality of life.

Liposomes are the archetypal, simplest form of a nanovector. They use the overexpression of fenestrations in cancer neovasculature to increase the drug concentration at tumor sites. Liposome-encapsulated formulations of doxorubicin were approved 10 years ago for the treatment of patients with Kaposi's sarcoma and are now also used against breast cancer and refractory ovarian cancer. Liposomes continue to be refined and applied to additional cancer indications. They are only the first in an ever-growing number of nanovectors under development for novel, more efficacious drug-delivery modalities.

Nanovectors in general have at least a tripartite constitution, featuring a core constituent material, a therapeutic or imaging payload, and biologic surface modifiers, which enhance the biodistribution and tumor targeting of the nanoparticle dispersion. A major clinical advantage sought by the use of nanovectors over simple immunotargeted drugs is the specific delivery of large amounts of therapeutic or imaging agents per targeting biorecognition event.

Targeting methods range from covalently linked antibodies to mechanisms based on the size and physical properties of the nanovector. Nanovector formulations are designed:

- To reduce the clearance time of small peptide drugs
- To protect active agents from enzymatic or environmental degradation
- To avoid obstacles to the targeting of the active moiety

An example is protective exclusion by the blood–brain barrier or the vascular endothelium; the augmented osmotic pressure states in cancer lesions results in outward convection of the therapeutic moiety and nanoparticle sequestration by the reticuloendothelial system.

23.4.7   Systems to Provide Real-Time Assessments of Therapeutic and Surgical Efficacy

Biologic systems are known to be highly transparent to 700- to 1100-nm NIR light. The strong optical absorbance of SWCNTs in this special spectral window, an intrinsic property of SWCNTs, can be used for optical stimulation of nanotubes inside living cells to afford multifunctional nanotube biologic transporters. For oligonucleotides transported inside living cells by nanotubes, the oligonucleotides may translocate into the cell nucleus upon endosomal rupture triggered by NIR laser pulses. Continuous NIR radiation may cause cell death because of excessive local heating of SWCNTs in vitro.

Selective cancer cell destruction can be achieved by functionalization of SWCNTs with a folate moiety, selective internalization of SWCNTs inside cells labeled with folate receptor tumor markers, and NIR-triggered cell death, without harming receptor-free normal cells. Thus, the transporting capabilities of CNTs combined with suitable functionalization chemistry and their intrinsic optical properties may lead to new classes of novel nanomaterials for drug delivery and cancer therapy.

Weismann's (2003) work has added a nondestructive optimal method that was developed to selectively and sensitively detect CNTs in biologic surroundings. This method provides a valuable tool for tracing the locations of nanotubes in cells, tissues, and organisms, and it could play a critical role in studying how nanotubes become distributed in the body. This type of data on nanotube biodistribution will not only aid researchers in fine tuning nanotube-enabled cancer therapeutics and imaging agents, but will also help provide some of the data needed to use such agents in humans.

23.5 Challenges Put Forth by This Novel Technology

In an ideal scenario, the onset of the transformational processes leading toward malignancy should be detected early, as a matter of routine screening, by noninvasive means such as proteomic pattern analysis from blood samples or the in vivo imaging of molecular profiles and evolving lesion contours.

The major challenge is to accurately determine and dictate choices for targeting and barrier-avoiding strategies for an intervention plan. Another approach is to treat cancerous cells without collateral effects on healthy tissues. Repeating this process in routine a number of times is another challenge that nanotechnologists are facing. After these challenges are met, treatment efficacy will surely be monitored in real time, and therapeutics will be supplanted by personalized prevention. If fully integrated with the established cancer research enterprise, nanotechnology might help this vision become reality. Some of the principal challenges along this path are discussed in the next sections.

23.5.1 Developing Approaches for the In Vivo Detection and Monitoring of Cancer Markers

The effective early detection of precancerous and neoplastic lesions remains an elusive goal. Clinical cancer imaging technologies do not possess sufficient spatial resolution for early detection based on lesion anatomy. To identify malignancies based on their molecular expression profiles, all imaging technologies require contrast agents made up of a signal-amplifying material conjugated to a molecular recognition and a targeting agent such as an antibody. Nanoparticle technologies are under development and testing as candidate multifunctional, molecularly, or physically targeted contrast agents for all clinical imaging modalities. The objectives are to detect smaller and earlier stage cancer tumors, identify molecular expressions of neoplasms and their microenvironment, and provide improved anatomic definition for lesions. Nanoprobes may be a good option for the detection of precancerous cells.

Nanoparticle probes, which have molecularly targeted recognition agents, might provide information on the presence, relative abundance, and distribution of cancer signatures and markers associated with the tumor microenvironment.

Sustained angiogenesis is an important marker for use in the early detection of cancer because it is found in premalignant lesions of the cervix, breast, and skin, and it might be expected to be an early- to midstage event in human cancers.

A different approach to molecular detection in vivo involves the use of implantable sensors that are equipped with technology to relay sensed information extracorporeally. Despite many years of research toward this vision, the unsolved challenge for the clinical deployment of implantable molecular sensors remains an unwanted, non-specific adsorption of serum proteins on the sensing surfaces. This phenomenon, which is known as *biofouling*, results in a rapid loss of the ability of the sensor to detect the protein of interest over the background signal. A challenge is to develop surface nanostructures that will prevent nonspecific adsorption. More realistically, however, nanotechnology might be expected to yield novel, biofouling-indifferent sensing strategies (e.g., on the measurement of physical properties) from which the contributions of the fouling molecules might be systematically decoupled by appropriate mathematical algorithms.

23.5.2   Refining Technology Platforms for Early Detection of Cancer Biomarkers Ex Vivo

Serum markers for the early detection of most cancers are not available. The markers that are in clinical use, such as PSA and carcinoembryonic antigen, are nonspecific and have widely different baseline expressions in the population, so they are of limited effectiveness for early detection. The goal of developing reliable early detection approaches from serum, other biologic fluids, or any sample obtained through minimally or noninvasive procedures remains of paramount importance.

Several nanotechnologies are realistic candidates for early detection platforms, starting with surface patterning approaches, including firmly established technologies such as DNA microarrays, and serum surface-enhanced laser desorption/ionization time-of-flight

(SELDI-TOF) mass spectroscopy for proteomics. For these, the transition from the micron to the nanoscale dimensional control on surface features translates into increases in information quality, quantity, and density.

Nanocantilever, nanowire, and nanotube arrays might enable the transition from single- to multiple-biomarker cancer diagnostic, prognostic, and treatment selection. However, the areas of concern and the current limitations of these approaches include the need for covalent binding of different antibodies or other biologic recognition molecules to the devices and the deconvolution of noise from the signal, especially in regard to biofouling. For the analysis of proteomic signatures, a major challenge will be the identification of signatures from low-concentration molecular species in the presence of extremely high concentrations of nonspecific serum proteins. Issues that pertain specifically to the cantilever arrays include the need to develop further mathematical models for the determination of stresses and biologic identification signatures from the beam curvatures.

When CNTs are irradiated with a light beam of a single wavelength, a precise map of the distribution of many molecular markers in a single cell, cell population, or tissue is generated. This approach offers the potential advantages of readily identifying the conjugate markers; yielding specific information on their tissue distribution; and introducing new protocols that include cell surface, endocellular, and microenvironmental antigens in the same test.

The use of CNTs as selective, enriching harvesting agents for serum proteomics has been proposed. The emphasis for this approach is on low-molecular-weight proteolytic fragments, which are found in trace quantities in ovarian and other cancers. The use of nanoparticles for this approach has two objectives: the maintenance of fragments in the circulation that otherwise would be rapidly cleared and the selectivity of the uptake of the desired molecular signals over the "noise" of the most abundant serum proteins. This approach raises the possibility, used in SELDI-TOF proteomics, that appropriate surface treatment can significantly increase protein uptake per unit area and help prefractionate the sample to focus on the spectral domains of interest.

Some of the critical issues and research needs that are relevant in the developing field of nanotoxicology, especially as they relate to CNMs, include:

- **The need for detailed material characterization:** The toxicity of carbon materials may depend on byproducts or residues of complex carbon structures as much or more than on the primary carbon structures.
- **The need for realistic exposure scenarios:** Risk is the product of hazard and exposure, and little is currently known about realistic exposure levels, especially for lung exposure.

- **The need for methods to track nanomaterials in biologic experiments:** Sensitive methods of detection are required to quantify the extent of systemic transport and persistence at distant organs after dermal exposure, inhalation, ingestion, injection, or implantation.

- **The issue of adsorptive interferences with fluorescent assays:** The question needs to be addressed whether CNMs may interfere with fluorescent probes through adsorption or other means.

- **Dose metrics:** Sensitive detection methods are required to determine the dose metrics required in toxicology studies. However, it is possible that lipophilic nanomaterials (e.g., fullerenes) may interact with plasma membrane lipids and exert toxicity directly in the absence of cellular uptake.

- **The most important indicators of toxicity:** CNMs may elicit additional types of pathologic reactions and not be limited to the usually used short-term indicators of toxicity, altered cellular function, or inflammation.

23.6 Future Perspectives for Carbon Nanomaterials for Cancer Treatment

One futuristic application for CNTs that researchers have started to envision is using branched nanotubes as tweezers for manipulating molecules within a cell. Other researchers are starting to use CNTs as nanoscale electron sources that could form the heart of a fiberoptic X-ray source. Imagine snaking a tiny X-ray machine into the body directly to a tumor and blasting the tumor with a lethal dose of energy without damaging the surrounding tissue.

Of course, only time will tell which of the many proposed uses of CNTs will ultimately impact cancer research and clinical oncology. The future submissions in cancer nanotechnology will demand the following minimal materials characterizations:

- Complete bulk chemical composition (specifically including metals and heteroatom content >0.1%)

- Specific surface area

- Detailed descriptions of morphology (aspect ratios, secondary carbon forms, metals location) by electron microscopy examination of multiple fields

Further desirable characterization would include:

- Surface chemical composition (by energy dispersive or X-ray photoelectron spectroscopies)

- Texture (spatial arrangement of graphene layers)
- The degree of crystallinity or perfection of the graphene layers

A realistic long-term goal for toxicologists as well as materials scientists is the development of "green" nanomaterials formulations (those co-optimized and surface engineered for both function and minimal health impact).

Indeed, the majority of work with CNTs is still in the early stages of development, and it will be several years before any of these applications will begin human clinical trials. But given that this field is only a few years old, the future of CNT–based cancer research seems as bright as the light given off by an irradiated nanotube itself. Although there is universal consensus among scientists that significantly more work is needed on all of the new CNMs to adequately assess their toxicity and health risks, it is promising that research into the toxicity of nanomaterials has begun in earnest.

However promising nanovector delivery systems might be, the eagerness for them must be placed against the backdrop of the proper considerations of safety for patients and health care workers and in the context of stringent regulatory approval perspectives. The relevant issues go well beyond considerations of biocompatibility of the carriers, their biodistribution, and the reliability of their production protocols, which of course remain central concerns. By their very tripartite nature, nanoparticles arguably fall under the purview of the three branches of regulatory agencies such as the Food and Drug Administration (FDA): drugs, medical devices, and biologic agents. Therefore, they might have to be examined from these three perspectives accordingly. The main advantage of nanoparticles resides in their multifunctionality; they can incorporate multiple therapeutic, diagnostic, and barrier-avoiding agents.

With current regulations, it could be expected that regulatory approval will have to be issued for each agent and then for their combination. The time required for ascertaining their suitability for clinical use might therefore be quite long, perhaps unnecessarily so. The establishment of faster, safe regulatory approval protocols would ameliorate concerns about the length of time it takes for agents to be assessed by the FDA. Nanotechnology might significantly contribute to realizing this goal. The development of approaches for the real-time assessment of the efficacy of therapeutic regimens would substitute for the direct observation of tumor size, molecular expression, and efficacy in targeting the desired signaling pathways over (or in parallel with) conventional endpoint analysis, such as length of remission and extension of life.

Research in this direction is steadily progressing using the technology for molecular assessment both in vivo and ex vivo, as described earlier. The development of agents for in vivo molecular imaging, the establishment of dual therapeutic and imaging nanovector technologies, and the promise of in vivo microscopy (with fluorescent multiphoton imaging reaching single-cell resolution) all have the potential to transform regulatory processes. Therefore, nanotechnology might be expected to accelerate and render more accurate the regulatory approval process for all drugs (both nanoencapsulated and conventional) and assist in the determination of preferred therapeutic options.

Carbon nanowire and nanotube arrays might contain several thousand sensors on a single chip and therefore offer even greater multiplexing advantages. For both nanowires and microcantilevers, the nanofabrication protocols afford very large numbers of identical structures per unit area and therefore massive multiplexing capabilities. The many similarities that these protocols share with the fabrication of microelectronic components indicate that they will be comparably suitable for production scale-up at low cost and with high reliability.

Thus, cancer nanotechnology is a complete area of study because the present therapies have not fully succeeded. But with the help of CNMs, cancer nanotechnology will eventually succeed.

23.7   Summary

Painless needles in the form of CNMs are the near-future syringes that have started their existence on a pilot scale. These syringes have eased the work of doctors because they are already drug loaded and will replace capsules and tablets. Moreover, when cancer is discussed, it is necessary to stress the important issue of identification of suitable early markers followed by the development of a technology for biomarker-targeted delivery.

Naturally existing nanomachines, such as hemoglobin, proteins, and liposomes, can be harvested and used as media to carry drugs and target them to specific sites. Thus, the biodegradation factor is satisfied because the drug is loaded in the tissue and then the carrier is eluted out of the system.

The complexity of cancer technology and nanotechnology go hand in hand. Hence, answers give rise to new questions. One such critical problem is the toxicity of CNMs. But this problem can be overcome by functionalization. Unpurified CNMs, especially CNTs, affect human epidermal kerotinocyte cells and human kin fibroblasts by triggering the proteins involved in response to oxidative stress. The toxicity of CNMs may also be due to structural manipulations. Thus, to nullify this toxicity, CNT surfaces are functionalized with different organic compounds.

Early detection of precancerous cells is helpful to diagnose the disease and treat it at an early stage. Another interesting tool is the

404 Applications of Carbon Nanomaterials in Biosystems

fabrication of protein-coated nanocantilevers. Because CNTs are target-specific drug-delivery tools, they can also act as nanobombs. Due to thermal energy confinement in SWCNTs and subsequent vaporization of liquid, such as alcohol and de-ionized water, between SWCNTs in bundles, pressure is created, eventually causing an explosion.

Apart from all the ideas contributed to date by scientists, the main challenge lies in providing a system for real-time assessment of therapeutics and surgical efficacy. When this is accomplished, then nanotechnology will succeed in a true sense to cure cancer.

CHAPTER 24

Carbon Nanomaterial As a Nanosyringe: A Near-Future Reality

Seema Parihar

Nanotechnology Research Centre, Birla College, Kalyan, Maharashtra, India

Madhuri Sharon

MONAD Nanotech Pvt. Ltd., Powai, Mumbai, India

24.1 Introduction

The application of nanotechnology has tremendous potential in health care, particularly for the development of better pharmaceuticals. Nanotechnology-enabled drug delivery has already been successful in delivering drugs to specific tissues within the body, with the promise of capability that will enhance drug penetration into cells, as well as other means to improve drug activity. It is known that the efficacy of a drug can be increased if it is delivered to its target selectively and if its release profile is controlled.

In the past decade, two blossoming technologies have been hot research topics internationally. The powerful utility of concerted application of nanotechnology and biotechnology has been recently exemplified by breakthroughs in bio-directed nanosynthesis, nanoassembly, and nano-aided biologic recognition. Efforts are currently focused on searching for methods to mimic or exploit the unique capabilities of bioagents in producing nanostructures.

Nanodevices have shown capability in performing clinical functions such as detecting cancer at its earliest stages, pinpointing cancer's location in the body, and delivering anticancer drugs specifically to malignant cells. Nanoscale devices can control the spatial and temporal release of therapeutic agents or drugs. Nanoscale devices are much smaller than cells and even many cell organelles. Most animal cells are 10,000 to 20,000 nm in diameter. This means that nanoscale devices smaller than 50 nm can easily enter most cells, and those smaller than 20 nm can transit out of blood vessels. As a result, nanoscale devices can readily interact with biomolecules on both the cell surfaces and within the cells. Hence, a nanoscale device may contribute to cancer therapy by being a drug carrier.

Nanoscale devices attached with antibodies and loaded with drugs can serve as a targeted drug-delivery vehicle that can transport chemotherapeutics or even therapeutic genes into diseased cells while sparing the loading of healthy cells with drugs. Targeting a drug to its site of action would not only improve the therapeutic efficacy but also enable a reduction in the total dose of the drug that must be administered to achieve a therapeutic response, thus minimizing unwanted toxic effects of the drugs. Dendrimers, silica-coated micelles, ceramic nanoparticles, and cross-linked liposomes have already been shown to have potential as drug carriers. Moreover, carbon nanomaterials (CNMs) that have been extensively used for various bioapplications also show the possibility of being used for drug delivery (Parihar et al., 2006).

One of the primary objectives in the development of a drug-delivery system is the controlled delivery of drugs to its site of action at an optimal rate (Kreuter, 1991) and in the most efficient way possible. Nanoparticles, chiefly because of their small particle size, offer many advantages for many medical applications (Kreuter, 1983b; Marty and Oppenheim, 1977). The particle size enables intravenous (IV) and intraarterial injection because particles of this size can easily traverse even the smallest blood capillaries with an inner diameter of 3 to 8 μm (Thews et al., 1999). A small size also minimizes possible irritant reactions at the injection site (Little and Parkhouse, 1962; Kreuter, 1994a & b).

Many nanoscientists fantasize that as soon as CNMs are introduced into a living system along with drugs, they can act as a self-driven syringe and deliver the medicine to the site of requirement. Scientists are even planning to make CNM a disposable syringe that can be removed from the system either by degradation or excreting it from the system. There have been reports by Wang et al. (2003) about carbon being a cytotoxic material. Another school of thought believes that CNMs' damaging effect on living cells is exhibited only when they are exposed to light (Sharon et al., 2000). On the brighter side, researchers from Rice University have found that the toxicity of water-soluble carbon nanotubes (CNTs) to human skin cells decreased as the functionalization of the CNTs increased. They have also shown that CNTs were generally less toxic than fullerenes.

24.2   Definition of Nanoparticles for Pharmaceutical and Medical Purposes

Nanoparticles for pharmaceutical purposes are defined by Kreuter (1994b) in the *Encyclopedia of Pharmaceutics Technology* as "solid colloidal particles ranging in size from 1–1000 nm (1 μm). They consist of macromolecular materials and can be used therapeutically as drug carriers, in which the active principle (drug or biologically active material) is dissolved, entrapped or encapsulated, or to which active principle is adsorbed or attached."

24.3   History of the Development of Carbon Nanomaterial

The use of different types of nanomaterials for drug delivery was envisaged more than 4 decades ago. In the late 1960s and early 1970s, Peter

Speiser at ETH (Swiss Federal Institute of Technology, Zurich) realized Paul Ehrlich's idea and developed the first nanoparticle for drug-delivery purposes (Birrenbch and Speiser 1976; Kopf et al., 1976, 1977). These nanoparticles were produced by emulsion polymerization of acrylamide cross-linked with N,N'-methylene bis acrylamide in hexane.

Zolle et al. (1970) used another process, denaturation of albumin dissolved in water and emulsified in hot cottonseed oil, to produce nanoparticles. $^{99m}TcO^-_4$ was bound to the nanoparticles, and these particles were used for radioimaging of the lungs after IV injection. Later, the albumin nanoparticles were used for drug delivery. Kramer (1974) incorporated the anticancer drug mercaptopurine into these particles using heat denaturation.

Widder et al. (1978) performed the first successful drug targeting with nanoparticles by incorporation of magnetite particles into similar albumin nanoparticles and the use of a magnetic field. They substituted a more efficient anticancer drug, doxorubicin, for mercaptopurine.

The latest important input in the improvement of drug targeting with nanoparticles is the development of Long circulating nanoparticles by the covalent linkage of polyethylene glycol (PEG) chains to poly lactic-co-glycolic acid (PLGA) (Gref et al., 1994) or to poly-alkyl cyanoacrylate nanoparticles (Peracchia et al., 1998) and delivering the drugs to the brain across the blood–brain barrier (Kreuter et al., 1995).

In 2005, scientists at the University of Trieste, University of Ferrara, International School for Advanced Studies, and National Consortium of Material Science and Technology used CNTs to boost neural signaling. They grew nerve cells from the brain's hippocampus region on substrates coated with networks of CNTs and found a large increase in neural signal transfer between cells (Lovat et al., 2005). Because CNTs are similar in shape and size to nerve cells, they could help to structurally and functionally reconnect injured neurons. Hippocampal neurons grown on CNTs display a sixfold increase in the frequency of spontaneous postsynaptic currents.

Because CNTs could act as supportive devices for bridging and integrating a functional neuronal network, scientists have also started viewing carbon as a possible tool for drug delivery.

24.4 Drug Delivery Promises of Carbon Nanomaterial

To use CNMs for drug delivery, it is important to understand their morphology and properties. Many characteristics of CNMs are discussed in detail in previous chapters.

24.4.1 Morphology of Carbon Nanomaterial

Carbon nanoparticles exhibit tubular, fibrous, and bead-like structures named CNTs, carbon nanofibers (CNFs), and carbon nanobeads

(CNBs), respectively. Although many physical and chemical parameters and characteristics of CNMs make them a suitable material for drug delivery, so far they have not been considered seriously, and very few successful results are available.

Carbon Nanotubes

CNTs are concentric shells of graphite formed by one sheet of conventional graphite rolled up into a cylindrical form. The lattice of carbon atoms of graphite sheets remains continuous around the circumference of the CNTs. Hence, CNTs are fullerene-related structures closed at both ends with caps containing pentagonal rings. They were discovered in 1991 by the Japanese electron microscopist Iijima, who was studying the material deposited on the cathode during the arc-evaporation synthesis of fullerenes. He found that the central core of the cathodic deposit contained a variety of closed graphitic structures, including nanoparticles and CNTs, of a type that had never previously been observed.

CNTs are of two types: single-walled CNTs (SWCNTs) and multi-walled CNTs (MWCNTs).

In SWCNTs, there are only tubules and no graphitic layers around them. The diameter of an SWCNT is up to 2 nm, and the length varies as per production procedures from 3 to 10 μm. The arrangement of carbon in an SWCNT can be of the arm-chair, zigzag, or chiral pattern. SWCNTs are mostly produced in a bundle and are then separated by chemical or physical methods.

MWCNTs are stacks of graphene sheets rolled up into concentric cylindrical structures. Their diameter is in the range of 10 to 50 nm, and their length can be up to or more than 10 μm. The individual graphene sheets are separated by about 0.34 nm.

Carbon Nanofibers

CNFs are filaments without a lumen. They can be produced in various shapes, including straight, coiled, cactus, cauliflower, octopus, stacked, or fish-bone (herring-bone) shaped.

Carbon Nanobeads

CNBs are spherical, hollow structures. When five to seven beads are covered by a broken graphene sheet, the bead is called a *spongy bead*. The thickness of each graphene sheet is 8 to 10 nm and the total diameter of the beads is around 250 to 800 nm.

24.5   Synthesis and Purification of Carbon Nanomaterial

The synthesis, opening, and thinning of SWCNTs, MWCNTs, and nanomaterials are topics of immense interest owing to their potential applications in catalysis, nanoelectronic devices, drug delivery, and the nanoscale chemistry of CNTs (Ajayan, 1999; Harris, 1999; Heer and Martel, 2004; Service, 1998).

For drug delivery, it is important to use highly pure CNMs. Carbon nanostructures are generally synthesized by three processes: arc discharge, laser ablation, and catalytic chemical vapor deposition. Details of purification of synthesized CNMs are given in Chapters 2 to 5.

Synthesis of CNM involves use of catalysts, which often remain along with the finally produced CNM. To get pure CNM, it is necessary to remove the catalyst. Moreover, amorphous carbon often also gets synthesized during CNM production. Hence, during purification, care is taken to remove the amorphous carbon.

Various purification procedures have been reported in the literature (Monthioux et al., 2001). Most of them are done by using acids such as HCl or H_2SO_4 or other oxidizing reactants such as H_2O_2. These purification procedures are performed with the main goal of removing the catalyst particle and impure carbon phases such as amorphous carbon, polyaromatic shells, and graphite particles.

Removal of catalysts, which are usually metals, is normally done by HCl treatment; oxidizing acids are used for removal of amorphous carbon deposits.

Recently, Hirsch and Vostrowsky (2005) proposed a method to obtain a pure homogenous dispersion of MWCNTs by functionalizing with pyrrolidine groups to enhance their solubility in the organic solvent dimethyl-formamide, which is evaporated and then heated at 350°C to defunctionalize them in the process, leaving purified nanotubes on the substrate.

To optimize the structural features of carbon nanostructures for drug delivery, heat treatment or treatment in activating or passivating gas atmospheres is done.

24.6 Properties of Carbon Nanotubes

Some important properties of CNTs and their molecular background that may affect drug delivery include their chemical reactivity, electrical conductivity, optical activity, and mechanical strength.

24.6.1 Chemical Reactivity

The chemical reactivity of CNTs is usually more than that of the graphene sheet. CNT reactivity is directly related to s-orbital mismatch caused by an increased curvature. Therefore, a distinction must be made between the sidewall and the end caps of CNTs for the same reason. A smaller CNT diameter results in increased reactivity. A covalent chemical modification of either the sidewall or end caps is possible.

Although CNMs are usually insoluble material, the solubility of CNTs in different solvents can be controlled or enhanced by creating chemical bonds.

24.6.2   Electrical Conductivity

The electrical conducting properties are caused by the molecular structure. The conductance is determined by quantum mechanical aspects and was proven to be independent of the length of the CNTs. However, depending on their chiral vector (which controls the diameter of the CNTs), CNTs are either semiconducting or metallic in nature.

24.6.3   Optical Activity

Theoretical studies have revealed that the optical activity of CNT disappears as the CNTs become larger. It is, therefore, expected that other physical properties could also be influenced by the size of the CNTs.

24.6.4   Mechanical Strength

CNTs have very high tensile strength, so they exhibit very large young modulus in their axial direction. CNTs are also highly flexible; therefore, they are potentially suitable for applications in composite material that need an isotropic property.

CNMs have diverse tunable physical properties as a function of their size and shape due to a strong quantum confinement effect and large surface-to-volume ratio. CNTs are hollow, tubular, caged molecules Because of these properties, they have been proposed as lightweight packing material for hydrocarbon fuels and as nanoscale containers for molecular drug delivery.

24.7   Carbon Nanomaterial for Drug Delivery

The minimum diameters of SWCNTs are similar to the diameter of a molecule of DNA, so SWCNTs can easily traverse through cells. However, the length of CNM varies according to the method of production, so it is important that they are tailored to the right size for drug delivery.

Filling of opened CNTs has already been successfully done with DNA and proteins. This has given impetus to consider CNM for drug delivery. Depending on whether the drug is to be adsorbed into the CNM surface or filled into the lumen of CNTs, the treatment of CNM for drug loading varies.

CNBs have recently attracted attention for use as a drug-carrying vehicle. The advantage of using CNBs would be that their smaller and desired sizes can be synthesized by controlling the pyrolysis conditions and the precursor (Sharon et al., 1998).

Another form of CNM that is also being investigated in our laboratory is CNFs. As mentioned, CNFs exist in various forms (e.g., straight, coiled, spiral, branched, bamboo-like, octopus). It has been found that the activation of CNFs with KOH creates pores on the CNF surface, thus increasing the available surface area for functionalization or drug attachment, which is a necessary requirement for drug loading onto CNM.

A detailed survey of the cytotoxicity and degradability of CNM inside living cells needs to be worked out before attempting the use of CNM for drug delivery. This is discussed in Chapters 19 and 24.

24.7.1   Steps Involved In the Delivery of Drugs Using Carbon Nanomaterial

The steps considered for drug delivery are discussed in the next sections.

Opening the Closed Ends of Carbon Nanotubes

CNT opening has been demonstrated to be a side effect of the various acid-based purification procedures (as mentioned in Sec. 24.5). Oxidative cleavages of C=C bonds and the presence of greater angular strain due to geometry (pentagonal rings) are considered to be some of the possible reasons for initiation of oxidation at the tips or caps (Hwang, 1995), thus opening the closed ends.

Cutting or Tailoring Carbon Nanomaterial to the Desired Size

Shorter tubes have found many applications, such as in field emission, composites, and medicines. Various approaches have been developed for either opening or cutting the tubes of both SWCNTs and MWCNTs (Ajayan et al., 1993; Tsang et al., 1993, 1994). The most common methods are abrasive treatment and ball milling.

Abrasive Treatment   Abrasive treatment is a physiochemical method used for reducing the length of CNTs. The MWCNTs are dispersed in organic solvent, and using ultrasound and magnetic stirring at ambient temperature, they are abrased into smaller segments. The alcohol in the homogeneous suspension of the CNTs is then evaporated completely. Evaporated fine carbon powders are smaller in length (Maurin, 2001). However, this method has its limitations in yield and often creates important structural damage.

Ball Milling   Ball milling is a physical method of reducing the length of CNMs (CNFs and CNTs). Different-sized CNMs can be obtained by controlling the frequency and time of ball milling (Fig. 24-1). The drawback of this method is that uniform-sized CNMs are not produced, and even the structure of the CNMs is damaged.

Super Critical Water Treatment   Water has unique properties above its critical point (T_c, 374°C; P_c, 3200 psi), where T_c is critical temperature and P_c is critical pressure. Lowering of the dielectric constant (from 80 at ambient conditions to less than 5 above the critical point) and reduction of hydrogen bonds make super critical water (SCW) an efficient nonpolar solvent.

Nonpolar organics are completely soluble in SCW in the presence of oxygen and could be rapidly and efficiently oxidized to carbon

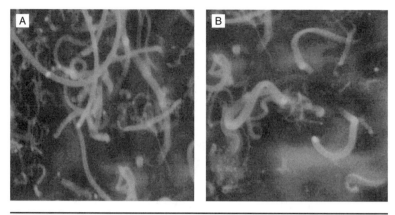

FIGURE **24-1**   Scanning electron micrograph of carbon nanofibers before (*A*) and ball milling after (*B*).

dioxide and water in a single homogeneous fluid phase with no interphase mass transfer limitations. Because the extent of hydrogen bonding in water is lowered under super critical conditions, it tends to be more reactive.

The use of the SCW system for opening and thinning of MWCNTs (Figs. 24-2 and 24-3) has been found to be an easy and rapid method that avoids the use of strong acids. In addition, it is pollution free.

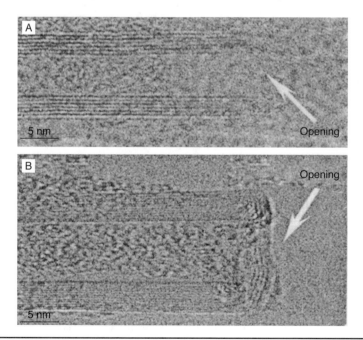

FIGURE **24-2**   Opening of multiwalled carbon nanotubes in super critical water in the absence (*A*) and the presence (*B*) of oxygen. (Courtesy of Chang et al., 2002.)

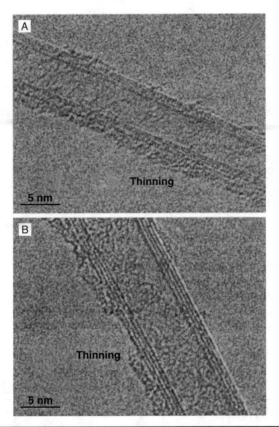

FIGURE 24-3   Thinning of multiwalled carbon nanotubes by super critical water in the absence (*A*) and the presence (*B*) of oxygen. (Courtesy of Cheng et al., 2002.)

Chemical Functionalization of Carbon Nanotubes

Because CNMs are insoluble in both polar and nonpolar solvent, they exhibit a very low process ability. Chemical functionalization of CNTs has been done to overcome the problem of process ability.

Several laboratories have investigated the surface chemistry of CNTs, focusing most frequently on the ends and the outer shell of the CNTs. When the ends of the CNTs are opened, which is a general consequence of purification of the as-produced nanotubes, the interior of the nanotubes becomes ready to accommodate guest materials. The carbon shell is closed by various functional groups, most frequently by $-NH_2$, $-COOH$, $-OH$, and $>C=O$ groups (Timea et al., 2004). For MWCNTs, the outer shell often contains discontinuous spots and imperfections. These local vacancies are also closed by functional groups mentioned above. However, for these local applications in which nanotubes are used to strengthen polymers, the presence of functional groups is advantageous to ensure chemical bonding between polymers and fillers or to enhance the solubility of CNTs in various solvents.

A team of scientists from Rice University has come up with a new technique for attaching amino groups to the sidewalls of SWCNT. They have produced functionalized CNTs by reacting flouronanotubes with terminal diamines. Attaching the amino functionality to the sidewalls of the tubes provides multiple sites for creating covalent bonds to monomers or polymers. This opens up an opportunity for covalent binding of DNA or drugs to the functionalized tubes.

Drug Loading

When organic polymers are used as a drug carrier, drugs are incorporated or loaded during preparation of the polymer. Similarly, there have been reports of CNMs being loaded with different metals during synthesis of CNM (Seraphin et al., 1993). But drug molecules cannot be loaded in this way because CNMs are prepared at a very high temperature (i.e., ~750°C onward), which may be damaging the drug molecules to be loaded. Before loading the drug onto the CNM, a detailed study of the characteristics of both the drug and the CNM is required.

Drugs may be bound to a nanoparticle by:

- Incorporation in to the interiors of a nanocapsule (Soppimath, 2001)
- Covalent binding (Kopf et al., 1976; Langer et al., 2000a)
- Electrostatic binding (Hoffmann et al., 1997; Langer et al., 1997)
- Surface adsorption (Berg et al., 1986; Vora et al., 1993)

In most cases, as per the physicochemical properties of the drug, the method of drug binding and the type of nanoparticles to be used are decided upon (e.g., thermolabile drugs cannot be incorporated into a nanoparticle produced by involving heating). Similarly, hydrophilic drugs cannot be incorporated into a hydrophobic nanoparticle without difficulties and vice versa. Surface adsorption in general is governed by a Langmuir type of interaction (Berg et al., 1986; Vora et al., 1993). Surfactants generally reduce adsorption. In some cases, however, they improve adsorption (Harmia et al., 1986a & b).

Although loading of a drug onto CNM has not been standardized, methods that have shown promises in fullerenes and metal insertion can be considered and tried for drug loading. Some of them are:

1. Collisions of accelerated atoms [with KE (kinetic energy) ≤ 150 eV for alkali metals] on SWCNTs (Farajian et al., 1999), which is used for filling endofullerenes with metal and other atoms

2. Opened SWCNTs along with the drug to be inserted are taken in glass ampoule, sealed under vacuum conditions, and then heated beyond the drug sublimation temperature. This method is specifically adapted for drugs with low sublimation temperatures and thermal stability. This method is quite successful; the filling rate can reach 100% (Hirahara et al., 2000) provided the SWCNT surfaces are adequately clean.

3. In situ filling: SWCNTs are synthesized while the filling material or drug is sublimed, typically during the electric arc process. For example, using various carbon anodes doped with C_{60} to form peapods is being tried (Monthioux et al., 2001). However, yields were found to be very low by this route, probably because of the high speed of the transient phenomena and the restricted volume while SWCNT formation occurs in the plasma. This provides little chance for the potential filler to actually enter the SWCNT cavity before the closing of the tubule while it grows, specifically considering that a closed tip growth model generally accepted mechanism for the SWCNT formation.

4. Filling via liquid phase: SWCNTs, along with the filler in molten state, are put together in a sealed ampoule. In this case, the capillary effect occurs despite the nanometric diameter of the SWCNTs. It is necessary to have a molten compound with a sufficiently low surface tension and low melting temperature to avoid undesirable effects, such as early closing of the previously opened carbon structure caused by an excessive melting temperature (Govindaraj et al., 2001). One way is to fill the tubes with molten salts and to reduce the salt using hydrogen gas, although it remains to be ascertained whether the reduction rate is 100%. Samarium oxide has been filled into MWCNTs by this method.

5. Solution phase chemistry has the advantage that without heat treatment, the drug can be filled at room temperature. SWCNT material is soaked in a solution of the compound to be inserted using a suitable solvent, depending on the filler's chemical composition and the specific requirement of the process. As mentioned previously, the elemental state is obtained by reducing the filler by hydrogen gas. Using this method, Sloan et al. (1998) have filled SWCNTs with ruthenium.

6. Nanoextraction and nanocondensation: Most drugs neither evaporate nor sublime, but rather degrade at elevated temperatures. Therefore, a new method of incorporating drugs into SWCNTs at room temperature is nanoextraction or nanocondensation.

- For nanoextraction: SWCNTs are heat treated at 1780°C in a vacuum for 5 hours and then further heated in an oxygen atmosphere at 570°C for about 10 minutes. This heat treatment enlarges the diameter of the SWCNTs from 1 nm or less to 1 nm or more [in a study by Yudasak et al. (2003), about 50% of them had diameters >2 nm].
- For nanoextraction, guest molecules must have a poor affinity to the solvent, but a strong affinity to the CNTs. Also, the solvent, must have a poor affinity to CNTs. To demonstrate nanoextraction, C_{60} crystallites (1 mg) were added to ethanol (10 mL) and ultrasonicated for 3 minutes and then SWCNTs (1 mg) were added. This mixture of SWCNTs, C_{60}, and ethanol was kept for 1 day at room temperature. The solubility of C_{60} in ethanol is about 0.001 mg/mL (Kimata et al., 1993); hence, C_{60} crystallites could hardly dissolve and remained at the bottom of the ethanol solution or suspended in it. After 1 day, the SWCNTs were taken out of the mixture and air dried at room temperature. Transmission electron microscopy (TEM) showed that C_{60} molecules were incorporated inside the SWCNTs $(C_{60})_n$.
- Nanocondensation: To prepare (C_{60}) SWCNTs through nanocondensation, 10 µ mL of C60-toluene standard solution (2 to 8 mg/mL) is dropped onto SWCNTs placed on a grid disk (a TEM sample holder) kept on a filter paper to soak the excess solution as quickly as possible (Ruoff et al., 1993). The grid disks are usually about 3 mm in diameter and about 0.035 mm thick and are made up of copper with carbon.
- Peapods can be applied to drug-delivery systems by replacing C_{60} with molecules having medicinal effects.

7. Opening of MWCNTs in a SCW medium creates an alternative possibility for filling. However, for filling of MWCNTs, the very important criterion is the surface tension threshold value of 100 to 200 mN/m. Also, there must be a route for escape of gas or air trapped in the MWCNTs. Thus, if the MWCNTs are opened inside a liquid with a low surface tension, the liquid should be pulled unhindered by capillarity (Ebbesen, 1996; Ugarte et al., 1998). If opening were achieved in SCW, it would be a useful mode to fill the desired compounds soluble in SCW.

Because most of the modified compounds are soluble in super critical fluid, it will be an efficient medium for drug delivery by filling the nanotubes with desired drugs.

So far, CNTs have been filled with halides, oxides, C_{60}, heavy metals, gases such as hydrogen, carbohydrates, DNA, enzymes, and other proteins. But none of the drugs has yet been loaded or filled in CNTs.

Drug Release

Drug releases from nanoparticles to the site of action and subsequent biodegradation are important for developing a successful formulation. Moreover, the drug should be released from the nanoparticles at the desired rate and cycle (Kumaresh et al., 2001; Mathiowitz et al., 1997). The release rates of nanoparticles depend on:

- Desorption of the surface-bound or adsorbed drug
- Diffusion through the nanoparticle matrix
- Diffusion (in case of nanocapsules) through the polymer wall
- Nanoparticle matrix erosion
- A combined erosion and diffusion process (Hu et al., 2003; Kreuter, 1994a & b).

Thus, diffusion and biodegradation govern the process of drug release. After the release of the drug, it is important that:

- The particles inserted into the living system should preserve and protect the drug from any degradation until they reach the site of action.
- The drug should not get released until it reaches the sites of action.
- The drug should be released at a rate that achieves the desired therapeutic effect on a continuous basis.
- It is of utmost importance to decide the type of drug release cycle to be applied (e.g., constant, cyclic), depending on the environmental conditions.
- The particles should recognize the site of action; this mostly depends on the choice of antibodies attached to the nanoparticle along with the drug.
- If desired, the nanoparticles should have the ability to get bound or associated with the sites of action.

An *in vitro* drug release studies may give a detailed insight into the problems and parameters associated with the release of drugs. However, in many cases, the *in vitro* release cannot be correlated with the *in vivo* situation (Park, 2002). In vitro drug release may be studied with a variety of methods (Hu et al., 2003b).

Because of the small size of the nanoparticles, the separation of the releasing particles from the rest of the sample represents a major problem. In principle, this separation can be achieved either by ultracentrifugation or by membrane separation using artificial or biologic membranes, dialysis bags, ultrafiltration, and centrifugal ultrafiltration.

All of these techniques have a common disadvantage that a significant time lapse exists between the immediate release from the particle and the time of sampling because ultrafiltration, membrane diffusion, ultracentrifugation, and so on, are time-consuming processes. During this time lapse, the release continues; therefore, it is not possible to obtain real-time release rates.

A very good alternative to study the release is the use of substances that change their analytical profile—for instance, color—by moving from a hydrophobic to a hydrophilic environment when traversing from the nanoparticle into the release medium.

Biodegradation of Drugs

Finally, when the drug is released, the nanoparticles should get degraded or removed from the body. Biodegradation of nanoparticles is a very important requirement for most therapeutic uses. Biodegradation has a profound influence on the drug release rate. In addition, with the possible exception of vaccines, the nanoparticle or the carrier material has to be eliminated rather rapidly to avoid accumulation in the body. However, degradation of nanocarbon in living systems still remains an illusive system that has not yet been worked out.

24.8 Progress Made So Far in Using Carbon Nanomaterial for Drug Delivery

One of the important questions that scientists working with CNM to be used as a drug-delivery vehicle is whether it can safely traverse through cells. Bianco et al. (2005) showed that CNTs are adept at entering the nuclei of cells; hence, they can be used as nanodelivery vehicle. They modified the CNTs by heating them for several days in di-methyl formamide, which enabled short linking chains of tri-ethylene glycol (TEG) to be attached. Then a small peptide was bonded to the TEG molecule. When the modified CNTs were mixed with cultures of human fibroblast cells, they rapidly entered and migrated toward the nucleus. At low doses, the CNTs appeared to leave the cells unharmed, but as the concentration was increased, cells began to die. Although the use of CNM is in its infancy in the delivery of drugs, researchers have started to believe that one day they may be able to use CNTs to deliver drugs and vaccines. A wide range of different molecules could be attached to the CNTs, increasing the possibility of an easily customized way of ferrying molecules into the cells.

The basic concept for using CNTs in vaccine delivery is to link the antigen to the CNTs while retaining their conformation and thereby inducing an antibody response with the correct specificity. In addition, CNTs should not trigger a response by the immune system (i.e., they should not possess intrinsic immunogenicity) (Pantarotto et al., 2003a).

One can imagine that in the distant future, instead of receiving a vaccine shot with a syringe, a patient may lick a lollipop coated with functionalized CNTs acting as a vaccine-delivery system. CNTs have already been used to build a computer, so the day is not far off when they can be applied to improved drug-delivery systems or can travel to the brain after being inhaled.

24.9 Possible Drawbacks in the Use of Carbon Nanomaterial for Drug Delivery

The use of carbon in therapeutic systems has begun with many apprehensions. Its suitability has been a big question mark in the mind of many scientists. One of the apprehensions is that use of nanosized material in medicine poses unique problems; they can be cleared out of the body before they complete their mission. Moreover, they also have a large surface area relative to their volume, which may allow unwanted friction.

As far as the breakdown products of nanocarbon are concerned, our knowledge at the moment is almost negligible, and it needs further research. The brighter side is that looking at the possibilities and applicability of CNMs in drug delivery, scientists have gotten actively involved in solving these issues.

24.10 Summary

In this chapter, the possibility of using CNMs as tools or vehicles to deliver drugs in desired amounts to sites of action have been discussed. Various properties that make CNMs a possible drug carrier have also been explored. Finally, different processes involved—from tailoring CNMs to the desired size, loading and releasing the drug, and the biodegradation of CNMs—have been discussed.

CHAPTER 25

Antimicrobial Effects of Carbon Nanomaterials

Arvind Gupta and Sejal Shah

Nanotechnology Research Centre, Birla College, Kalyan, Maharashtra, India

Madhuri Sharon

MONAD Nanotech Pvt. Ltd., Powai, Mumbai, India

25.1   Introduction

Carbon nanomaterials (CNMs) are nanometer-sized particles. After reading the earlier chapters, this statement must be known by heart by the readers. Still, an essential property of CNMs is left to be discussed. CNMs can have a semiconducting nature depending on their internal diameter. In living systems, this property of CNMs supports photocatalytic activity. Because of this property of carbon at the nanolevel, some microbiologists call CNMs a "killer tool." CNMs have played a role as a drug-delivery tool, and now CNMs are able to kill cells.

Nanoparticles have unique physicochemical properties that are not found in their bulk materials. In general, they have a much higher reactivity, and because of their ultra-small size, they can easily penetrate skin and cells, rapidly distribute in the human body, and even directly interact with organelles within cells. Their huge ratio of surface area to mass increases their chemical activities and therefore allows them to become efficient catalysts. These increased chemical and biologic activities have resulted in many engineered nanoparticles being designed for specific purposes, including diagnostic and therapeutic medical uses and environmental remediation, requiring introduction of these novel materials into the human body or the environment.

Carbon nanoparticles are available in several geometric forms, including spherical fullerenes, cylindrical nanotubes, and planar nanoplatelets. Although these nanoparticles all share the same graphene structure that imparts semiconductor properties, their diverse geometries afford a spectrum of unique chemical, electrical, magnetic, and optical properties. Composite materials containing carbon nanoparticles electrically coupled to biomolecules could yield an array of high-performance technologies. Inclusion of a biologic recognition element along with a necessary cofactor or mediator allows the composite electrode to serve as an integrated, bioactive electrode unit (i.e., biocarbon network). This work includes:

- Development of simple and rapid methods to modify and process carbon nanoparticles for self-directed assembly
- Incorporation of the nanoparticles into biocarbon networks
- Characterization of the network's fundamental properties

Entry of CNMs into cells is not a problem; nanoparticles can easily enter the cell membrane because the cell membrane has a much larger pore size. Nevertheless, it would be interesting to know the mode of entry because cell membranes have different designs depending on whether the cell is an animal, plant, or bacterial cell. As shown in Fig. 25-1, a plant cell generally has a cell membrane protected by a cell wall; therefore, a CNM has to pass through a double entrance to react with the cell organelles. Bacterial and animal cells have a single

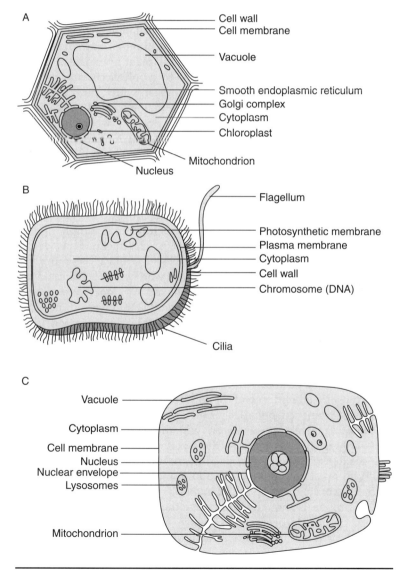

Figure 25-1   Schematic diagrams of plant (*A*), microbial (*B*), and animal (*C*) cells.

entrance gate in the form of a cell membrane, but with a different cell composition.

25.1.1   Plant Cells

One of the most important distinguishing features of plant cells is the presence of a cell wall. The relative rigidity of the cell wall renders plants sedentary, unlike animals, whose lack of this type of structure allows their cells more flexibility, which is necessary for locomotion. The plant cell wall serves a variety of functions. Along with protecting the intracellular contents, the structure bestows rigidity to the plant; provides a porous medium for the circulation and distribution of water, minerals, and other nutrients; and houses specialized molecules that regulate growth and protect the plant from disease.

Growing plant cells are surrounded by a polysaccharide-rich primary wall. This wall is part of the apoplast, which itself is largely self-contiguous and contains everything that is located between the plasma membrane and the cuticle. The primary wall and middle lamella account for most of the apoplast in growing tissue. The symplast is another unique feature of plant tissues. This self-contiguous phase exists because tube-like structures known as plasmodesmata connect the cytoplasm of different cells.

Cell walls are significantly thicker than plasma membranes and were visible even to early microscopists, including Robert Hook, who originally identified the microscopic structures of a sample of cork and then coined the term *cells* in the 1660s. The thickness, as well as the composition and organization, of cell walls can vary significantly. Many plant cells have both a primary cell wall, which accommodates the cell as it grows, and a secondary cell wall developed after the cell has stopped growing. The primary cell wall is thinner and more pliant than the secondary cell wall and is sometimes retained in an unchanged or slightly modified state without the addition of the secondary wall, even after the growth process has ended.

The main chemical component of a primary plant cell wall is cellulose, a complex carbohydrate made up of several thousand glucose molecules linked end to end. In addition, the cell wall contains two groups of branched polysaccharides, the pectins and cross-linking glycans. Organized into a network with the cellulose microfibrils, the cross-linking glycans increase the tensile strength of the cellulose. The coextensive network of pectins provides the cell wall with the ability to resist compression. In addition to these networks, a small amount of protein can be found in all plant primary cell walls. Some of this protein is thought to increase the mechanical strength, and part of it consists of enzymes, which initiate reactions that form, remodel, or break down the structural networks of the wall. Such changes in the cell wall directed by enzymes are particularly important for fruit to ripen and leaves to fall in the autumn.

25.1.2   Bacterial Cells

A bacterial cell wall is a unique structure that surrounds the cell membrane. Although not present in every bacterial species, the cell wall is very important as a cellular component. Structurally, the wall is necessary for:

- **Maintaining the cell's characteristic shape:** The rigid wall compensates for the flexibility of the phospholipid membrane and keeps the cell from assuming a spherical shape.

- **Countering the effects of osmotic pressure:** The strength of the wall is responsible for keeping the cell from bursting when the intracellular osmolarity is much greater than the extracellular teichoic osmolarity.

- **Providing attachment sites for bacteriophages:** Teichoic acids attached to the outer surface of the wall are similar to landing pads for viruses that infect bacteria.

- **Providing a rigid platform for surface appendages:** Flagella, fimbriae, and pili all emanate from the wall and extend beyond it.

The two major types of bacteria are gram positive and gram negative. The cell walls of gram-positive bacteria consist of many polymer layers of peptidoglycan connected by amino acid bridges. The peptidoglycan polymer is composed of an alternating sequence of N-acetylglucosamine and N-acetyl-muraminic acid. Each peptidoglycan layer is connected or cross-linked to the next by a bridge made of amino acids and amino acid derivatives.

The cell wall of gram-negative bacteria is much thinner, being composed of only 20% peptidoglycan. Gram-negative bacteria also have two unique regions that surround the outer plasma membrane: the periplasmic space and the lipopolysaccharide layer. The periplasmic space separates the outer plasma membrane from the peptidoglycan layer. It contains proteins, which destroy potentially dangerous foreign matter present in this space. The lipopolysaccharide layer is adjacent to the exterior peptidoglycan layer. It is a phospholipid bilayer construction similar to that in the cell membrane and is attached to the peptidoglycan by lipoproteins.

25.1.3   Animal Cells

Animal cells are typical of eukaryotic cells, which are enclosed by a plasma membrane and containing a membrane-bound nucleus and organelles. Unlike the eukaryotic cells of plants and fungi, animal cells do not have cell walls. This feature was lost in the distant past by the single-celled organisms that gave rise to the kingdom Animalia. Most cells of microbes, animals, and plants range in size between 1 and 100 μm; therefore, they are visible only with the aid of a microscope.

The lack of a rigid cell wall allowed animals to develop a greater diversity of cell types, tissues, and organs. Specialized cells that formed nerves and muscles gave these organisms mobility. The ability to move around with the use of specialized muscle tissues is a hallmark of the animal world, although a few animals, primarily sponges, do not possess differentiated tissues. Protozoa locomote via nonmuscular means, in effect using cilia, flagella, and pseudopodia.

25.2   Killing of Microbes

Certain chemical agents are commonly used for fighting microbes. For surface sanitization, two inorganic approaches are commonly used. The first is based on photocatalytic activity. The second exploits the toxicity of certain metallic cations such as silver. Silver has long been known for its excellent antimicrobial effect due to the release of silver ions, which exert a toxic effect when taken up by microbes. Use of ultrafine silver nanoparticles has increased the antimicrobial activity. The extreme increase in surface area enhances silver's natural sanitizing ability. However, the photocatalytic activity of CNMs has been found to be much more effective in killing microbes.

25.2.1   Photocatalytic Action

Had there not been a control system for growth of microbes, then our planet would not be governed by humans; rather, it would be ruled by microbes because of their high multiplication rate. The photocatalytic process perhaps may be one of the processes controlling the growth of pathogens in the water and atmosphere in a manner that their growth does not become a menace. For the photocatalytic process to occur spontaneously in nature, one needs the presence of fine oxide material such as iron oxide, solar rays, and moisture. These three things are abundantly available in nature. In the photocatalytic process, if a particle of a semiconductor is in contact with another material, it can form a depletion region (e.g., a p:n junction). When its interface is illuminated with a light of photon energy greater than the band gap of the materials, formation of electron–hole pairs takes place in a similar fashion as one gets when a p:n junction is illuminated. These photon-generated electron–hole pairs are highly reactive and can initiate an oxidation/reduction process with any organic material that comes into physical contact with them.

25.2.2   Photocatalytic Mechanism

Any organic material or water present near the interface of such semiconducting particles would get quickly oxidized. For example, phospholipids are expected to be oxidized to give some phosphorous salts, CO_2, and a few degraded phospholipid products. The photocatalytic mechanism involves three steps i.e., at interface semiconductor and

1. At interface semiconductor and solution.

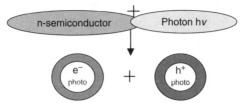

2. Water gets oxidized to h^+_{photo} to hydroxyl radicals (OH$^\bullet$)

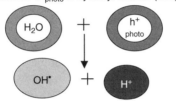

3. These hydroxyl radicals or e^-_{photo} and h^+_{photo} can perform the following reactions with organic materials that are physically in contact;

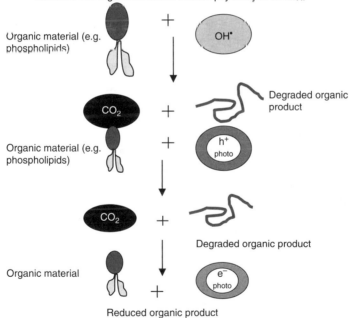

FIGURE 25-2   Photocatalytic mechanism of organic material or water present at the interface of semiconducting particles.

solution water gets oxidized to h^+_{photo} to hydroxyl radicals (OH$^\bullet$); then these hydroxyl radicals or e^-_{photo} and h^+_{photo} reacts with organic material which are physically in contact and degrade then producing reduced organic product (Fig. 25-2). Considering this logic, it would be interesting to study the antimicrobial effects of CNMs on microbial cells.

25.3   Antimicrobial Effects of Carbon Nanomaterial

25.3.1   In Vitro Action of Carboxyfullerene

Teo et al. (2003) have studied in the vitro action of carboxyfullerene in detail. Fullerene compounds have avid reactivity with free radicals and are regarded as "radical sponges." The trimalonic acid derivative of fullerene is one of the water-soluble compounds that has been synthesized and found to be an effective antioxidant both in vivo and in vitro. Carboxyfullerene has been shown to be effective in the treatment of both gram-positive and gram-negative infections, although its mode of action is poorly understood. In this study, all gram-positive species were inhibited by 50 mg/L of carboxyfullerene, but gram-negative species were not inhibited, even at 500 mg/L of carboxyfullerene. Bactericidal activity was demonstrated only for gram-positive species, particularly for *Streptococcus pyogenes* A-20, which was killed rapidly. Intercalation of carboxyfullerene into the cell wall of staphylococci and streptococci was demonstrated by transmission electron microscopy (TEM) and the anti-carboxyfullerene binding assay. Damage to the cell membrane in gram-positive (but not gram-negative) bacteria was confirmed by the membrane perturbation assay. These findings indicate that the action of carboxyfullerene on gram-positive bacteria is achieved by insertion into the cell wall and destruction of the membrane integrity.

25.3.2   In Vivo Studies of Fullerene-Based Materials Using Endohedral Metallofullerene Radiotracers

Cagle et al. (1998) have demonstrated the feasibility of using water-solubilized metallofullerene [$^{166}Ho_x@C_{82}(OH)_y$] radiotracers to monitor the fate of fullerene-based materials in animals. They showed that 20% of it was excreted within 5 days, suggesting that water-solubilized fullerene materials, in general, may be useful components in drug design.

25.3.3   Cellular Localization of a Water-Soluble Fullerene Derivative

Fullerenes are a new class of compounds with potential uses in biology and medicine. Many insights have been made in the knowledge of their interaction with various biologic systems. However, their interaction with organized living systems as well as the sites of their potential action remains unclear. A fullerene derivative could cross the external cellular membrane and localize preferentially to the mitochondria. (Foley et al., 2002). Because their structure mimics that of clathrin, which is known to mediate endocytosis, the potential use of fullerenes as drug-delivery agents is being envisaged.

25.3.4 Ecotoxicology of Carbon-Based Engineered Nanoparticles

To more fully assess the toxicity of water-soluble fullerenes (nC_{60}), acute toxicity assays have been performed (Oberdörster et al., 2005) on several environmentally relevant species, including freshwater crustaceans *Daphnia magna* and *Hyalella azteca*; a marine *Harpacticoid copepod*; and two fish species, the fathead minnow *Pimephales promelas* and the Japanese *Oryzias latipes*. The latter two species have been used to assess the sublethal effects of fullerene exposure by assessing mRNA and protein expression in the liver.

It was found that solubilization of fullerenes (either by sonication or by using tetrahydrofuran) increased the toxicity of nC_{60}. For the invertebrate studies, nC_{60} could not be prepared at high enough concentration levels to cause 50% mortality (LC_{50}) at 48 or 96 hours. The maximum concentrations tested were 35 ppm for freshwater and 22.5 ppm for full-strength (35 ppt) seawater because at higher concentrations, the nC_{60} precipitated out of solution. Daphnia 21 day exposures resulted in a significant delay in molting and significantly reduced offspring production at 2.5 and 5 ppm nC_{60}, which could possibly produce impacts at the population level. In the case of the fish, it was found that neither the mRNA nor protein-expression levels of cytochrome P450 isozymes CYP1A, CYP2K1, and CYP2M1 were changed. The peroxisomal lipid transport protein PMP70 was significantly reduced in the fathead minnow, but not in the medaka, indicating potential changes in acetyl-CoA pathways.

25.3.5 Cytotoxicity of Water-Soluble Fullerenes in Vascular Endothelial Cells

Nanoscale materials are presently under development for diagnostic, nanomedicine, and electronic purposes. In contrast to the potential benefits of nanotechnology, the effects of nanomaterials on human health are poorly understood. Nanomaterials are known to translocate into the circulation and could thus directly affect vascular endothelial cells (ECs), causing vascular injury that might be responsible for the development of atherosclerosis.

To explore the direct effects of nanomaterials on endothelial toxicity, human umbilical vein ECs were treated with 1 to 100 µg/mL of hydroxyl fullerene (C60[OH]24; mean diameter, 7.1 ± 2.4 nm) for 24 hours. C60(OH)24 induced cytotoxic morphologic changes such as cytosolic vacuole formation and decreased cell density in a dose-dependent manner. The lactate dehydrogenase assay revealed that a maximal dose of C60(OH)24 (100 µg/mL) induced cytotoxic injury. The proliferation assay also showed that a maximal dose of C60(OH)24 inhibited EC growth. C60(OH)24 did not seem to induce apoptosis, but caused the accumulation of polyubiquitinated proteins and facilitated autophagic cell death. Chronic treatment with low-dose C60(OH)24 (10 µg/mL for eight days) inhibited cell attachment and delayed EC growth. Only maximal doses of fullerenes caused cytotoxic injury or death and inhibited cell growth. EC death seemed to be caused by activation of ubiquitin-autophagy cell death pathways, although exposure to nanomaterials appears to represent a risk for cardiovascular disorders.

25.3.6   Photocatalytic Degradation of *Escherichia coli* and *Staphylococcus aureus* Microbes by Multiwalled Carbon Nanotubes

Carbon nanotubes (CNTs) can be either metallic or semiconducting in nature, depending on their diameter. Their photocatalytic behavior has given an impetus to use them as antimicrobial agents. More than 95% of *Escherichia coli* and *Staphylococcus aureus* bacteria were killed when exposed to CNT for 30 minutes in the presence of sunlight (Sharon et al., 2007). CNTs are supposed to have smooth surface onto which they accumulate positive charges when exposed to light. The surface, which is exposed to light, accumulates positive charges at the surface that is nonilluminated and has negative charges. At the cellular level, microorganisms produce negative charges on the cell membrane. Therefore, the damaging effect of multiwalled CNTs (MWCNTs) exposed to light on the microorganisms is possible. Sharon et al. (2007) have reported the photocatalytic killing of microbes by MWCNTs. Killing was due to damage in the cell membrane, as seen in scanning electron micrographs.

Moreover, biochemical analysis of membrane as well as total cellular proteins by SDS-PAGE (sodium dodecyl sulfate polyacrylamide gel electrophoresis) showed that there was denaturation of membrane proteins as well as total proteins of both microbes studied. The killed microbes that showed a decrease in the number of protein bands (i.e., due to the breakdown of proteins) also showed an increase in the level of free amino acids in the microbes. This further confirmed that proteins got denatured or broken down into shorter units of amino acids. Increased levels of free amino acids were recorded in both the microbes treated with MWCNTs and sunlight (Figs. 25-3 and 25-4).

Thus, not only microbes but also animal cells are being tried for maintaining asepsis in foodstuffs. But the applications of antimicrobial activity is not only limited to food items or water purification; there are also other applications, which are mentioned in the following sections.

25.4   Application of the Antimicrobial Properties of Carbon Nanomaterial

25.4.1   Water Purification

Water purification, or drinking water treatment, is the process of removing contaminants from surface water or groundwater to make

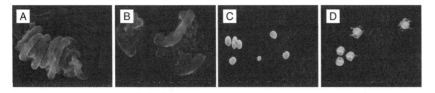

FIGURE 25-3   Scanning electron micrograph (SEM) of *Escherichia coli*. (A) Control SEM. (B) After treatment with 100 ppm of MWCNT and sunlight, showing ruptured cell membranes of *E. coli* and some disintegrated cells. (C) Healthy control cells. (D) Cells with ruptured cell membranes of *Staphylococcus aureus*.

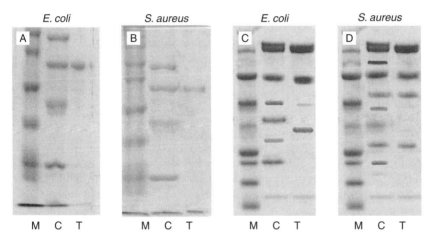

Notes: C = Control; M = Marker protein; T = Treated.

FIGURE 25-4   SDS-PAGE (sodium dodecyl sulfate polyacrylamide gel electrophoresis) of membrane proteins of *Escherichia coli* (A) and *Staphylococcus aureus* (B) showing only one surviving band of membrane protein and of total cellular proteins of *E. coli* (C) and *S. aureus* (D) showing only five surviving bands of membrane protein after denaturation due to the photocatalytic activity of multiwalled carbon nanotubes.

it safe and palatable for human consumption. A wide variety of technologies may be used, depending on the raw water source, contaminants present, standards to be met, and available financial resources. Many techniques are used for purification of water. The most common ones are carbon filtering, solar disinfection, and chlorination or iodination.

Carbon Filtering

Charcoal, a form of carbon with a high surface area due to its mode of preparation, adsorbs many compounds, including some toxic compounds. Water is passed through activated charcoal to remove such contaminants. This method is most commonly used in household water filters and fish tanks. Household filters for drinking water sometimes also contain silver; trace amounts of silver ions have a bactericidal effect.

Activated carbons are created from wood, coal, or coconut shells that have been ground up and activated by heating at a controlled temperature and pressure to promote the active sites where pollutants can be adsorbed. Adsorption is the physical process in which certain water pollutants are attracted to the surfaces of the carbon rather than to the water. The pollutants are locked into the carbon and are removed from the water. Activated carbons are also able to filter out sediment through mechanical filtration by trapping particles in the spaces between the carbon granules. Activated carbon has been used for hundreds of years to treat taste, odor, and color problems. Major water utilities and water treatment manufacturers have found activated carbon to be an excellent medium to produce better-tasting water and remove harmful water contaminants at a reasonable cost.

Solar Disinfection

Microbes are destroyed through temperature and ultraviolet radiation provided by the sun. Water is placed in a transparent plastic bottle, which is oxygenated by shaking, followed by topping up. It is placed on tile or metal for six hours in full sun, which increases the temperature and gives an extended dose of solar radiation, killing any microbes that may be present. The combination of these two techniques provides a simple method of disinfection for tropical, developing countries.

Chlorination or Iodination

Chemical additives include chlorine dioxide and iodine solutions. Iodine, in solution, crystallized, or in tablets, is added to water. Iodine kills off many, but not all, of the most common pathogens that may be present in natural fresh water sources such as lakes, rivers, and streams. Carrying iodine for water purification is a lightweight but imperfect solution for those in need of field purification of drinking water. Chlorination is done and microbes are killed effectively.

Each of these techniques for filtration of contaminated water has a drawback. Activated carbon can only filter the microbes, but killing of microbes is still a question. If we consider sodis treatment, then the required exposure time to sunlight is minimum 6 hours; in today's high-tech world, 6 hours is truly a long time; hence, killing of microbes has to be done at faster rate. Also, microbes are not completely killed in the iodination technique, so again this method not very useful. Moreover, chlorination is harmful, considering environmental pollution. There is one solution to this problem—the development of carbon nanofilters.

Nanofilters using CNTs have been developed. So far, carbon nanobeads (CNBs) have not being worked out for this purpose, which has been developed by Sharon's group in India. (Mukhopadhya et al., 1998) There are many challenges to be faced while using CNBs (e.g., their properties, such as the band gap and required sized, have to be determined).

The advantages of using CNBs are that they are still smaller and have more surface area compared with CNTs, so they attempt to develop nanofilters. CNBs will be a better option for making nanocolumns, which resemble the column filtration assembly with a unique feature that has the capacity to filter and kill microorganisms.

The basic challenge in water purification is to minimize the microbial load in the sample. For this purpose, *E. coli* can be used as an indicator organism. The principle is very simple—the photocatalytic properties of semiconductor carbon are used. The idea is that when a semiconductor is in contact with any other material, such as a solution of some inorganic salt or a buffer solution, a depletion region is formed (see Chap. 18). When this depletion region is illuminated with a light source of energy equivalent to the band gap of the material, the interface gets accumulated with photo-generated electrons and holes. These photo-generated carriers are highly reactive. Because of their high reactivity, if they come in contact with any organic material, the material is completely decomposed. *E. coli* or any other living organism has a membrane made out of organic materials such as lipoproteins. When these membranes come in contact with the photo-generated electron–holes, they get decomposed, killing the bacteria.

The various reactions that take place at the interface of the semiconductor and the electrolyte are shown in the following equations. The illumination of the interface of the semiconductor and the solution produces the following reaction:

$$\text{n-Semiconductor} + h\nu \rightarrow e^-_{\text{photo}} + h^+_{\text{photo}} \qquad (25.1)$$

Water gets oxidized by h^+_{photo} to form hydroxy radicals (OH•):

$$H_2O + h^+_{\text{photo}} \rightarrow OH^\bullet + H^+ \qquad (25.2)$$

These hydroxy radicals or h^+_{photo} and e^-_{photo} can perform reactions with the cell membrane:

$$\text{Cell membrane} + OH^\bullet \rightarrow CO_2 + \text{Degraded organic product} \quad (25.3)$$

$$\text{Cell membrane} + h^+_{\text{photo}} \rightarrow CO_2 + \text{Degraded organic product} \quad (25.4)$$

$$\text{Cell membrane} + e^-_{\text{photo}} \rightarrow \text{Reduced organic product} \qquad (25.5)$$

In this fashion, pathogenic bacteria in water can be killed, making the water free of bacteria.

This type of approach was used with CNTs with a band gap 1.2 eV. Because of the low band gap, visible light was used to study this reaction.

E. coli exposed to	No. of E. coli per mL	% Survival of E. coli
Light (30 min)	3.5×10^6	87.50
CNTs	3.60×10^6	90.00
Light (30 min) + CNTs	1.6×10^6	03.75

CNT = carbon nanotube.

TABLE 25-1 Photocatalytic Killing of *Escherichia coli* by Carbon Nanotubes

In other words, solar radiation can be used for this purpose. The results of this exercise are given in Table 25-1, which suggests that in 30 minutes of exposure of water containing bacteria and CNTs, only 3.75% of the total bacteria would survive. This experiment also suggests that CNTs are not toxic to the growth of the bacteria.

25.4.2 Effect of Carbon Nanomaterials on Plant and Algal Cells

The effects of CNMs on plant cells have also been studied. Unlike microbes, plants have rigid cell walls made up of cellulose and lignins. It has been observed that the cell walls of plant cells also get ruptured when exposed to CNTs in the presence of sunlight. However, compared with microbes, higher concentrations of CNTs and exposure to sunlight are required.

This property will have an application in killing undesired algal cells that accumulate during rainy seasons on floors and walls. Maybe paints will have CNMs incorporated into them to stop algal accumulation on walls.

25.4.3 Carbon Nanotubes as Sterilants

Hydrogen-free CNT composite films have been created that possess unusual free radical generation properties (Narayan, 2004). These films have the ability to nonspecifically kill bacterial cells. They can be used to sterilize a surface. These composites have been formed by simultaneous pulsed laser ablation of carbon and bombardment of nitrogen ions generated by a Kaufman ion source at high temperatures (600° to 700°C). The results suggest the creation of a new form of carbon that is predominantly trigonally coordinated, with small fractions of carbon–nitrogen bonds. TEM studies have allowed us to conclude that the material consists of sp^2-bonded ribbons wrapped approximately 15 degrees normal (perpendicular) to the surface. Plain-view, high-resolution TEM specimens demonstrate layers with curvatures similar to those seen in MWCNT structures. In addition, the interlayer order extends to approximately 15 to 30 nm. These novel structures result from the use of energetic ions that create nonequilibrium conditions that alter the growth mode of graphitic planes. In vitro testing revealed significant antimicrobial activity against *S. aureus* bacteria. Possible applications include use on the functional surfaces of dialysis equipment, scalpels, and other sterile equipment.

25.4.4   Nanostructured Bandages, Surfaces, and Textiles

The development of products that will encourage cell growth and reduce infection are in the offing. Thus, the challenge for antimicrobial activity will take more consideration in the use of CNMs because all the biologic systems are not the same . But in the near future, we will surely benefit greatly from this technology.

25.5   Summary

CNTs are endowed with either a metallic or a semiconducting property depending on the diameter of the CNT. Their photocatalytic behavior has given an impetus to use CNTs as antimicrobial agents.

The antimicrobial property (i.e., the ability to kill microbes) has been found to be caused by disrupting the cell membrane. Moreover, CNTs have also been able to disrupt the plant cell wall and membrane; hence, various applications are being envisaged such as killing unwanted algae and as a sterilant.

CNTs' antimicrobial capacity and their use as a filter material can solve and improve water purification and filtration. CNTs are being thought of as a better alternative to existing water purification and filtration technology. Along with this, the basic application of CNM is water purification, as a sterilant, and for use in bandages to clean surfaces of the skin that will be bacteria proof.

CHAPTER 26

Food, Cosmetics, and Carbon Nanotechnology

Suprama Datta

Nanotechnology Research Centre, Birla College, Kalyan, Maharashtra, India

Madhuri Sharon

MONAD Nanotech Pvt. Ltd., Powai, Mumbai, India

The real power of nanoscale science is the potential to converge disparate technologies that can operate at this scale. With applications spanning in all the industry sectors, technological convergence at the nanoscale is poised to become the strategic platform for global control of manufacturing, food, agriculture and health in the immediate years ahead.

H. Shand and K. Wetter (2006)

26.1   Introduction

Nanotechnology is the "ability to remake the world from the atom up." Nanotechnology controls the building blocks of both living and nonliving things, and nanotechnology now embodies the dream of its indomitable monopoly in every corner of human civilization. From the multifarious application of nanosciences, the cosmetic industry has not remained untouched, and even food sectors are poised for benefits from the technology.

26.1.1   Cosmetic Industry

It is difficult to estimate the number of cosmetics, sunscreens, and personal care products containing nanoparticles that are now commercially available. The list of cosmetics and toiletries that now contain nanomaterials include sunscreens, antiaging and antiwrinkle creams, lipsticks, face powders, moisturizers, toothpaste, soap, deodorants, shampoos, aftershave lotions, and many more. The present market is endowed with products that are more readily available and cheaper and yet without compromising the product quality, which is possible only because of the convergence with nanotechnology.

26.1.2   Food Industry

During the past 5 years, nanotechnology has been integrated into a number of food and food packaging products. It is the idea of "cooking at the-bottom." Dr. Heini Menzi, vice president for European R&D of Givaudan, a Swiss company) says:

> The public assumes that natural means healthy, that NI (nature identical) means dubious—and that's not true. . . . All food is only chemistry, after all—but that sort of language is scary to the general public. So people become concerned, "Oh, this is artificial. . . ." And to some extent, yes, it is. But the chemist can get to the same molecule, whether he takes

it physically out of the raw plant or via synthesis. From a taste point of view, you could use either.

   Would it not feel great having a grape juice without natural extracts but made from just artificially synthesized and engineered chemical molecules? An optimal amount of trans-2-hexenal can enhance a sweet's apple flavoring. A slight change in the recipe with added heptyl acetate can turn the apple flavor to that of a pear and make it cheaper, but safely processed. Moreover, naturally sourced flavors may include all sorts of undesirable residues left over from the farm, including herbicides, pesticides, microorganisms, and even levels of plant toxins that could be harmful.

   Production of flavorings is just one example of the application of nanotechnology in the food flavor and color improvement sector of the food industry. Other segments of the food industry that can benefit from nanotechnology are:

- Food production
- Food processing
- Food preservation
- Food safety
- Food packaging

   Food products containing nanomaterials now include canola oil, chewing gum, meal-replacement milkshakes, vitamins, and color additives. Most of the big food companies have nanoresearch and development projects underway, and the use of nanotechnology in food production and packaging is expected to expand significantly over the next 10 years. With the technology of manipulating the molecules and the atoms of food, the future food industry has a powerful method to design food with much more capability and precision, lower costs, and sustainability.

26.2   History

Several thousands of years ago, Indians, Egyptians, Greeks, and Romans used nanotechnology to dye their hair without being aware that the particles they were using were of nanosize. Dr. Philippe Walter from the Centre de Recherche et de Restauration des Musées de France in Paris in the recent *Nanoweek* news release "Nanotechnology in Cosmetics—2000 Years Ago," stated:

> For thousands of years, cosmetics have been used and were made by the judicious combination of naturally available minerals with oils, various creams, or water. Since the Greco-Roman period, organic hair dyes obtained from plants such as henna have been used, but other unusual

formulas based on lead compounds, such as the recipes describing several methods to dye hair and wool black, were also common. It is remarkable that these Greco-Roman techniques have been used up to modern times.

In recent experiments, Walter et al. (2006) showed that an ancient dyeing process for coloring hair black is a remarkable illustration of synthetic nanoscale biomineralization.

26.3   Commercial Impact of Nanomaterials

Nanotechnology is a multibillion dollar industry, which is expected to grow to 1 trillion US$ by 2015. The majority of nanomaterial manufacturing and use occur in the United States (49%), with the European Union responsible for 30% and the rest of the world accounting for the remaining 21%. Within the European Union, the United Kingdom accounts for nearly one third of the market (Aitken et al., 2005).

The cosmetic and food industries are among the greatest revenue-producing areas for nanoparticles (although not carbon nanoparticles in particular).

26.3.1   Nanocosmetic Market

Nowhere are nanomaterials entering manufacturing and reaching the consumer faster than in personal care products and cosmetics. In 2004, the United Kingdom's Royal Society noted that of the engineered nanomaterials in commercial production, the majority were being produced for use in the cosmetic industry.

According to the report "Nanomaterials, Sunscreens and Cosmetics Are Small Ingredients and Big Risks" by Friends of Earth, which focuses on the use of nanoparticles in the personal care industry; this sector is one of the primary early adopters of nanomaterials. A survey by the Woodrow Wilson Center for International Scholars, which included 116 products, found that 71 cosmetics, 23 sunscreens, and 22 personal care products incorporate nanomaterials. The Australian Therapeutic Goods Administration has stated that there are close to 400 sunscreen products alone that contain nanoparticles and certain antiaging face creams and moisturizers, particularly containing carbon fullerenes, that are currently commercially available in Australia. These data represent a small fraction of personal care products containing nanomaterials that are currently on the market and may not reflect the overall pattern of nanoparticles used across these sectors.

26.3.2   Nanofood Market

The nanofood market has soared from $2.6 billion in 2003 to $5.3 billion in 2005 and is expected to reach $20.4 billion by 2010. The nano-featured food packaging market will grow from $1.1 billion in 2005 to $3.7 billion in 2010. More than 400 companies around the world are active in research, development, and production. The United States is the leader followed by Japan and China. According to Helmut Kaiser, more than 200 companies worldwide are engaged in nanotechnology research and development related to food. Among the 20 most active companies are five that rank among the world's 10 largest food and beverage corporations. By 2010, Asia, with more than 50% of the world's population, will become the biggest market for nanofood, with China in the leading position.

Any food company that wants keep its leadership in food industries is set up to work with nanotechnology as the immediate response. The impact of nanotechnology is huge, ranging from basic food to food processing, from nutrition delivery to intelligent packaging. It is estimated that the nanotechnology and nano-bio-info convergence will influence more than 40% of food industries by 2015. The risk for the food companies lies not in entering the nanotechnology field, but in entering it too late.

26.4 Nanomaterials Used in the Cosmetic and Food Industries

26.4.1 Nanoparticles and Nanopowder

Nanoparticles and nanopowders, such as ZnO_2 and TiO_2, are used in sunscreens, and silicate nanoparticles are used in plastic food packaging by Bayer.

26.4.2 Nanocapsules

Nanocapsules are novel carriers that can be incorporated into nutraceuticals and in food systems. These vehicles transmit nutrients, such as lycopene, lutein, coenzyme Q10, phytosterol, and vitamin D, in food products. The addition of nanocapsules in cosmetic formulations increases the bioavailability of the product.

26.4.3 Nanofibers

Nanofibers are also being used in cosmetics.

26.4.4 Fullerenes

Carbon fullerenes, Buckminster fullerene, and buckyballs (C60) are used as antioxidants in many face creams. Fullerene serves as the most predominant use of carbon nanomaterials in the cosmetic industry.

26.4.5 Single-Walled Carbon Nanotubes and Multiwalled Carbon Nanotubes

Single-walled and multiwalled carbon nanotubes (SWCNTs and MWCNTs) aid in the food packaging industry. They are being used as fillers in food packaging products and additives in foods and drinks and are on the verge of being used in cosmetics.

26.5 Use of Carbon Nanomaterials in Cosmetics

As discussed earlier, the CNMs mostly used in the cosmetic industry are fullerenes, SWCNTs, and MWCNTs.

26.5.1 Fullerenes

Fullerenes or buckyballs are compounds composed solely of an even number of carbon atoms, which form a cage-like fused-ring polycyclic system with 12 five-membered rings and the rest six-membered rings. The archetypal example is C_{60} fullerene, in which the atoms and bonds delineate a truncated icosahedron.

The most remarkable claimed property of the fullerenes used in the cosmetic industry is their antioxidant (elimination of free radicals linked to aging and cell death) properties. The antioxidant properties of fullerenes, which serve to "mop up" free radicals, are exploited in the cosmetic and food industries.

According to multiple research studies on the antioxidant action of fullerenes, fullerenes act like a sponge for free radicals (due to their structure of the cage, bonding configuration, and high number of available bonding sites), effectively neutralizing the power of the free radicals to damage tissue. Fullerenes have shown to be up to 100 times more effective than commonly used compounds as vitamin E and coenzyme Q10. Additional factors, such as promising results in preliminary toxicity tests and their suitability for water-soluble and lipid-soluble derivatives, reveal fullerenes as a very promising antioxidant material for health and personal care products (e.g., antiwrinkle and antiaging skin creams, and burn creams). Moreover, their nanoscale size renders efficient transport to the needy cells and hence achieves appropriate targeting. Mitsubishi Corp. and ITO announced a joint venture in 2003 focused on antiaging and antiwrinkle creams, with a fullerene-based product planned for commercialization. Scientists are also working on using C_{60} in fullerene-based after-burn creams (Table 26-1).

26.5.2 Single-Walled Carbon Nanotubes and Multiwalled Carbon Nanotubes

CNTs are small cylinders of carbon atoms. Unlike other carbon materials such as diamond and carbon black, CNTs have a completely different structure that gives them interesting and very promising properties. There are also different types of structures related to CNTs. When combined into macrostructures, they are called *nanotube ropes*. When they resemble Greek scrolls, they are referred to as *nanoscrolls*. Nanohorns, on the other hand, take the shape of a cone instead of a

Product	Manufacturer	Website reference
Defy: Age Management Exfoliator	Bellapelle Skin Studio	http://www.bellapelle.com/ products/products_skin_nourish.php
EGF Complex Cocktail	Bellapelle Skin Studio	http://www.bellapelle.com/ products/products_skin_nourish.php
Nourish	Bellapelle Skin Studio	http://www.bellapelle.com/ products/products_skin_nourish.php
Dr. Brandt New Lineless Cream	Dr. Brandt	http://www.drbrandtskincare.com
Revitalizing Night Cream	MyChelle Dermaceuticals LLC	http://www.mychelleusa.com/ revital_night_cream.htm
Sircuit Addict/ Firming Antioxidant Serum	Sircuit Cosmeticeuticals	http://manstouch.com/skinbody/ sircuitskin.html
White Out/Daily Under Eye Care	Sircuit Cosmeticeuticals	http://manstouch.com/skinbody/ sircuitskin.html
Zelens Fullerene C-60 Day Cream	Zelens	http://www.zelens.co.uk/
Zelens Fullerene C-60 Night Cream	Zelens	http://www.zelens.co.uk/

TABLE 26-1   Cosmetic Companies Using Fullerenes in their Products

tube. CNTs can be either "closed" or "open" ended. CNT possesses some remarkable properties, which make them a suitable material to be used in the cosmetic industry.

Structural Properties of Carbon Nanotubes
The structural property of CNTs makes them of potential use in the cosmetic industry. Because CNTs are hollow cylinders and can be open-ended, closed-ended, or capped, they can serve as containers for biologic molecules and ferry the molecules to the target cells.

Chemical Properties of Carbon Nanotubes
The chemical property of CNTs aids in carrying out reactions at the nanoscale level, where the nutrient (vitamin D, coenzyme Q10) is precisely targeted to the desired cells.

26.6   Use of Carbon Nanomaterials in Food Processing and Packaging

26.6.1   Fullerenes in the Food Industry

Because fullerenes are free radical scavenging molecules (i.e., antioxidants) and their preliminary toxicity tests have approved their usage in certain food products, they are now being tried in food-related industries.

26.6.2   Carbon Nanotubes in the Food Industry

CNTs' tensile and compressive strength are being made use of in the food packaging industry. Researchers have found that adding CNTs to polyethylene increases the polymer's elastic modulus by 30%. In addition, it provides a much slicker surface than Teflon, which is waterproof. Mitra and Zafar (2005) have found that by heating CNTs in a closed-vessel microwave oven, they can be chemically modified without damaging their essential structure. This novel method of functionalization has improved the solubility of CNTs, leading to new types of packaging. Potential applications of functionalized CNTs are believed to include the following.

Antimicrobial Packaging

CNTs will scavenge the oxygen molecules and other microorganisms from entry into food products, thereby keeping it fresh.

Embedded Sensors in Nanotube "Electronic Tongue" Technology

In "electronic tongue" technology, the sensors could detect substances in parts per trillion and would trigger a color change in the packaging to alert the consumer if a food has become contaminated or if it has begun to spoil, hence detecting the entry of foodborne pathogens.

Intelligent Food Packaging

The CNTs in the packaging material will release a preservative if the food inside begins to spoil. This preservative release is operated by means of a bioswitch, again developed by nanotechnology.

Super Sensors

CNTs will lead to instantaneous detection of certain bacteria and hence will play a crucial role in the event of a terrorist attack on the food supply.

Future Possibilities of Using Carbon Nanotubes in the Food Industry

Scientists are envisaging that the use of CNMs will enable the food industry to predict, control, and improve agricultural production.

Field of food packaging	Institutes and companies
Antimicrobial packaging and active packaging	Kodak
Electronic tongue technology	Kraft, Rutgers University, University of Connecticut
Intelligent food packaging	Researchers in The Netherlands
Super sensors	BioVitesse, a start-up company supported by Purdue University

TABLE 26-2   Institutes and Companies Involved in the Application of Nanotechnology in the Food Packaging Industry

With the technology of manipulating the molecules and the atoms of food, the future food industry will have a powerful method of designing food with much more capability and precision, lower costs, and sustainability.

Moreover, the combination of DNA and nanotechnology research is expected to generate new nutrition delivery systems, which brings the active agents more precisely and efficiently to the parts of human bodies and cells that need them. Functional food will benefit from the new technology in food preservation. Future nanosensors would tell us when the food starts to be spoiled; the packaging will change color to alert the consumer.

Food manufacturers will produce food with less waste and use less energy, resulting in more production efficiency and environment safety (Table 26-2).

26.7   Toxicity Concerns of Carbon Nanomaterials in the Cosmetic Industry

The primary concerns about the health and safety impact of CNMs arise both from consideration of the inherent nature and novel properties of CNMs and from the surprising results seen in many of the relatively small number of nanotoxicity studies conducted to date. In some cases, the very properties that make CNMs uniquely useful in commercial applications (e.g., in the cosmetic and food industries) also raise the potential for novel mechanisms and targets of toxicity.

The effect of C_{60} fullerenes are not largely studied in mammalian models. A recent study of fullerenes found that although individual fullerenes do not dissolve well in water. They have a tendency to form aggregates that are both very water soluble and bactericidal; a property that raises strong concerns of ecosystem impacts because bacteria constitute the bottom of the food chain in many ecosystems.

C_{60} fullerenes are also capable of being transported via the gills from water to the brains of fish (Oberdörster et al., 2005), where they can cause oxidative damage to brain cell membranes. In experiments with human cultured cell lines, fullerenes show high toxicity, causing oxidative damage to cell membranes that leads to cell death.

Studies in which CNTs were instilled into the lungs of rodents have consistently demonstrated that within 30 days, CNTs cause unusual localized immune lesions (granulomas) and other signs of lung inflammation. SWCNTs caused dose-dependent fibrosis in lung at sites far removed from the sites of particle deposition. One study of MWCNTs showed similar lung toxicity. Oxidative stress may be part of the mechanism behind the damage to lung tissue that has been observed in these studies of CNTs.

SWCNTs and MWCNTs have also been shown to induce oxidative stress in skin cells. These studies raise concern for potential toxicity at the beginning or end of the lifecycle of products containing CNTs. The possibility of a toxic effect is also envisaged through workplace exposures of CNT-containing products that undergo weathering, erosion, or grinding during recycling or disposal. Moreover, CNTs are reported to stimulate blood clotting by interacting with platelets either by aggregating themselves or by activating a receptor on platelet–integrin receptors.

Although scientists and toxicologists have not yet reached any consensus on the level of toxicity of CNTs, there are indications that under certain circumstances, CNTs might be hazardous. If CNTs are to be used in novel products, then they are likely to come in contact with the human body through the skin and access the inner organs through the respiratory and digestive tracts. During manufacture, CNTs may enter the respiratory airways of the workers and accumulate in their lungs. If CNTs are used as fillers in food packaging products, then they could reach the stomachs and intestines of consumers. If cosmetics and biofilters are developed using nanotubes, then they will certainly come in contact with human skin.

Various potentials of CNMs that have been used in the food and cosmetic industries that may cause concerns include:

- **CNMs may penetrate cell membranes.** This property, although exploited as a means of carrying the nutraceuticals to the skin cells or the digestive tract, suggests that they can cross physiologic barriers.

- **CNMs may cross physiologic barriers (e.g., lung–blood and blood–brain) and enter the systemic circulatory system.** Hence, they can enter the body compartments, which are not readily accessed, by larger particles and other smaller molecules. Carbon nanoparticles of different size gain entry into the body cells via various mechanisms:

- >500 nm: By active endocytosis
- <200 nm: By variety of active and nonactive mechanisms
- ≤20 nm: Directly pass through the cell membrane without requiring specific transport mechanisms.

After they are inside the cell, CNMs are distributed in the cytoplasm and appear to bind to a variety of cell structures. The manner in which different individual and aggregated carbon nanoparticles may interact with critical cell structures is poorly understood.

- **Translocation of inhaled carbon nanoparticles from the lung to the brain or into the systemic circulation may take place.** CNMs can deposit throughout the respiratory tract when inhaled. Some of the particles settle in the nasal passages, where they have been shown to be taken up by the olfactory nerves and carried past the blood–brain barrier directly into brain cells. Smaller CNMs have been shown not only to penetrate deeply into the lungs, but to also readily cross through lung tissue and enter the systemic circulation. These and other studies suggest that some CNMs can evade the lung's normal clearance and defense mechanisms. This potential for rapid and widespread distribution within the body offers the promise of a new array of diagnostic and therapeutic applications for these substances, but it also heightens the importance of having a full understanding of their toxicity.

- **CNMs may stimulate blood clotting.** Researchers from The University of Texas Health Science Center at Houston and at Ohio State University examined the impact of various forms of carbon nanoparticles in laboratory experiments on human platelets (blood's principal clotting element) and in a model of carotid artery thrombosis, or blockage, using anesthetized rats. Carbon nanoparticles interact with platelets and also aggregate on their own, which may enhance blood clotting (Radomski, et al., 2005). Medical evidence has been accumulating mainly from epidemiologic studies that exposure of humans to particulate matter and to very small particles increases the risk of cardiovascular disease. Except for C_{60} fullerenes, mixed carbon nanoparticles, SWCNTs, and MWCNTs had a provoking impact on human platelet clumping and thrombosis in rats. Mixed carbon nanoparticles showed the greatest impact, SWCNTs had the second greatest impact, and MWCNTs were third. These nanoparticles were also shown to activate a receptor—the glycoprotein integrin receptor—on platelets that is vital to their aggregation. Although the mechanism is not yet clear, each particle used a different molecular pathway to activate the receptors. This research suggests a risk of thrombosis from airborne pollution in addition to the risk of atherosclerosis and heart attack.

- CNMs may evade the body's usual metabolic and immune defense mechanisms.

- CNMs may directly interact and possibly interfere with cellular components.

- CNMs may deliver secondary molecules to intracellular targets or reach nontarget cells or organs.

- CNMs may persist and accumulate in the body or environment.

There is currently little understanding of the mechanisms that lead to the biologic effects that have been observed in toxicity studies. Scientists are beginning to examine the extent to which these behaviors can result in significant toxicologic impacts and at what levels of exposure toxicity occurs. Such effects include the potential to kill skin cells in culture; damage brain tissue in mammals and fish; impair lung function and generate unusual granulomas in the lungs of rodents; and kill microorganisms, including ones that may constitute the base of the food chain.

Irrespective of the immense utilization of fullerenes and CNTs as the sole source of CNMs in the cosmetic and food industries, toxicity studies of both have raised concerns regarding the consumption of nanofood and nanocosmetics.

C_{60} fullerenes have been less studied in mammalian models. A recent study of fullerenes found that although individual fullerenes do not dissolve well in water, they have a tendency to form aggregates that are both water soluble and bactericidal. This property raises strong concerns for ecosystem impacts because bacteria constitute the bottom of the food chain in many ecosystems. C_{60} fullerenes are also capable of being transported via the gills from water to the brains of fish, where they can cause oxidative damage to brain cell membranes. In experiments with human cultured cell lines, buckyballs show a high toxicity, causing oxidative damage to cell membranes that leads to cell death.

Although there is a lack of data on chronic toxicity of CNMs after ingestion, they are thought to readily enter the human body and gain access to the bloodstream via ingestion and crossing the blood–brain barrier. As soon as they cross the physiologic barriers and enter the highly protected body compartments, they are likely to invade the cells by penetrating their membranes and binding with various cell structures. Further study regarding the aggregated carbon nanoparticles formed is not yet clear.

26.8   Toxicity Concerns of Carbon Nanomaterials in the Food Industry

These carbon nanoparticles are also much more mobile than larger-sized particles. It appears likely that such carbon nanoparticles can penetrate the skin directly as stated by the NANODERM Project report (dated 2008) that a possible risk factor to skin penetration is revealed by tape stripping studies.

Skin has a strong external barrier, the stratus corneum, that protects the internal organs from environmental exposure. Although healthy skin is generally impervious to particle exposure, such particles in liposomal formulations penetrate the horny layer with large concentrations at the stratum granulosum, thus developing a possible risk for further penetration into vital tissue, especially for skin with an impaired barrier function.

26.9   Summary

The introduction of carbon nanotechnology in the cosmetic and food industries is waiting for a new era of revolution. It is claimed that with the convergence of both, the manufacture of health care and food products has become easier and cheaper as well, yet with an increased bioavailability.

Despite the potential of these tiny particles, their very novel properties, which have been made use of in cosmetic and food products, pose risks for consumers and the environment as well.

Before both industries are allowed to introduce CNMs into personal care products and food products, there is a need to define the substantiation of safety means for nanomaterials, identify the presence of nanomaterials in all products, and obtain information on particle size from all personal care product manufacturers.

Above all, it is for the benefit of humankind that nanotechnology has been made to go hand in hand with several such industries, so the positive aspects of nanotechnology should be used while evading its adverse side effects.

CHAPTER 27

Toxicity versus Biocompatibility of Carbon Nanotubes

Shailesh Khatri, Prashant Ambilwade, Amruta Phatak and Maheshwar Sharon

Nanotechnology Research Centre, Birla College, Kalyan, Maharashtra, India

Madhuri Sharon

MONAD Nanotech Pvt. Ltd., Powai, Mumbai, India

It is a mistake for someone to say nanoparticles are safe, and it is a mistake to say nanoparticles are dangerous. They are probably going to be somewhere in the middle. And it will depend very much on the specifics.

V. Colvin, Director of Center for Biological and Environmental Nanotechnology, Rice University, quoted in Technology Review.

27.1   Introduction

The discovery, characteristics, properties, synthesis methods, and some of the applications of carbon nanomaterials (CNMs) have been discussed in previous chapters. In this chapter, the toxicologic and biocompatibility issues related to the potential use of carbon nanotubes (CNTs) in pharmaceutical products and their features that affect toxicity are presented.

CNMs are the most notable products of nanotechnology. CNMs can enter the human body because of inhalation, skin contact, ingestion, or intentional injection, and they may affect the environment if released in significant quantities. Early studies of CNM toxicity have yielded conflicting results, which have generated more questions than answers. Although the toxicity and biocompatibility of fullerenes have been widely investigated, data concerning the biologic properties and biotoxicity of CNTs are insufficient and ambiguous.

As the number of products containing CNTs entering the market increases, the chances of free CNTs escaping during production or disposal processes or even through accidental spillage and entering the environment and living beings have increased. A dramatic increase in environmental, occupational, and public exposure to engineered nanoparticles in the near future is expected because of extensive use of nanoparticles for consumer and industrial products. The extent of future exposure to nanoparticles associated with these new products is still unknown. So far, limited data are available regarding CNT toxicity. Hence, the exact impact of CNTs on biologic systems is poorly assessed. Discussions regarding the potential benefits and risks associated with their widespread use have just begun (Fig. 27-1).

The properties of CNTs are being explored and exploited to improve the performance and efficacy of biologic detection, imaging,

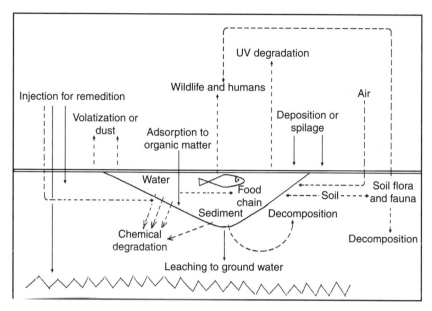

Notes: UV = Ultraviolet.

FIGURE 27-1 Probable routes through which carbon nanotubes may enter the environment.

and therapy applications. In many of these envisaged applications, CNTs will have to be injected or implanted in the body. After CNTs are inside the body and after they discharge their medical payloads, their fate, at present, is not accurately predicted. Cell culture studies by Shvedova and Castranova (2003) have shown evidence of cytotoxicity and oxidative stress induced by single-walled CNTs (SWCNTs), depending on whether and to what degree they are functionalized or oxidized. A recent report (Muller et al., 2005) also found that inhaled SWCNTs can cause damage to the lungs of animals. On the other hand, another study by Zhuang et al. (2008) reported that CNTs leave the body without accumulating in the organs and without observable toxic effects.

27.2   Nanotoxicology

Toxicology is an interdisciplinary field concerned with the study of the adverse effects of chemicals and materials on living organisms. It applies knowledge, methods, and techniques from fields such as chemistry, physics, material sciences, pharmacy, medicine, and molecular biology.

Particle toxicology, as a subdiscipline, has developed in the context of lung disease arising from inhalation exposure to dust particles of workers in the mining industry. With the rapid development of nanomaterials and nanotechnology, such as particle toxicology, a new subdiscipline, nanotoxicology, is emerging. Current toxicologic knowledge about engineered nanoparticles is extremely limited, and traditional toxicology does not allow for a complete understanding of the size, shape, composition, and aggregation-dependent interactions of nanostructures with biologic systems. An understanding of the relationship between the physical and chemical properties of nano-structures and their in vivo behavior would provide a basis for assessing toxic responses. The development of nanotechnology has raised questions as to the effects of nanomaterials on human health. The main route of exposure to the public is envisaged to be nanoparticles coming in the form of air pollution (Maynard and Howard, 1999), thus mainly affecting the lungs, digestive tract, or skin. However, nanoparticles may also get incorporated into the body through epidermal creams (sunscreens) or therapeutic and diagnostic applications (e.g., drug delivery) that will result in exposure to the digestive tract and possibly the bloodstream (Donaldson et al., 1992).

27.3   Toxicology of Carbon Nanotubes

Factors that can determine the potential harm of CNM include:

- **The ratio of surface area to mass of CNMs:** The larger the CNM's surface area, the greater the area of contact with the cell membranes causing increased absorption and transport of toxic substances (Donaldson et al., 2004). Hence, absorption studies of CNM are important to accurately assess the risk posed.

- **The particle retention time and biopersistence:** A long contact duration of the CNMs with the cell membranes increases the chances for damage. Hence, clearance and migration rates also need to be considered.

- **The reactivity or inherent toxicity of the chemical(s) contained within the particle:** CNMs may have their inherent toxic effects enhanced by virtue of the tendency to absorb the residue of carbon and the catalyst used during their production. These metals catalyst and amorphous carbon may also be toxic to cells. A higher surface area-to-mass ratio may also contribute to CNMs' inherent toxicity.

- **The dispersion properties dictate the toxicity of CNTs:** Agglomerates are expected to be less toxic than free CNTs because in agglomerates, the surface area in contact is relatively less than for free CNTs. Thus, dispersion studies are also critical in the study of CNT toxicity.

Because systematic and thorough quantitative analysis and phys-icochemical analysis such as pharmacokinetics (absorption, distribution, metabolism, and excretion) of nanoparticles have not been conducted, the overall behavior of nanostructures can be summed up only in general terms:

1. Carbon nanostructures enter the body via six principle routes: intravenous, dermal, subcutaneous, inhalation, intraperitoneal, and oral.

2. Absorption may occur where the nanostructures first interact with biologic components (proteins, cells).

3. Afterward, nanostructures can distribute or disperse to various organs in the body and may remain unaffected structurally, may be modified, or may be metabolized or excreted.

4. Nanostructures enter the cells of the organ and reside in the cells for an unknown period of time before leaving or moving to other organs or to be excreted.

Some of the ways that CNTs enter body systems are presented in Fig. 27 2. Their presence may be deleterious in some of these body systems, such as the neuronal and circulatory systems. The implications of CNMs on humans' vital systems are yet to be finely viewed.

Studies involving lung toxicity, skin irritation, and cytotoxicity have given some clarity about the toxicity of CNMs (Fig. 27-3).

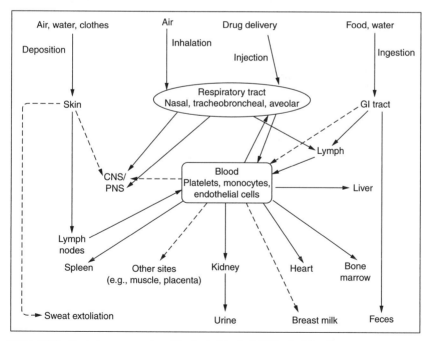

Notes: CNS = Central nervous system; GI = Gastrointestinal; PNS = Peripheral nervous system.

FIGURE 27-2   Considerations for nanomaterial toxicity and biocompatibility studies.

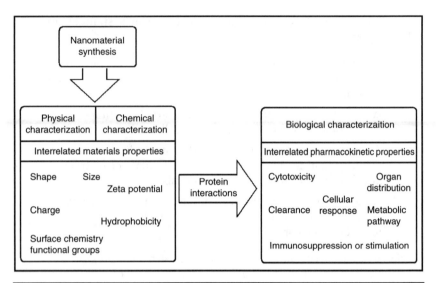

FIGURE 27-3 Probable routes through which carbon nanotubes may enter the body.

27.3.1   Lung Toxicity

Entry of CNTs through respiration into the lungs is more detrimental than entry into other organs. In general, the characteristics of respirable fibers that determine their pathogenicity (Donaldson et al., 2002b) include:

- **Fiber dimensions and length**
- **Biopersistence:** A high biopersistence increases the toxicity of all fibers. Insoluble long fibers are capable of persisting in living systems. A low biopersistence decreases the toxicity of even very long fibers (>20 μm). It is not clear how the inherent chemical stability of CNTs (Kam et al., 2007) influences their biopersistence.
- **Reactivity or inherent toxicity:** Toxicity depends largely on the chemical components.

Unpurified and pristine CNTs have been found to cause lung inflammation and granuloma formation (Lam et al., 2004, Warheit et al., 2004; Huczko et al., 2005; Muller et al., 2005). Lam et al. (2004) found that SWCNT products induced dose-dependent lung lesions characterized by interstitial granuloma regardless of the levels of metal impurities. Similarly, Muller et al. (2005) observed dose-dependent inflammation and granuloma formation by multiwalled CNTs (MWCNTs). Warheit et al. (2004) investigated the lung toxicity of SWCNTs in rats via intratracheal instillation. The study, which exposed rats to unrefined SWCNTs, showed a mortality rate of approximately 15% in rats exposed to SWCNTs. However, this was later attributed to mechanical blockage of the upper airways causing asphyxiation rather than any inherent toxicity of the SWCNT soot. Hence, although initial evidence urges considerable caution in the handling of CNTs, the actual toxicology of CNTs cannot be verified until advances have been made in CNT-delivery methods.

One such study of instillation of unrefined CNTs intratracheally in guinea pigs showed that working with soot containing CNTs is unlikely to be associated with a health risk (Huczko et al., 2001). Moreover, C_{60} fullerenes have not demonstrated significant toxicity (Jenson et al., 1993) despite evidence that intravenous injection in rats leads to rapid distribution among many tissues (including the brain) and accumulation in the liver and spleen (Yamago et al., 1995).

Nevertheless, it cannot be denied that the light weight and small size of CNTs is a matter of concern because they pose a potential occupational inhalation hazard. This particulate matter is thought to be present in many industries involving combustion of various chemicals. Several rodent studies (Warheit 2004) in which test dusts were administered intratracheally or intrapharyngeally to assess the pulmonary toxicity of manufactured CNTs showed that regardless of the process by which CNTs were synthesized and the types and amounts of metals they contained, CNTs were capable of producing inflammation, epithelioid granulomas (microscopic nodules), fibrosis, and biochemical and toxicologic changes in the lungs. Comparative toxicity studies (Lam, 2004) in which mice were given equal weights of test materials showed that SWCNTs were more toxic than quartz, which is considered a serious occupational health hazard if it is chronically inhaled; ultra-fine carbon black was shown to produce minimal lung responses.

Environmental fine particulate matter is known to form mainly from combustion of fuels and has been reported to be a major contributor to the induction of cardiopulmonary diseases by pollutants. Given that many adverse effects of CNT have been noted:

- SWCNTs and MWCNTs were found to elicit pathologic changes in the lungs.
- SWCNTs administered to the lungs of mice produced impairment of respiratory function.
- CNTs were found to damage the mitochondrial DNA in the aorta, increase the percent of aortic plaque, and induce atherosclerotic lesions in the brachiocephalic artery of the heart.
- It is speculated that exposure to combustion-generated MWCNTs may play a significant role in air pollution–related cardiopulmonary diseases.
- High-purity carbon black with an average primary particle diameter of 14 nm increased oxidative stress in human type II

alveolar epithelial cells in vitro and increased murine alveolar macrophage migration in fetal calf serum nearly twofold compared with high-purity carbon black with an average primary particle diameter of 260 nm.

- Long-term, subchronic inhalation of carbon black with an average diameter of approximately 16 nm can cause the development of pulmonary tumors in rats.

27.3.2 Skin Toxicity

Only a few dermatologic tests have been performed to assess the safety levels and hazards in handling CNT. Huczko et al. (2001) performed two set of tests:

1. Volunteers with allergy susceptibilities were exposed to a patch test for 96 hours.
2. A modified Draize rabbit eye test (using a water suspension of unrefined CNTs) was conducted with four albino rabbits monitored for 72 hours after exposure.

Both tests produced negative results compared with a CNT-free soot control. Thus, the researchers concluded that "no special precautions have to be taken while handling these carbon nanostructures." But certain studies on human epidermal keratinocytes have presented some concerns that need to be further examined (Shvedova et al., 2003; Monteiro-Riviere et al., 2005).

27.3.3 Cytotoxicity

Shvedova et al. (2003) have studied the effect of unrefined SWCNTs with high levels (~30%) of iron catalyst present in them on human epidermal keratinocytes by incubating the cells for 18 hours in CNTs. The incubated cells lost viability, produced morphologic alterations, and increased oxidative stress. These effects were believed to be due to iron impurity. The authors warned of possible dermal toxicity in handling unrefined CNTs, but stressed the role of SWCNT particle size and structure in these findings.

Muller et al. (2005) evaluated the inflammatory potential of CNTs on peritoneal macrophages. Cytotoxicity was probed by measuring lactate dehydrogenase release during 24 hours of incubation. The inflammatory potential was assessed by measuring mRNA expression of tumor necrosis factor (TNF)-α (a proinflammatory cytokine). The CNTs used for incubation of the peritoneal and alveolar macrophages were ground MWCNTs (having fewer agglomerates) and ungrounded MWCNTs (relatively agglomerated). The ungrounded particles were relatively less cytotoxic and showed proinflammatory responses because increased agglomeration decreases their availability to the macrophages.

Tamura et al. (2004) found that purified CNTs significantly increased superoxide anion and TNF-α production by neutrophils isolated from human blood, and cell viability decreased. Donaldson et al. (2006) used transmission electron microscopy and confocal microscopy to image the translocation of SWCNTs into human cells. The CNTs were seen to enter the cytoplasm and localize within the cell nucleus, causing cell mortality in a dose-dependent manner.

27.3.4   Toxicity Assessment of Carbon Nanomaterial

Toxicity assessments of CNMs used for biologic purposes require:

- Detailed characterization of materials because the toxicity of CNMs may depend on byproducts or residues of complex carbon structures as much or more than on the primary carbon structures

- Consideration of realistic exposure scenarios

- Sensitive methods to detect and quantify nanomaterials in biologic experiments

- The extent of systemic transport and persistence in distant organs after CNM exposure by various routes

- Answers to the issue of absorptive interferences caused by CNMs with fluorescent probes in assays

- Sensitive detection methods to determine dose metrics in toxicologic studies (however, it is possible that lipophilic nanomaterials [e.g., fullerenes] may interact with plasma membrane lipids and exert toxicity directly in the absence of cellular uptake)

- Exploration of important indicators of toxicity because CNMs may elicit additional types of pathologic reactions and not be limited to the usually used short-term indicators of toxicity, altered cellular function, or inflammation

Considering the above points, toxicity studies should be assessed at various hierarchical levels as shown in Fig. 27-4.

27.4   Biocompatability of Carbon Nanotubes

As per the standard definition, "biocompatibility is a characteristic of some materials that when they are inserted into the body they do not produce a significant rejection or immune response." (Williams, 1999)

There is a common assumption that the small size of nanostructures allows them to easily enter tissues, cells, organelles (i.e., mitochondria, DNA, ribosomes), and functional biomolecules. Because the actual physical size of an engineered nanostructure is similar to those of many biologic molecules (e.g., antibodies, proteins) and

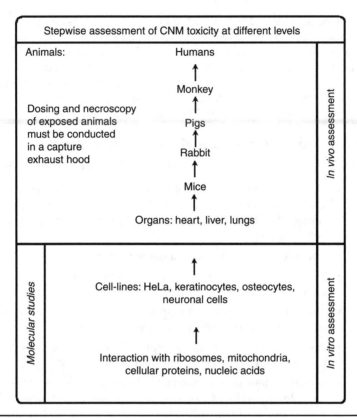

FIGURE 27-4   A schematic representation of toxity studies.

structures (e.g., viruses), the entry of nanostructures into vital bio-logic systems could cause damage and subsequently cause harm to human health. However, several recent studies (Hu et al., 2004; McKenzie et al., 2004; Webster et al., 2004; Gabar et al., 2005) have demonstrated that despite the size of nanostructures, they do not freely go into all biologic systems, but are instead governed by the functional molecules added to their surfaces.

Existing carbon-based biomaterials are already well known. Pyro-lytic carbon has been used for many years in biomedical implants and coatings, particularly in the manufacture of heart valve prostheses (Haubold et al., 1979), which were found to have excellent biocompat-ibility properties (Cenni et al., 1995; Ma et al., 2000). Studies demon-strated excellent blood compatibility of pyrolytic carbon heart valves with good adherence of endothelial cells and minimal adherence and activation of platelets (Haubold, 1983).

Development in biomedical carbons includes diamondlike car-bon (DLC). This dense meta-stable form of amorphous carbon has several properties that make it suitable for biomedical applications.

The most notable properties are its high hardness, low coefficient of friction, chemical inertness, and good corrosion and wear resistance (Cui et al., 2000; Grill et al., 2003; Sheeja et al., 2004). Many studies have investigated the biocompatibility of DLC coatings for orthopedic and cardiovascular applications. Biocompatibility studies reported no inflammatory response from macrophage cells in vitro (Thomson et al., 1991, (Allen et al., 1994; Linder et al., 2002), and no cytotoxic effects on fibroblasts or osteoblast cells have been observed (Allen et al., 1994; Allen et al., 2001).

The exploration of CNTs in biomedical applications is just underway but has significant potential. Because a large part of the human body consists of carbon, it is generally thought of as a very biocompatible material. Cells have been shown to grow on CNTs, so they appear to have no toxic effect. The cells also do not adhere to the CNTs, potentially giving rise to applications such as coatings for prosthetics and surgical implants. The ability to functionalize the sidewalls of CNTs also leads to biomedical applications such as vascular stents and neuron growth and regeneration. It has also been shown that a single strand of DNA can be bonded to a CNT, which can then be successfully inserted into a cell; this has potential applications in gene therapy.

27.4.1   Functionalization of Carbon Nanomaterial to Improve Biocompatibility

CNTs are insoluble in water and, thus, are incompatible in the water-based environments of living systems. Scientists at Rice University examined the cytotoxicity of water-soluble forms of CNTs and found that water-soluble CNTs are significantly less toxic and can be rendered nontoxic with minor chemical alterations. Moreover, pristine CNTs are unfortunately insoluble in many other liquids such as resins and most solvents. This means they are difficult to disperse evenly in a liquid matrix, complicating efforts to use their outstanding physical properties in the manufacture of nanocomposite materials, as well as in other practical nanotechnology applications that require preparation of uniform mixtures of CNTs with many different organic, inorganic, and polymeric materials. To make CNTs more easily dispersible in liquids, it is necessary to physically or chemically attach certain molecules or functional groups to their sidewalls without significantly changing the desirable properties of CNTs. This process is called *functionalization*.

Chemical functionalization of CNTs is commonly used to improve their solubility (Tagmatarchis et al., 2004). The production of robust composite materials requires strong covalent chemical bonding between the filler particles and the polymer matrix rather than the much weaker van der Waals physical bonds that occur if the CNTs are improperly functionalized. Functionalization methods such as chopping, oxidation, and "wrapping" of the CNTs in certain polymers can create more active bonding sites on the surface. Typically, this involves covalent attachment with appropriate molecules such as peptides (Tagmatarchis et al., 2004; Georgakilas et al., 2002; Pantarotto et al., 2003a and b; Pantarotto et al., 2004; (Jiang et al., 2004), acids, amines, polymers (Peng et al., 2003; Lin et al., 2003), and poly-L-lysine (Zhang et al., 2004), to the walls of CNTs. This is most commonly achieved either by amidation or esterification of the -COOH groups after CNT purification (Jiang et al., 2004; Peng et al., 2003; Lin et al., 2004; Zhang et al., 2004) or by 1,3-dipolar cyclo-addition (Tagmatarchis et al., 2004; Georgakilas et al., 2002; Pantarotto et al., 2003a and b; Pantarotto et al., 2004). Because of their reactive nature, many of these methods are associated with uneven surface coverage or inter-CNT coupling. In the context of biomedical applications, the chemical functionalization of CNTs or adsorption of biomolecules (e.g., lipids, proteins, biotins) is the most promising dispersion technique because no CNT-specific cytotoxicity was observed in studies into drug or vaccine delivery (Tagmatarchis et al., 2004; Pantarotto et al., 2003a and b; Pantarotto et al., 2004). To obtain a pure, homogenous dispersion, the MWCNTs were functionalized with pyrrolidine groups. This enhanced their solubility in the organic solvent dimethylformamide, which was evaporated by heating at 350°C to defunctionalize and leave purified CNTs on the substrate (Fig. 27-5).

27.4.2 Dispersion of Carbon Nanotubes to Improve Biocompatibility

One of the major and recurring problems encountered by researchers in the investigation of CNT toxicity and biocompatibility is the tendency of CNTs to aggregate in large bundles and ropes. The manipulation and characterization of large numbers of individual CNTs is a difficult task because high molecular weights and strong intertubular forces (both van der Waals and electrostatic) promote the formation of such bundles and ropes (Tagmatarchis et al., 2004). This is especially true for the saline, media, and serum solutions commonly used in toxicology testing. Hence, it is likely that a large portion of the discrepancies in

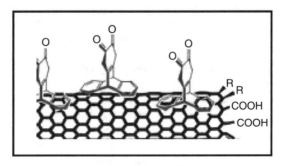

FIGURE 27-5   Chemical functionalization of carbon nanotubes.

toxicity and biocompatibility data are due to differences in CNT dispersion, the factor that ultimately dictates the presentation of CNTs to cells. Because many applications of CNTs (biomedical or otherwise) require their dispersion in a variety of solvents (e.g., organic solvents for polymer interactions and aqueous solvents for drug delivery), there have been many investigations into improving CNTs' solubility. Some of the successful and popular methods are sonication, stabilization with surfactant, and covalent functionalization.

Sonication

Sonication is a commonly used method for separating CNT aggregates in solution because it quickly disperses CNTs without the need for any chemical modification (Hilding et al., 2003). The two main methods of sonication are the ultrasonic bath and the ultrasonic probe (also known as a horn or wand), both of which use a bubble nucleation and collapse mechanism (Hilding et al., 2003). Although both the probe-style and bath-style ultrasonic systems can be used for dispersing CNTs, it is widely believed that the probe-style ultrasonic systems work better for dispersing CNTs.

Surfactant

It is also widely known that adding a dispersing reagent (surfactant) into the solution accelerates the dispersion effect. The reagents polyvinyl pyrrolidone (PVP), sodium dodecylbenzene sulfonate (SDBS), and polyvinyl alcohol (PVA) are good dispersion agents. Biomolecules, such as DNA, carbohydrates (Numata et al., 2004; Lii et al., 2003; Stobinski et al., 2003; Dodziuk et al., 2003), and peptides, are also studied as surfactants to solubilize CNTs in aqueous solution. The wide variety of synthetic and biologic surfactants makes them attractive for tailoring solubility toward specific applications.

It is critical to understand the cytotoxicity of CNT–surfactant conjugates because these reagents are increasingly being used in industry and laboratories. The toxicity of a surfactant may not only cause health issues for those working with CNTs but may also prove toxic to cells when used as nonviral transporters for biomolecules, including DNA, siRNA, and proteins, for therapeutic purposes. The cytotoxicity of SWCNTs suspended in various surfactants shows that the conjugates sodium dodecyl sulfate (SDS)–CNT and sodium dodecylbenzene sulfonate (SDBS)–CNT are toxic to astrocytoma cells solely because of the toxicity of the SDS and SDBS molecules.

Addition of Functional Groups

Achieving a stable dispersion that will last for days, weeks, or months with little to no settling requires other agents in the solution to prevent the CNTs from falling out of the solution over time. Emulsifier T-60 (also known as Tween 60) is commonly used with deionized water or isopropyl alcohol. Organic titanates can be used with acetone and xylene. The specific application determines whether these agents remain in the solution with further processing or if they need to be removed. Some organic titanates can be removed by heating the solution above 250°C. The addition of the OH and COOH functional groups assists the CNTs dispersing in DI water and other solvents as well as the chemical bonding to other materials during further processing.

27.4.3   Studies Using Neuronal Cells to Confirm the Noncytotoxicity and Biocompatibility of Carbon Nanomaterial

Carbon nanofibers (CNFs) with diameters smaller than 100 nm showed a reduction in neural scar tissue formation (astrocyte proliferation and function). McKenzie et al. (2004) have also shown that both pristine and chemically functionalized CNTs affect neural generation. MWCNTs chemically functionalized with carboxylic acid, ethylenediamine, or poly-m-aminobenzene sulfonic acid provide a substrate for neurite extension. This observation suggests that neurite extension is loosely based on the ionic charge, with positively charged ethylenediamine–MWCNT producing the most neurite extension. Neither the CNT-patterned surfaces nor the functionalized MWCNTs demonstrated cytotoxicity toward neuronal cells.

27.4.4   Studies Using Osteoblast Cells to Confirm the Noncytotoxicity and Biocompatibility of Carbon Nanomaterial

CNTs and nanocomposites are being considered as favorable bone biomaterials and are being investigated for use in osteoblast regeneration (Webster et al., 2004; Price et al., 2003). Elias et al., (2002) found that CNFs smaller than 100 nm (nanophase) increased osteoblast proliferation. Alkaline phosphatase activity, intracellular protein synthesis, and deposition of extracellular calcium all increased in the smaller-diameter CNFs compared with CNFs with diameters exceeding 100 nm. The authors concluded that CNFs did not induce a cytotoxic response and that the nanophase CNFs demonstrated potential as orthopedic materials.

Similarly, Price et al., (2003) also found that nanosized CNFs promoted osteoblast adhesion. But fibroblasts, chondrocytes, and smooth muscle cells were unaffected by the diameter of the CNFs. In certain nanocomposites, increasing concentrations of CNFs led to increased osteoblast and decreased fibroblast adhesion. None of the nanocomposite formulations differed significantly from control materials. It was concluded that CNFs did not show any cytotoxic effects and are promising candidates for use in orthopedic materials.

Supronowicz et al. (2002) fabricated conductive polylactic acid (PLA)–MWCNT nanocomposites. Osteoblast cells were seeded onto the surface and then exposed to alternating current stimulation. Control samples (PLA–MWCNT nanocomposite films) were run without electrical stimulation. The results showed an increase in osteoblast proliferation and extracellular calcium deposition on the nanocomposite compared with the control samples.

27.4.5 Studies Using Carbon Nanotubes as Drug- and Vaccine-Delivery Vehicles to Confirm the Noncytotoxicity and Biocompatibility of Carbon Nanomaterial

Applications of CNTs for drug-delivery operations require their functionalization to increase their solubility in aqueous milieu. Pantarotto et. al. (2003a) demonstrated the use of functionalized SWCNTs to create a vaccine-delivery device by linking a small peptide sequence from the foot and mouth disease virus (FMDV) to the sidewall of purified SWCNTs. This SWCNT–FMDV peptide complex elicited a specific antibody response in vivo, suggesting that vaccine delivery is a viable application for CNTs. Pantarotto et al. (2004) also demonstrated that the translocation of functionalized SWCNT complexes behaved like a cell-penetrating peptide across human 3T6 and murine 3T3 cell membranes.

27.4.6 Other Efforts to Confirm the Noncytotoxicity and Biocompatibility of Carbon Nanomaterial

To date, researchers have focused on the potential biologic applications of fullerenes and CNTs, but other CNMs, especially nanodiamonds, are beginning to emerge as alternative candidates for similar and many different applications. Researchers have assessed the cytotoxicity of nanodiamonds ranging in size from 2 to 10 nm. Assays of cell viability, such as mitochondrial function and luminescent adenosine triphosphate production, showed that nanodiamonds were not toxic to a variety of cell types. Furthermore, nanodiamonds did not produce significant reactive oxygen species (ROS). Cells can grow on nanodiamond-coated substrates without morphologic changes compared with controls. Nanodiamonds that are 2 to 10 nm in size with and without surface modification by acid or base have been found to be biocompatible with a variety of cells of different origins, including neuroblastoma, macrophage, keratinocyte, and PC-12 cells. Although the cell types used may have different mechanisms of internalization of the nanodiamonds and the long-term effect of the internalized nanodiamonds on the cells needs to be further investigated, the resultant retention of the mitochondria membrane along with low levels of ROS suggests that after they are inside the cell, the nanodiamonds remain nonreactive.

27.5   Some Unanswered Questions and Challenges

Several unanswered questions and challenges prevail regarding CNTs, including:

1. Overall, the relationship between the size, shape, and surface chemistry of nanostructures and their correlation to intracellular and in vivo biodistribution as well as metabolic fate are unknown.

2. Another considerable challenge is the variability in manufacturing methods, associated raw materials, and reaction scaling necessary to produce adequate volumes of uniform nanostructures.

3. To qualify observed in vivo results, an understanding of the nanostructure–protein interface is vital because it can potentially dictate behavior in vivo.

4. For medical applications based on free nanostructures, as with any new medicine, safety issues such as the systemic distribution, kinetics, variations depending on the route of administration, accumulation phenomena, dose response by tissues and organs involved, effects on cellular metabolism, conformational changes in proteins, and impact on promote tumor formation are important and should be addressed.

5. Basic scientific questions need to be answered, including:

   - How do cells interact with nanoparticles, and is this interaction similar to or different from the reaction to microparticles?

   - What is the mechanism of cellular uptake?

   - Does subcellular compartmentalization occur?

   - What determines the intracellular accumulation of CNT?

   - Does nanomaterial act as a primary or secondary messenger or induce a cascade of signal transduction pathways when in contact with the cell surface or at the intracellular level?

   - What is the relative importance of the size, shape, and chemistry of nanoparticles?

   - What are the mechanisms for transcellular transportation? This is an essential question because, as in imaging and targeted delivery, it is of great significance whether the nanoparticles stay within a biologic barrier (e.g., blood–brain barrier, air–blood barrier in the lungs or skin) or are able to cross it. Such scientific considerations should lead to specific research.

- Development of suitable in vitro models to study nanobiology (e.g., how nanoparticles interact with cells, especially of human origin).

- Search for suitable cell and molecular biologic parameters, which could indicate deleterious effects of nanoparticles in different cell and tissue systems.

- Which animal models are suitable to study nanobiology, and how can they be reduced to a minimum?

- A need for comparison of in vivo and in vitro models.

27.6   Conclusion: Are Carbon Nanotubes Safe?

Based on the data available so far in which one school of thought suggests that CNMs are nontoxic and biocompatible and the other that finds CNMs to be toxic and nonbiocompatible, it can be concluded that a comprehensive physicochemical characterization of CNMs is essential before they are used for biocompatibility and toxicity studies. Until we are able to fully extract and unravel all the relevant information on CNMs' toxicity, an immediate implementation of preventive measures should be carried out to get reliable results.

The toxicity of CNMs needs to be understood in the framework of material characterization. However, before successful use of CNMs in biomedical implants, drug- and vaccine-delivery vehicles, biosensors, and so on, their biocompatibility must be established. The surface coating, the presence of metal catalysts or graphite, the exposure to ultraviolet radiation, the dispersion properties, and the tendency to deposit as aggregates because of high van der Waals forces all greatly influence the behavior of these CNT particles. Only after all of these issues have been evaluated can the selection of appropriate doses and concentrations related to the cell types and tissue species be considered.

CHAPTER 28

Carbon Nanotubes: A Paradigm Scaffold Fabricating Tissues

Madhuri Sharon

MONAD Nanotech Pvt. Ltd., Powai, Mumbai, India

Sunil Pandey

Nanotechnology Research Centre, Birla College, Kalyan, Maharashtra, India

Goldie Oza and Avinash Kadam

Indian Institute of Technology Bombay, Mumbai, India

28.1 Introduction

28.1.1 Tissue Engineering

Tissue engineering is a discipline that is in its infancy, but has emerged rapidly and received extensive attention (Miller and Patrick, 2003; Takezawa, 2003; Vats et al., 2003). It involves replacement of anatomic structure of damaged, injured, or missing tissue or organs by agglomerating biomaterials, cells, and biologically active molecules (Atala, 2000; Ibarra et al., 2000; Lanza et al., 2000).

Biomaterials are three-dimensional scaffolds that provide mechanical support to the growing cells or tissue, guiding them in a microenvironment mimicking the biologic system. These can then be transplanted into the system (Fig. 28-1). The principal function of a scaffold is to direct cell behavior, such as migration, proliferation, differentiation, maintenance of phenotype, and apoptosis, by facilitating sensing and responding to the environment via cell matrix

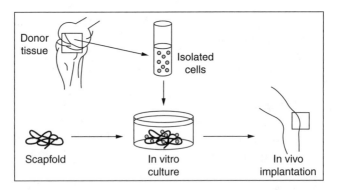

Figure 28-1 One approach to tissue engineering is to use a biomaterial scaffold to direct growth of the body's own cells in a laboratory and then implant the cells back into the body. (*Courtesy of Christine Schmidt, Professor, Dept. of Biomedical Engineering CPE 4.418, BME 4.202l, The University of Texas, Austin.*)

communications and cell–cell communications (Vats et al., 2003). Such scaffolds should have the capacity to give highly porous surfaces, allowing seeding of cells at high densities as well as facilitating proper cell–cell interaction through a regulated molecular myriad (Cima et al., 1991; Mikos et al., 1993). In addition, the scaffolds provide spatial signals to actuate the organization of the cells as well as that of the extracellular matrix (ECM) derived from them.

The ECM secreted by cells is responsible for supporting tissue development and providing the microenvironment needed for cellular growth. The ECM is composed of an array of proteinaceous nanofibers conferring physical and chemical homeostasis of the tissue. For example, collagen is a natural ECM molecule with a fibrillar structure found in many tissues, such as bone, skin, tendons, ligaments, and other connective tissues. Earlier tissue engineers were mostly focused on investigations of macrolevel structures. However the present-day need is to engineer functional units of the subcellular scale structures and nanostructures to control cell-molecular and cell–cell interactions (Wen et al., 2005). Thus, engineering tissue for miniaturization at the nanolevel to precisely design mimicry of the native tissue structures is one of the most promising directions for tissue fabricators (MacDonald et al., 2005).

The four basic components in tissue engineering are biomaterial scaffold, functional cells, biomolecules (e.g., growth factors, ECM, other functional molecules), and dynamic forces. This chapter deals with the most useful biomaterial scaffold, carbon nanotubes (CNTs).

28.1.2 Why Carbon Nanotubes Are Versatile Scaffolds

It is clear that tissue engineering needs smart biomaterials that can support cell growth without altering the biologic functions required for tissue growth. A marriage between nanotechnology and microengineering could be the best option for designing biomaterials to generate scaffolds for tissue engineering.

Nanotechnology, the science that deals with material properties at the nanoscale (1 to 100 nm), focuses on constructing designs with either a bottom-up or top-down approach. The ability of nanotechnology to form materials that can act as biologic machineries, such as molecular motors, actuators, and optical tweezers, has made it a reliable methodology for tissue engineering. Interestingly, nanomaterials can also be multifunctional, capable of both targeting and imaging (Gao et al., 2004).

One nanomaterial that has the potential for multiple uses in tissue engineering is CNTs. The field of biomaterial fabrication was revolutionized soon after the discovery of CNTs by Iijima (1991). CNTs are cylindrical tubes 2 to 100 nm in diameter, resulting in a very large ratio of surface to volume. These are made up of carbon atoms arranged in a series of condensed benzene rings rolled up into a tubular structure.

This smart material belongs to family of fullerenes, the third allotropic form of carbon along with graphite and diamond. Single-walled CNTs (SWCNTs) are constructed from a single sheet of graphite, and multiwalled CNTs (MWCNTs) consist of multiple concentric cylinders of graphite. Both SWCNTs and MWCNTs possess very interesting properties such as ordered structures with a high aspect ratio, ultra lightweight, high mechanical strength, high electrical and thermal conductivity, metallic or semimetallic behavior, and high surface area (Sharon et al., 1995). These properties have made CNTs the favorite material for nanofabrication of compatible biomaterials that can act as scaffolds for engineering tissues (Martin et al., 2003).

CNTs have significant relevance for tissue engineering in four areas:

- Cell tracking and labeling
- Monitoring cellular physiology
- Regulating cellular behavior
- Structural support for tissue engineering

28.2   Cell Tracking and Labeling

It is important to trace implanted cells and monitor the advancement of tissue formation in vivo in tissue-engineered fabrications of clinically relevant sizes. Tagging implanted cells would assess or evaluate the viability of engineered tissue and help in understanding the biodistribution and migration pathways of transplanted cells. For tagging, contrasting agents in vivo need to have good biocompatibility and stability. The literature thus far has shown that CNTs are feasible as imaging contrast agents for optical labeling, magnetic resonance, and radiotracer modality.

28.2.1   Optical Labeling

Optical detection can be a simple and direct diagnostic method for biomedical imaging. CNTs reflect many characteristic properties desirable for optical detection. They possess optical transitions in the near-infrared (NIR) region. The infrared spectrum between 900 and 1300 nm is an important optical window for biomedical applications because of the lower optical absorption (greater penetration depth of light) and small autofluorescent background (Kam et al., 2005c). In addition, CNTs display good photostability. Raman scattering and fluorescence spectroscopy may be promising methods for tracking CNTs in cells over long durations of time. For instance, SWCNTs dispersed in a pluorinc surfactant incubated with mouse peritoneal macrophage-like cells can be imaged using fluorescence techniques. Approximately 70,000 CNTs were reported to be ingested per cell at the rate of one CNT per second (Cherukuri et al., 2004). Excitation at 660 nm is the signature absorption of SWCNTs. SWCNTs are very stable when they are inside biologic cells. Nucleic acid–encapsulated SWCNTs have been shown to be stable for weeks when ingested by 3T3 fibroblasts and murine myoblast stem cells (Hellar et al., 2005). CNTs remain in the cells during repeated cell divisions, suggesting that such probes could be used for studying cell proliferation and stem cell differentiation.

CNTs remain stable in cells because of their hydrophobic nature. They are internalized by a myoblast stem cell. The concentration of such CNTs within the cell can be determined using the signal intensity. Because Raman spectroscopy is very sensitive to functional groups in a molecule, it may provide useful information about the microenvironment of the cell. This is one of the most promising methods for using CNTs as optical biosensors in vivo and may serve as a cornerstone technique (Hellar et al., 2005).

The problems associated with this method are tissue absorbing or scattering light, autofluorescence, and photobleaching. In many cases, the optical absorption bands are narrow, and the resulting emissions are broad. This can limit the simultaneous detection of multiple fluorophores; hence, this aspect has to be examined before CNTs can be used for optical labeling (Cherukuri et al., 2004).

28.2.2 Magnetic Resonance Imaging Contrast Agents

Magnetic resonance imaging (MRI) involves proton relaxation associated with the water molecules interacting with coordinated paramagnetic ions. MRI contrasting agents, such as gadolinium-based nanoparticles, iron oxide superparamagnetic nanoparticles, and quantum dots, have had a significant impact (Ballou et al., 2004; Gupta et al., 2005; Mornet et al., 2004). Fullerenes, such as buckminsterfullerenes and CNTs, are composed entirely of carbon, thus providing poor contrast in MRI. Fullerenes require functionalization to make them more readily detectable, which is done by attaching a heavy metal such as gadolinium to the surface of the fullerene or caging the heavy atom (Bolskar et al., 2003; Toth et al., 2005).

CNTs can be characterized by Raman spectrometry. Each part of the Raman spectrum (Fig. 28-2)—the radial breathing mode (RBM), the disorder-inducing mode (D mode), and the high-energy mode (HEM)—can be used to access different properties of SWCNTs. Of all the Raman modes, the RBM is unique to SWCNTs. In the high-energy range around 1600 cm-1, SWCNTs show a characteristic double-peak structure.

There has been demonstration of gadolinium loading into ultrashort (20 to 100 nm long) CNTs through sidewall defects, which display 40 times greater proton relaxivities than the standard gadolinium clinical contrast agent. With such a large contrasting capability, such

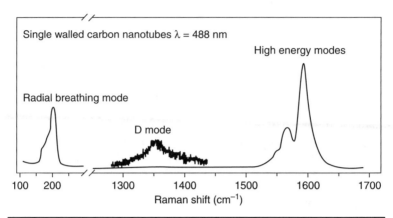

Figure 28-2   Raman spectra of single-walled carbon nanotubes (SWCNTs) showing the radial breathing mode (RBM), the disorder-inducing mode (D mode), and the high-energy mode (HEM) (Reich et al. 2004).

materials can be dispersed throughout an engineered tissue, allowing the complete process of its monitoring.

28.2.3   Radiolabeling

Radiolabeling involves the use of radioisotopes such as 111 In whose radioactivity can be measured by scintillation counters. CNTs are covalently bound to the radioisotope and are then administered into the biologic system to measure the biodistribution of CNTs. In a research study with BALB/c mice, such functionalized SWCNTs were shown not to accumulate in any specific organ, they got rapidly cleared from the blood system through the renal excretion route without any toxic side effects or mortality, and they could be visually observed in the urine (Singh et al., 2006).

28.3   Monitoring Cellular Behavior

Monitoring cellular physiology, such as solute transport, enzyme–cofactor interactions, secretion of protein and metabolites, and cellular behavior such as matrix adhesion, assists in fabricating better-engineered tissues. Continuous monitoring of the performance of the engineered tissues could be provided by biosensors. Tissue and cellular responses have already been observed using other nanoparticle contrast agents such as inflammation of pancreatic islets (Denis et al., 2004), apoptosis (Jung et al., 2004; Zhao et al., 2001; Schellenberger et al., 2004), and angiogenesis (Schmieder et al., 2005). One method of monitoring engineered tissues would be to use implantable sensors capable of relaying information extracorporeally. Such a sensor would provide real-time data related to the physiologic relevant parameters such as pH, pO_2, and glucose levels.

There are several advantages in using nanosensors for evaluating engineered tissues in vivo. First, because the sensing element is nano-sized, miniaturization of the probe and implanting throughout an engineered tissue would not adversely disturb the system. Second, because of their enormous surface area, the active area is large for immobilizing numerous biologic and chemical compounds, including DNA (Singh et al., 2005) and proteins (Chen et al., 2004), for improved sensitivity.

Biosensors are electrochemical devices that rely on the transduction of the molecular recognition event into an analytical output signal. Among the many molecular recognition biomolecules, SWCNTs are one of the most promising candidates for the development of nanoscale biosensors because of their capability to provide strong electrocatalytic activity and minimize surface fouling of the sensors (Lin et al., 2004a).

SWCNTs were reported to have a high ability to promote electron transfer reactions in electrochemical measurement (Wang, 2004). They have a high aspect ratio, so the total surface for electrode, on to which SWCNT is deposited, becomes significantly larger on the same site than the surface of the bare electrode. Electrochemical biosensors based on a CNT array–modified electrodes have been reported by Kerman et al. (2004, 2005a, 2005b).

Among the many biosensors, the one that is responsible for detecting electrochemically total prostate specific antigens with SWCNT-modified platinum array microelectrodes was monitored using some linkers such as dimethyl formamide solution. Label-free electrochemical immunosensors in connection with a linker molecule that has a high affinity for graphite structure exhibited high selectivity with successful suppression of the nonspecific adsorption (Okuno et al., 2006).

CNTs have a very strong potential to address the needs of even label-free sensing for protein arrays (Ramachandran et al., 2004). They are functionalized, and their conductance is modulated as the target molecules bind to the functionalized CNTs. This sensing technology has been used to detect the binding of small virus particles (Patolsky et al., 2004), small molecules binding to proteins (Wang et al., 2003), streptavidin (Cui et al., 2001), and antibiotin immuno-globulin binding to immobilized biotin (Cui et al., 2001).

Coupling CNTs with macromolecules has paved the way for various applications in fields such as molecular electronics, drug delivery, and pharmaceuticals. Because of their unique electronic structures, CNT electrochemical sensors can potentially simplify the analysis of redox-active proteins and amino acids, allowing cell monitoring in engineered tissues.

28.4   Regulating Cellular Behavior

CNTs can make an impact in tissue engineering in controlling the production or delivery of tissue-inducing substances such as growth factors. These are very well known for a number of cell-altering applications, including localized drug delivery (Martin et al., 2003) and transfection (Singh et al., 2005) and as ion channel blockers (Joseph et al., 2003). Many of these methods take advantage of the large aspect ratio and ease of functionalization of CNTs. Thus, CNTs could be components of drug-delivery systems. A significant advantage of CNTs over spherical nanoparticles is that they can be heterogeneously functionalized. The ends and sidewalls of CNTs possess different chemical reactivities that can facilitate a dual-functionalized drug-delivery platform. For instance, amine-containing targeting agents such as antibodies could be attached to the ends of the nanotube. Drugs can then be linked with biodegradable linkers to the sidewall of the CNT, and a targeting or imaging moiety can be attached to the ends. In another instance, experiments with starch have revealed that although CNTs are not soluble in an aqueous solution of starch, they are soluble in an aqueous solution of the starch–iodine complex (Fig. 28-3). The reversible water solubilization of SWCNTs in water using starch may provide the means for developing fully integrated biologic nanotube devices. These observations suggest that iodine preorganizes the backbone of the amylose in starch into a helical conformation and makes its hydrophobic cavity accessible to a single CNT or bundles of CNTs. The formation of such starch-wrapped SWCNT complexes is driven by simultaneous enthalpic and entropic gains that result from creating favorable van der Waals interactions and from expelling the many small iodine molecules located inside the helix out into the solvent by a "pea-shooting" type of mechanism. This result has led to a simple protocol for cleaning up SWCNTs with starch.

CNTs facilitate delivery of DNA or any bioactive agent to cells. They can be functionalized to attach either electrostatically or covalently to DNA and RNA, and the remaining unfunctionalized and

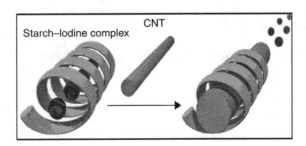

FIGURE 28-3   Starch–iodine complex with carbon nanotubes (CNTs).

hydrophobic portions of the nanotubes can be attracted to the hydrophobic regions of the cells. This hydrophobicity explains nonspecific binding between nanotubes and proteins (Chen et al., 2003). In addition, CNTs can be functionalized with bioactive proteins to help cross the cellular membrane. For example, biotin-functionalized CNTs bound to fluorescent dyes are capable of intercellular transport of fluorescent streptavidin (Kam et al., 2004). Nanotubes can also be functionalized with cationic components capable of penetrating human and murine cell types. Such nanotubes can be complexed with plasmid DNA to facilitate transfection of cells (Singh et al., 2005). In addition, DNA or siRNA can be bound to nanotubes with cleavable disulfide bonds and delivered to mammalian cells (Kam et al., 2005a).

In addition to heterogeneous functionalization, CNTs could provide localized delivery of therapeutic agents triggered by external sources. It has been shown that CNTs absorb NIR light at wavelengths that are optically transparent to native tissue. For example, irradiation with 880-nm laser pulses can induce local heating of SWCNTs in vitro, thereby releasing their molecular cargo without harming cells, or can be internalized within a cancer cell and kill the cell with sufficient heating (Kam et al., 2004). This could allow selective delivery of drugs to certain cell types, helping to control the distribution of such cells throughout the engineered tissue.

28.5   Carbon Nanotubes as Structural Support for Tissue Engineering

CNTs as structural support for tissue engineering structural support to the growing tissue is pivotal role played by CNTs. Structural support or matrix defines the space for engineered tissue and helps in the process of tissue development. Synthetic polymers such as (polylacticco-glycolic acid) PLGA and PLA (polylactic acid) lack the required mechanical strength, and such materials cannot be easily functionalized. The best support to the growing tissue is provided by CNTs. CNTs have the potential for providing the needed structural and mechanical reinforcement for tissue scaffolding (Harrison and Atala, 2007). CNTs can be grown on a patterned bed of catalysts, creating a three-dimensional array with complete control of the size, shape, and periodicity. This was exemplified by the growth of L929 mouse fibroblasts and neurons on such scaffolds (Correa-Duarte et al., 2004).

In a recent study, Ago et al. (2006) demonstrated mechanical immobilization of HeLa cells on an aligned CNT array through insertion of the nanotube tips into the cell. This type of an array has proven successful because it can fix a much larger number of cells compared with the flat surface of a silicon substrate, which indicates strong binding force of the aligned nanotubes toward the cell. This has been considered to be the first avalanche of mechanical cell immobilization, which offers future applications of the nanotube array for cell biology, tissue engineering, and biomolecular devices. Because CNTs are not biodegradable and are not readily dispersed in aqueous solutions, they are blended with polymers. For example, MWCNTs mixed with chitosan brings significant improvement in mechanical properties, Young's modulus, and tensile strength compared with neat chitosan (Wang et al., 2005).

An in vitro study showed several different types of cells successfully grown on CNTs and nanocomposites. For example, natural scaffolding agents such as collagen blend with SWCNTs, supporting smooth muscle cell growth (MacDonald, 2005) Because of their diameter (~100 nm), CNTs can possibly be used to mimic neural fibers for neuronal growth. Hippocampal neurons from Sprague-Dawley rats that were 0 to 2 days old have been grown on CNTs coated with 4-hydroxylneonetal (Hu et al., 2004).

Nanotubes in an array can be used to construct neural networks. Cortical tissue from 1-day-old Charles liver rats have been cultured on a nanopatterned substrate containing CNTs (Gabay et al., 2005). The dissociated neuronal cells were self-organized into an ordered, compactly wired network on the three-dimensional array of CNT islands. Neurons deposited on CNTs showed a significant increase in the frequency of spontaneous postsynaptic currents. This significant difference is because of the electrical conductivity of the CNTs, a property that can be useful for directed cell growth. For example, when an alternating current is applied to the substrate, nanocomposites of polylactic acid and MWCNTs have been shown to increase osteoblast proliferation; increase calcium production; and lead to an upregulation of collagen I, osteonectin, and osteocalcin. This has intrigued researchers to use nanocomposites as stimulators of bone formation (Supronowicz et al., 2002).

CNTs can also be functionalized to release bioactive factors, such as glucose oxidase, which can be attached to CNTs and still retain the enzymatic activity (Besteman et al., 2003). This has intrigued researchers to use CNTs with inherent electrochemical properties coupled with enzymatic or protein functionalization as a paradigm scaffold, providing mechanical integrity for cell growth and simultaneously monitoring the growth process.

28.6   Cytotoxicity of Carbon Nanotubes

When foreign substances are introduced into the body, understanding the organism's response is a crucial step (Harrison and Atala, 2007). There is a dilemma in the literature concerning the cytotoxicity of fullerene nanomaterial such as buckyballs and CNTs. Several in vitro studies report that CNTs are cytotoxic when incubated for six hours with alveolar macrophages (Jia et al., 2005). It is worth noting that CNTs that are cytotoxic are 90% pure; the remaining balance is amorphous carbon and residual catalyst. Interestingly, C_{60} was also analyzed and found to be relatively less toxic. Molecular characterization of cytotoxic mechanism of MWCNTs and multiwalled nano-onions on human fibroblast indicates that the effect of gene expression is dose dependent (Ding et al., 2005). At concentrations of 0.6 μg/mL, serious impact on cellular functions in maintenance, growth, and differentiation occurs. The observed effects on the expression of P38/ extracellular signal-regulated kinase and epidermal growth factor receptor from fibroblasts in contact with these fullerenes suggest that they may be useful nanomedicine for cancer therapy, especially those derived from epithelia. The reasons for the observed cytotoxicity are oxidative stress and activation of nuclear transcription factor-κ B in human keratinocytes (Manna et al., 2005).

Three important factors decide the potential of a particle to be toxic (Donaldson et al., 2002; Tran et al., 2000; Zhang et al., 2005):

- **The particle's ratio of the surface area to the mass:** A large surface area provides the particle a greater area for interaction with the cellular membrane, as well as great capacity for absorption and transport of toxic material.

- **The particle retention time:** The longer the particle stage with the cellular membrane, the higher the chance of damage.

- **The reactivity or the inherent toxicity of the chemical content within the particle**

The following properties of carbon are particularly more hostile:

- Highly pure carbon black with an average particle diameter of 14 nm increases oxidative stress in human type II alveolar epithelial cells in vitro (Barlow et al., 2005; Stone et al., 1998)

- Long-term inhalation of carbon black with an average diameter of approximately 16 nm can lead to development of pulmonary tumors in rats (Gallagher et al., 2003; Nikula et al., 1995).

- C_{60} fullerenes have not been proven to have significant toxicity (Jenson et al., 1996) but there are evidences of its rapid distribution among vital tissues like brain and accumulation in liver and spleen (Yamago et al., 1995).

Shvedova and Castranova reported the first cytotoxicity study on CNTs. They investigated the effects of impure SWCNT on immortalized human epidermal carenatocides (HaCaTs). HaCaT cells were incubated up to 18 hours in media containing unrefined SWCNTs (0.062 to 0.024 mg/mL). Exposure to SWCNTs resulted in accelerated oxidative stress, loss of cell viability, and morphologic alteration of the cellular structure. It was concluded that these effects were the results of a high level of iron catalyst present in the unrefined SWCNTs. Dermal toxicity was also a consequence of handling unrefined CNTs.

Similar dermal toxicity was reported in a study that found that MWCNTs initiated an irritation response in human epidermal keratinocytes (Monteiro-Riviere et al., 2005). Purified MWCNTs (synthesized via chemical vapor deposition) incubated at doses of 0.1 to 0.4 mg/mL with human embryonic kidney (HEK) cells for up to 48 hours were observed to localize within the cells, eliciting the production of proinflammatory cytokine (IL-8) release and decreasing cell viability in a time- and dose-dependent manner. Thus, the lack of a catalyst led to a toxicologic effect.

Jia et al. (2005) investigated the effects of carbonaceous nanoparticles on the cytotoxicity of alveolar macrophages, which showed that SWCNTs exhibited a higher cytotoxic response than MWCNTs. Cui et al. (2005) investigated the cytotoxicity of SWCNTs and showed that SWCNTs inhibited HEK 293 cells by inducing apoptosis and decreasing cellular adhesion ability. Both cell proliferation and adhesion ability decreased in a dose- and time-dependent manner. Whereas genes involved in apoptosis were upregulated, genes associated with the G1 phase of the cell cycle were downregulated along with the genes associated with adhesion.

28.7   Biocompatibility of Carbon Nanotubes

Biocompatibility is defined as "the ability of a material to perform with an appropriate host response in a specific application" (Williams, 1999). As CNTs are applied in different facets of technology, it has become imperative that their biocompatibility be properly investigated before their widespread use. Pyrolytic carbons were initially considered to be excellently blood compatible. Heart valves made of pyrolytic carbon were highly compatible, with good adherence of endothelial cells and minimal adherence and activation of platelets (Haubold, 1983). Recently, diamondlike carbon (DLC) has been demonstrated to be used as coatings for many orthopedic and cardiovascular applications (Allen et al., 1994; Thomson et al., 1991). DLC has been found to be biocompatible in a way that it can adhere to osteoblast and fibroblast cells and allow proliferation of these cells. DLC coatings on cardiovascular stents have also been found to reduce platelet and macrophage activation (Ball et al., 2004; Gutensohn et al., 2000).

The above DLC-based biomaterials have been found to be biocompatible; hence, CNTs are also considered to be the next probable candidates as a biomaterial. Various investigations dealing with interaction between CNT-based materials and neural cells, osteoblasts, fibroblasts, antibodies and the immune system, ion channels, and cellular membranes are discussed in the next sections (Fig. 28-4).

FIGURE 28-4 Biocompatibility of platelets and red blood cells with a single-walled carbon nanotube scaffold.

28.7.1 Carbon Nanotubes with Neuronal Cells

CNTs have been found to be biocompatible for neural applications (Gabay et al., 2005; Hu et al., 2004; McKenzie et al., 2004; Webster, 2004). The investigation done by McKenzie et al. (2004) showed effects of varying carbon nanofibers (CNFs) and surface energy on the proliferation and function of astrocytes. It was found that astrocytes preferentially adhered to and proliferated on CNFs with larger diameters and higher surface energies. Webster et al. (2004) simultaneously investigated interactions between nanocomposites (polyurethane and CNF) and astrocytes. But because of a lack of dispersion, CNFs have turned out to have poor mechanical properties. This nanocomposite also exhibits decreased astrocyte adhesion and retarded neurite in rat pheochromocytoma cells.

Both pristine and chemically functionalized CNTs have a positive impact on neuronal growth. Confluent layers of neurons were seeded onto lithographically patterned CNT surfaces, and axons formed an interconnected network that replicated the pattern of the CNT template. Similarly, purified MWCNTs that were chemically functionalized with carboxylic acid, ethylenediamine, or poly-m-aminobenzene sulfonic acid were observed to provide a substrate for neurite extension.

28.7.2 Carbon Nanotubes as Scaffolds for Bone Growth

Artificial bone scaffolds are from a wide variety of materials, such as polymers or peptide fibers. Their drawbacks include low strength and the potential for rejection by the body. Chemically functionalized SWCNTs have been used as scaffolds for the growth of artificial bone material (Zhao et al., 2005). The strength, flexibility, and light weight of SWCNTs enable them to act as scaffolds to hold up regenerating bone. Bone tissue is a natural composite of collagen fibers and crystalline hydroxyapatite, which is a mineral based on calcium phosphate. SWCNTs can mimic the role of collagen as a scaffold for inducing the growth of hydroxyapatite crystals. By chemically treating the nanotubes, it is possible to attract calcium ions and to promote the crystallization process while improving the biocompatibility of the nanotubes by increasing their water solubility. SWCNTs may lead to improved flexibility and strength of artificial bone and new types of bone grafts and to inroads into the treatment of osteoporosis and fractures. Bone cells can grow and proliferate on a scaffold of CNTs. Because CNTs are not biodegradable, they behave like an inert matrix on which cells can proliferate and deposit new living material, which becomes functional, normal bone (Zanello et al., 2006). CNTs carrying neutral electric charges sustained the highest cell growth and production of plate-shaped crystals. There was a dramatic change in cell morphology in osteoblasts cultured on MWCNTs, which correlated with changes in plasma membrane functions. CNTs hold promise in the treatment of bone defects in humans associated with the removal of tumors, trauma, and abnormal bone development and in dental implants.

Elias et al. (2002) investigated the proliferation and function of osteoblast cells seeded onto four variants of compacted CNFs (with diameters >100 nm or <100 nm). They concluded that there was an increased osteoblast proliferation on the nanophase (<100 nm) CNFs and increased alkaline phosphatase activity, intracellular protein synthesis, and deposition.

Osteoblast proliferation on nanocomposites of CNT (polylactic acid–MWCNT) has been investigated by Supronowicz et al. (2002). Osteoblast cells were seeded onto the surface and then exposed to alternating current stimulation, which showed an increase in osteoblast proliferation and extracellular calcium deposition on the nanocomposite (Fig. 28-5).

Figure 28-5   Osteoblast cells grown on a carbon nanotube scaffold.

28.7.3   Carbon Nanotubes with Antibody Interactions

The generation of fullerenes specific antibody was first determined in by Chen et al. (1998). This was achieved by the immunization of BALB/c mice with a C_{60} fullerene–thyroglobulin conjugate, yielding a polyclonal IgG antibody that could bind to both C_{60} and C_{70} fullerene derivatives. In a similar study by Erlanger et al. (2001), binding of C_{60}-specific monoclonal antibodies to SWCNTs was investigated. The study showed that the antibodies are bound to aqueous SWCNT ropes. Poly-l-lysine adsorbed on to CNTs was shown to enhance antibody and protein binding. Similarly, binding of monoclonal antibody is influenced by the hydrophilicity and surface disorders of the fibers (Naguib et al, 2005).

28.7.4   Ion Channel Interactions with Carbon Nanotubes

The interaction between SWCNTs, MWCNTs, fullerenes, and ion channels was investigated. It was found that SWCNTs blocked the channels formed from exp-2, kvs-1, and hERG (the human *Ether-à-go-go* Related Gene) potassium channels. Fullerenes were discovered to be less effective channel blockers than CNTs (Park et al., 2002).

28.8   Dispersion of Carbon Nanotubes

The most oblivious question raised about CNTs is their nonbiodegradability and nondispersibility, which lead to toxicity. The manipulation and characterization of large numbers of individual CNTs is a difficult task because high molecular weights and strong intertubular forces (both van der Walls and electrostatic) promote the formation of bundles and ropes of CNTs (Tagmatarchis and Prato, 2004). Because many applications of CNTs (biomedical or otherwise) require the dispersion in a variety of solvents, there has been much investigation into improving CNTs' solubility. The most successful and popular methods are sonication, stabilization with surfactant, and covalent functionalization.

28.8.1   Sonication

Sonication is the most efficient method for segregating CNT agglomerates in solution, thus leading to their dispersion. The two main methods of sonication are the ultrasonic bath and the ultrasonic probe. Both methods are involved in bubble nucleation and a collapse mechanism. Ultrasonic baths typically have higher operating frequencies (40 to 50 KHz) and no defined cavitation zone, and ultrasonic probes operate at frequencies around 25 KHz with a conical bubble nucleation zone extending from the probe tip (Jia et al., 2005; Lam et al., 2004; Muller et al., 2005).

28.8.2 Stabilization with Surfactant

Surfactants are used to enhance CNTs' solubility in aqueous solution. Sodium dodecyl sulfates, and triton X-100 are three of the most popular surfactants (Lin et al., 2004b; Numata et al., 2004). Likewise, biomolecules such as DNA (Barisci et al., 2004), protein, and carbohydrates (Lii et al., 2003) are commonly studied surfactants to solubilize CNT in aqueous solutions. Unfortunately, some processing conditions have been found to cause surfactant dissociation from CNTs.

28.8.3 Chemical Functionalization

CNTs can be solubilized using various covalently attaching molecules, such as peptides, acids, amines, or polymers, to the sidewalls of CNTs. This is most commonly achieved by amidation or esterificaion of the carboxyl groups present after CNT purification by the reactive nature. However, many of these methods are associated with uneven surface coverage or inter-CNT coupling (Lin et al., 2003; Peng et al., 2003; Zhang et al., 2004).

28.9 Summary

The principal function of a scaffold is to direct cell behavior, such as migration, proliferation, differentiation, maintenance of phenotype, and apoptosis, by facilitating sensing and responding to the environment via cell matrix communications and cell–cell communications. Such scaffolds should have the capacity to give highly porous surfaces allowing seeding of cells at high densities, as well as provide spatial signals to actuate the organization of the cells as well as that of the ECM derived from them. Both SWCNTs and MWCNTs possess very interesting properties as scaffolding agents, such as ordered structures with high aspect ratio, ultra light weight, high mechanical strength, high electrical and thermal conductivity, metallic or a semi-metallic behavior, and high surface area. These properties have made CNTs the favorite material for nanofabrication of compatible biomaterials that can act as scaffolds for engineering tissues.

CNTs have found important use in various aspects of tissue engineering, including cell tracking and labeling, monitoring and regulating cellular behavior, and acting as structural support for fabrication of a tissue. All of these applications will be rendered useless if CNTs are not safe and biocompatible. Hence, it becomes imperative to check for CNTs' cytotoxicity. Dispersion of bare CNTs is also a matter of relevance that requires proper functionalization.

CHAPTER 29

Carbon Nanotubes: Robust Nanorobots Interfacing Neurons to Fuel Up Neurogenesis in Neurodegenerative Diseases

Goldie Oza, Sunil Pandey and Shailesh Khatri

Nanotechnology Research Centre, Birla College, Kalyan, Maharashtra, India

Madhuri Sharon

MONAD Nanotech Pvt. Ltd., Powai, Mumbai, India

29.1   What Is a Neuron?

Neurons are the fundamental units of the central nervous system (CNS), made up of a cell body (soma), several dendrites, and a single axon (Fig. 29-1). More than one billion neurons exist in the CNS and are interconnected with sound precision. This meshwork of neurons has complete control on the thoughts, emotions, movements, and sensation that are a part of human experience. Neurons have given us the capacity to sense a needle pricking our skin; we can think and reciprocate to a person talking to us. When a mosquito bites, we can sense it and give it a response with a smack. Thoughts are aligned with movement. We think of moving out of town, and this thought is then flawlessly supported by the movement of the body. Mothers' care, love, and affection are the best examples of emotions seen not only in humans but also in animals.

The brain and spinal cord contain a plethora of distinct varieties of neurons, supportive cells, synaptic connections, chemical-regulating agents, and minuscule blood vessels, all of them working meticulously for the well-being of the body. Sensory neurons are responsible for many of the activities in the nervous system. Signals from the body surfaces traveling through the peripheral nerves reach the spinal cord

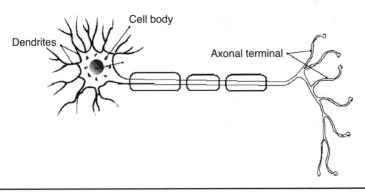

FIGURE 29-1   Structure of a neuron.

and are then transmitted throughout the brain. Incoming sensory messages are processed and integrated with information stored in various pools of neurons such that the resulting signal can be used to generate an appropriate motor response. The motor division of the nervous system is responsible for controlling a variety of bodily activities, including contraction of striated and smooth muscles and secretion by the exocrine and endocrine glands.

It is not surprising that the sustenance of such a complex circuit is difficult. Diseases such as Alzheimer's disease, schizophrenia, Parkinson's disease, and other neurodegenerative diseases erupt because of the erring of neural circuits. Early diagnosis and therapy of such diseases are very difficult. In all of the neurodegenerative diseases, natural regenerative capacity is thought to abate. Hence, an early diagnosis combined with some form of treatment that stops the pathogenic process is seen as the most promising way of battling these diseases.

29.2 Carbon Nanotechnology for Neurodegenerative Disorders

Nanobiotechnology has come to the rescue of such a problem. Nanobiotechnology is the convergence of engineering and molecular biology that is leading to development at the atomic, molecular, and macromolecular size range to create structures and to manipulate them to have novel properties. Nanomedicine, which is one of the prongs of nanobiotechnology, is a cutting-edge area of research that combines the concepts of nanotechnology and medicine, providing new hopes for research in both areas. The early genesis of the concept of nanomedicine sprang from the visionary idea that tiny nanorobots and related machines could be designed, manufactured, and introduced into the human body to perform cellular repairs at the molecular level (Freitas et al., 2005). Nanopolymers, metal nanoparticles, and carbon nanotubes (CNTs, which are the most robust materials) are the armamentariums of nanobiotechnologists. All of these materials play critical role in nanomedicine, but CNTs have a special role to play in the diagnosis and therapy of many neurologic disorders.

CNTs have intriguing electrochemical, mechanical, and chemical properties that make them excellent candidates for the improvement of neural interfaces, including:

- They have a very high mechanical stiffness, but at the same time they are very flexible, making them attractive for building penetrating electrodes in neural prostheses.
- Their very high aspect ratio and small size allow making tiny electrodes while maintaining a high electrical current density, an essential property for electrical stimulation.

- They have good electrochemical stability, reducing the possibility of damaging the electrodes and introducing abnormalities in neural function and cell structure.

- CNTs can be grown or assembled on a variety of surfaces and can give rise to structures with widely different morphologies, such as flat nanostructured continuous mats, sparse electrically conductive networks, localized three-dimensional nanoporous bushes, columnar closely packed forests, and spiked localized bundles or single fibers. This allows tailoring of the neural interface morphology to better mimic the neural tissue microenvironment and to enhance electrical coupling.

- Recent advances in CNT chemical functionalization have opened the route for designing appropriate functional electrode coupling down to the subcellular nanoscale.

This chapter explains how neuronal signaling mechanisms work normally and how they lead to diseases if errors occur. The chapter discusses different therapeutics used in the field and their associated difficulties, which are unanswered. Special attention is devoted to nanotechnology, which has the capability of answering some of these questions. The chapter also provides an overview of the state of the art of CNT research applied to neural interfaces and discusses the novel phenomena arising from their nanoscale properties that occur when they are coupled to in vitro cultured neural networks. Original results in the microfabrication of CNT electrodes for in vivo neuro-electrical activity recording and nerve cell stimulation are also presented, opening the route for a new generation of neural interfaces for long-term stable implantation.

29.3 Neuronal Signaling Mechanism for Excitatory and Inhibitory Postsynaptic Potential

Despite the varied signals carried by different classes of neurons, the form of the signal is always the same, consisting of changes in the electrical potential across the neuron's plasma membrane. Communication occurs because an electrical disturbance produced in one part of the cell spreads to other parts. Such a disturbance attenuates with increasing distance from its source unless energy is expended to amplify it as it travels. Thus, for large neurons, an active signaling mechanism is fruitful, using large amounts of energy for amplification of the signals traveling through it. This traveling wave of electrical excitation, known as an *action potential* or *nerve impulse*, can carry a message without attenuation from one end of a neuron to the other at speeds as great as 100 m/sec or more.

Action potentials are the direct consequence of the properties of voltage-gated cation channels. The plasma membranes of all electrically excitable cells (i.e., egg cells) contain voltage-gated cation channels, which are responsible for generating the action potentials. An action potential is triggered by a depolarization of the plasma membrane—that is, by a shift in the membrane potential to a less negative value. In nerve and skeletal muscle cells, a stimulus that causes sufficient depolarization promptly causes voltage-gated Na^+ channels to open, allowing a small amount of Na^+ to enter the cell down its electrochemical gradient. The influx of positive charge further depolarizes the membrane, thereby opening more Na^+ channels, which admit more Na^+ ions, causing still further depolarization. This process continues in a self-amplifying fashion until (within a fraction of a second) the electrical potential in the local region of the membrane has shifted from its resting value of about -70 mV to almost as far as the Na^+ equilibrium potential of about $+50$ mV. At this point, when the net electrochemical driving force for the flow of Na^+ is almost 0, the cell comes to a new resting state, with all of its Na^+ channels permanently open if the open conformation of the channel is stable. The cell is saved from such a permanent electrical spasm because the Na^+ channels have an automatic inactivating mechanism, which causes the channels to reclose rapidly even though the membrane is still depolarized. The Na^+ channels remain in this inactivated state, unable to reopen, until a few milliseconds after the membrane potential returns to its initial negative value.

In addition to the inactivation of Na^+ channels, voltage-gated K^+ channels open, so that the transient influx of Na^+ is rapidly overwhelmed by an efflux of K^+, which quickly drives the membrane back toward the K^+ equilibrium potential even before the inactivation of the Na^+ channels is complete. These K^+ channels respond to changes in membrane potential in much the same way as the Na^+ channels do but with slightly slower kinetics; for this reason, they are sometimes called *delayed K^+ channels*.

Synapse is a site where neuronal signals are transmitted from one cell to another. The cells are electrically separated by synaptic cleft-transmitting signals from a presynaptic cell to postsynaptic cells. Any alteration in electrical potential in the presynaptic cell stimulates exocytosis of small signaling molecules (also called *neurotransmitters*) in the cleft, which then diffuse inside the postsynaptic cell. The binding of neurotransmitter to the transmitter-gated ion channels fires up the postsynaptic cell. The residual neurotransmitters are removed from the cleft via enzymes or by a variety of Na^+-dependent neurotransmitter carrier proteins. Rapid removal ensures both spatial and temporal precision of signaling at a synapse; it prevents the neurotransmitter from influencing neighboring cells and clears the synaptic cleft before the next pulse of neurotransmitter is released so that the timing of repeated rapid signaling events can be accurately communicated to the postsynaptic cell.

Transmitter-gated ion channels are specialized for rapidly converting extracellular chemical signals into electrical signals at chemical synapses. The channels are concentrated in the plasma membrane of the postsynaptic cell in the region of the synapse and open transiently in response to the binding of neurotransmitter molecules, producing a brief permeability change in the ion channel. Unlike the voltage-gated channels responsible for action potentials, transmitter-gated channels are relatively insensitive to the membrane potential; therefore, they cannot produce a self-amplifying excitation. Instead, they produce local permeability changes, hence changes of membrane potential, that are graded according to how much neurotransmitter is released at the synapse and how long it persists there. An action potential can be triggered from this site only if the local membrane potential increases enough to open a sufficient number of nearby voltage-gated cation channels that are present in the same target cell membrane. Ion channels for neurotransmitters, such as acetylcholine, gamma-aminobutyric acid, serotonin, and glutamate, are useful for the stimulation or inhibition of action potentials in the postsynaptic cells (Fig. 29-2).

Neurotransmitters released at an excitatory synapse cause a small depolarization in the postsynaptic membrane called an *excitatory postsynaptic potential* (excitatory PSP). Neurotransmitters that are released at an inhibitory synapse, causing a small hyperpolarization, are called *inhibitory postsynaptic potentials*.

The number of presynaptic cells secreting the neurotransmitters at the synaptic site should be very large because individual presynaptic cells have very few voltage-gated ion channels. An individual presynaptic cell cannot trigger an action potential; hence, there need to be many cells at the synaptic site to give a threshold value for excitation of postsynaptic cells.

The summation of all synapses contributes to the grand postsynaptic potential (grand PSP). *Spatial summation* is the general term used for combining the effects of signals received at different sites on the membrane of postsynaptic cells, and *temporal summation* combines the effects of signals received at different times. Together, temporal and spatial summations provide the means by which the rates of firing of many presynaptic neurons jointly control the membrane potential (the grand PSP) in the body of a single postsynaptic cell. The final step in the neuronal computation made by the postsynaptic cell is the generation of an output, usually in the form of action potentials, to relay a signal to other cells. The output signal reflects the magnitude of the grand PSP in the cell body having action potentials of all or no type.

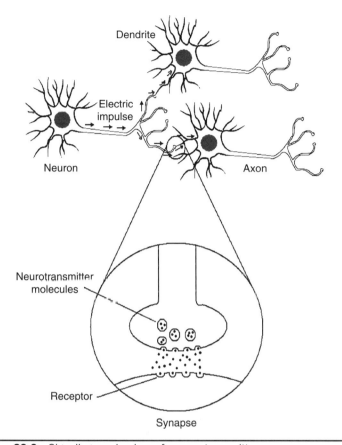

FIGURE 29-2   Signaling mechanism of a neurotransmitter.

29.4   Modus Operandi of Some Neurodegenerative Diseases

29.4.1   Parkinson's Disease

Parkinson's disease is a neurodegenerative disorder associated with the loss of dopaminergic neurons present in substantia nigra. These neurons transmit neurochemical signals via a tyrosine-derived substance called dopamine (L-DOPA). The crucial associations of these neurons are performing parkinsonian motoric functions of the various parts of the body. Degeneration of neurons, as mentioned earlier, is caused by an erring mitochondrial respiratory complex I and oxidative stress. Aggregation of proteins such as α-synuclein and others (Lewy bodies) is another characteristic property of Parkinson's diseases (Moore et al., 2005). Mitochondria play a master role in

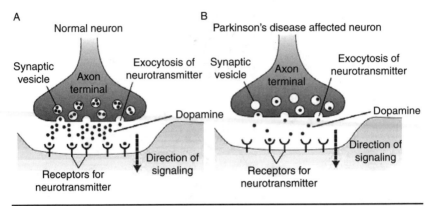

FIGURE 29-3   Neurotransmission of dopamine in a (*A*) normal and (*B*) Parkinson's disease–affected neuron.

generating reactive oxygen species (ROS) via respiratory chain complex I and hence is considered to be the main culprit in mediating the death (Fig. 29-3) of dopaminergic neurons (Kudin et al., 2004). Generation of ROS catalyzed by tyrosine hydroxylase and monoamine oxidase might activate preferential degeneration of neurons that function via firing dopamine at synaptic cleft (Fig. 29-4). The molecular mechanism explaining the mitochondrial dysfunction in dopaminergic neurons is speculated to be due to accumulation of a deleted mDNA molecule at a single cell level (Bender et al., 2006; Kraytsberg et al., 2006). The consequence of such a deletion is depleted residual amount of wild-type mDNA, leading to mitochondrial dysfunction by reduced expression of protein encoded by mDNA.

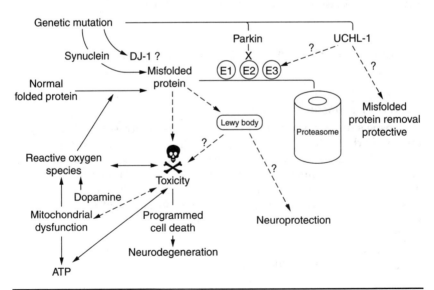

FIGURE 29-4   Reactive oxygen species (ROS) signaling leading to neurodegeneration.

29.4.2   Friedreich's Ataxia

Friedreich's ataxia (FRDA) is associated with a degeneration of Purkinjë neurons of the cerebellum that causes limb ataxia, loss of proprioception, dysarthria (problems with speech), skeletal abnormalities, hypertrophic cardiomyopathy, and a higher risk of diabetes. Individuals with human frataxia (FXN) are found to be deficient in FRDA. FXN is functional in several aspects of intracellular iron metabolism such as biosynthesis of haem (Yoon and Cowan., 2004) and iron–sulfur clusters (Stehling et al., 2004), iron binding and storage (Cavadini et al., 2002), and iron chaperone activity (Bulteau et al., 2004). Consequently, FXN-defective organisms, from unicellular yeast to humans, exhibit a plethora of metabolic disturbances caused by intramitochondrial iron accumulation, such as a loss of iron–sulfur cluster-dependent enzymes (Rotig et al., 1997), reduced oxidative phosphorylation (Ristow et al., 2000), and altered antioxidant defenses (Chantrel-Groussard et al., 2001). Deletion of FXN is due to expansion of GAA (triplets of phosphates called *codons*) with the first intron (noncoding sequence of a gene) of the FXN gene located on chromosome 9q13 (Campuzano et al., 1996). The hyperexpansion of GAA repeats determines the formation a triple-helix, non-BDNA stretch, resulting in reduced FXN mRNA transcription (Sakamoto et al., 2001).

29.4.3   Amyotropic Lateral Sclerosis

Amyotropic lateral sclerosis (ALS) is rare neurodegerative disorder caused by progressive deterioration of the anterior horn cells of the spinal cord and cortical motor neurons. The cause of neural death in ALS is under investigation. Glutamate-induced excitotoxicity is speculated to be one of the causes of neurodegeneration (Spencer et al., 1987). There is supportive evidence for enhanced oxygen radical–induced toxicity in the brain tissue of patients with ALS (Bowling et al., 1993). Some patients with autosomal dominant familiar ALS (fALS) have point mutations in the $Cu2+/Zn2+$ SOD1 (superoxide dismutase 1) gene (Rosen et al., 1993). Until now, more than 100 different mutations in SOD1 have been described (Bruijn et al., 2004). Most of these mutant SODs retain full enzymatic activity, and therefore a "toxic gain of function" as the cause for the disease has been postulated. However, the selective loss of motor neurons caused by mutation in SOD1 is still an unsolved mystery.

29.4.4   Temporal Lobe Epilepsy

Epilepsy is a neurologic disorder, affecting approximately 0.5% to 0.7% of the human population worldwide. The common symptoms of epilepsy are recurrent seizures consisting of synchronized discharges of large groups of neurons that alter normal function. Long-term seizures (status epilepticus), initiated in experimental models by kainic acid or pilocarpine, are known to activate neuronal cell death mechanisms in temporal lobe structures similar to other neurodegenerative disorders. This neuronal cell death is also observed in human temporal lobe epilepsy and is one of the most important aspects of epileptogenesis. The loss of CA1 and CA3 pyramidal neurons of the hippocampus, with relative sparing of the granular neurons of the dentate gyrus and some types of interneurons, is the histopathologic indicative of Ammon's horn sclerosis (Kunz et al., 2000).

29.4.5 Alzheimer's Disease

The most prevalent dementia-type disease is Alzheimer's disease constituting about two thirds of cases of dementia among other diseases such as Pick's disease and diffuse Lewy body dementia (Aronson et al., 1991; Helmer et al., 2001). Alzheimer's disease is characterized by an irreversible loss of neurons, particularly in the cortex and hippocampus regions (McKhann et al., 1984). This leads to progressive impairment in memory, judgment, decision making, orientation to physical surroundings, and language. The pathologic mechanism of Alzheimer's disease is neuronal loss, extracellular senile plaques containing the peptide amyloid, and neurofibrillary tangles containing the hyperphosphorylated microtubular protein tau (Fig. 29-5).

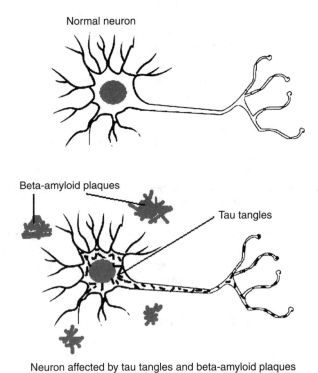

Normal neuron

Beta-amyloid plaques

Tau tangles

Neuron affected by tau tangles and beta-amyloid plaques

FIGURE 29-5   β-Amyloid plaques and tau tangles formed in neurons.

Amyloid in the extracellular senile plaques is a result of scissoring a big β-amyloid precursor protein (APP) via a series of proteases, α-, β-, and γ-secretases (Hutton et al., 1998). The cleavage product formed using γ-secretase is especially responsible for generating a β-amyloid peptide (Aβ42) that is 42 amino acids long, plays a critical role in the formation of insoluble toxic fibrils, and accumulates in the senile plaques isolated from the brains of patients with Alzheimer's disease (Esler et al., 2001; Iwatsubo et al., 1994). Neurofibrillary tangles are made up of paired helical filaments of abnormally filled and phosphorylated tau protein in the neuron and its dendrites. Many tau tangles are seen as the disease proceeds, leading to a reduction in synaptic density, loss of neurons, and degeneration in hippocampal neurons. A specific degeneration of neurons concerned with maintenance of specific transmitter cysteines results in deficits of acetylcholine, norepinephrine, and serotonin (Palmer et al., 1988). Although plaque formation arrests, plaque formation of tangles continues. This correlates with the progress and severity of dementia. A thorough genetic study done for familial Alzheimer's disease showed that mutations of genes of APP, presenilin-1, and presenilin-2 are responsible for the agglomeration of fibrillary tangles.

The most important evidence is based on the genetic observations from familial Alzheimer's disease (Hardy and Selkoe, 2002). This research showed that mutation of the genes of APP (Goate et al., 1991), presenilin-1, and presenilin-2 (Cruts et al., 1998) that cause inherited Alzheimer's disease lead to increased accumulation of fibrillary β-amyloid in the brain (Hardy and Selkoe, 2002; Saido, 2003). The other culprits associated with Alzheimer's disease are the apolipoprotein E gene, allele 4 (Levy-Lahad et al., 1995; Slooter et al., 1998), the endothelial nitric oxide synthase-3 gene (Dahiyat et al., 1999), and the α-2-macroglobulin gene (Blacker et al., 1998).

29.5 Nanomedicine Has Better Prospects for the Treatment of Neuronal Diseases

Because nanotechnology deals with materials of nanometric size, its applications in medicine are enormous. Nanomedicine has its impact in diagnosis, monitoring, and treatment of diseases. It is also useful in controlling and understanding biologic systems. Nanomedicine has especially contributed to the understanding of many neurologic diseases and has played a critical role in their diagnosis and therapy. Neuroprotective properties are exhibited by many hydroxyfullerenes, CNTs, dendrimers, and so on. A neuroregenerative capacity is also seen in CNTs acting as a conduit, which can then be mixed with stem cells to repair neurons. Many polymers, CNTs, and metal nanoparticles have the capacity to cross the blood–brain barrier and deliver drugs. The details are given in Fig. 29-1. Because CNTs have major applications in the treatment of neurologic disorders, we will

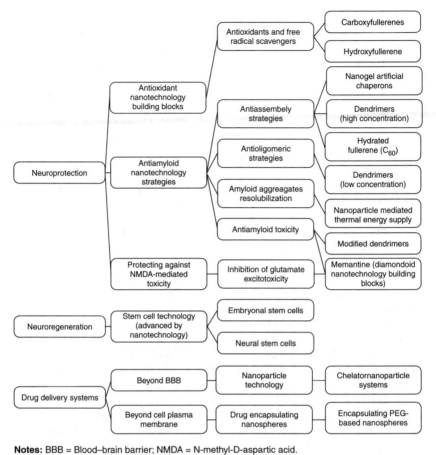

Notes: BBB = Blood–brain barrier; NMDA = N-methyl-D-aspartic acid.

FIGURE 29-6   Summary of applications of nanotechnology in medicine.

now deal with the properties of CNTs and their applications in the diagnosis and treatment of these disorders, making them a suitable candidate (Fig. 29-6).

29.6   Properties and Biocompatibility of Carbon Nanotubes

CNTs are well-ordered, high-aspect-ratio allotropes of carbon. The two main variants, single-walled CNTs (SWCNTs) and multiwalled CNTs (MWCNTs), both possess a high tensile strength, are ultra lightweight, and have excellent chemical and thermal stability. They also possess semimetallic and metallic conductive properties. This startling array of features has led to many proposed applications in the biomedical field, including biosensors, drug and vaccine delivery, and the preparation of unique biomaterials such as reinforced and conductive polymer nanocomposites.

The distinct electronic properties of CNTs have stimulated several studies to investigate the biocompatibility of CNTs for neural applications, focusing particularly on neurite extension (Gabay et al., 2005; Hu et al., 2004; McKenzie et al., 2004; Webster et al., 2004). McKenzie et al. (2004) investigated the effects of varying carbon nanofiber (CNF) diameter and surface energy on astrocyte proliferation and function. The study compared surfaces created from four different samples of CNFs (large MWCNTs with diameters of 60 to 200 nm synthesized via carbon vapor deposition [CVD]). Two CNF samples (with diameters of 100 and 200 nm) were used unrefined (classed as low-surface-energy CNFs), and the other two CNFs (with diameters of 60 and 125 nm) were subjected to pyrolytic stripping to remove the outer hydrocarbon layer. Astrocytes, the cells largely responsible for the scar tissue formation seen with current implantable neural devices, were then seeded onto the surfaces for adhesion, proliferation, and function studies. It was found that astrocytes preferentially adhered to and proliferated on CNFs with larger diameters and higher surface energies. The authors concluded that CNFs with diameters smaller than 100 nm show potential for neural applications because of a speculated reduction in neural scar tissue formation.

A follow-up study by Webster et al. (2004) investigated the interactions between a series of polyurethane (PU)–CNF nanocomposites and astrocytes. The PU–CNF composites were synthesized by solvent-casting techniques, after PU–CNF sonication in chloroform. The CNFs used in the study had an average diameter of 60 nm and were synthesized via CVD and purified by pyrolytic stripping. The mechanical properties of the nanocomposites were significantly poorer than those reported by other studies involving PU materials (Sen et al., 2004), suggesting that it was unlikely that the level of CNF dispersion was ideal. Neural cell seeding studies demonstrated that increased CNF loading in the nanocomposite resulted in both decreased astrocyte adhesion and retarded neurite growth in rat pheochromocytoma cells by a small but statistically significant amount.

The authors concluded that these findings, coupled with the ability to tailor the electrical resistance of PU–CNF nanocomposites, warranted further investigation into their use in neural probe applications. In contrast to the previous work, subsequent studies found that both pristine and chemically functionalized CNTs have a positive impact on neuronal growth (Gabay et al., 2005; Hu et al., 2004). Confluent layers of neurons were seeded onto lithographically patterned CNT surfaces (islands of CNTs grown via CVD on a quartz substrate). Over four days, neurons were observed to localize in CNT-rich regions, and their associated neurites and axons formed an interconnected network that replicated the pattern of the CNT template (Gabay et al., 2005). Similarly, purified MWCNTs that were chemically functionalized with carboxylic acid, ethylenediamine, or poly-m-aminobenzene sulfonic acid (each holding a different ionic charge at physiologic pH) were each observed to provide a substrate for neurite extension (Hu et al., 2004). It was concluded that neurite extension was loosely based on the ionic charge, with positively charged ethylenediamine–MWCNT producing the most neurite extension. Neither the CNT-patterned surfaces nor the functionalized MWCNTs demonstrated cytotoxicity toward neuronal cells.

29.7   Application of Carbon Nanotubes in Neurodegenerative Disorders

If you have seen the movie *The Matrix*, then you are familiar with "jacking-in," a brain machine neural interface that can connect the human brain to a computer network. It is still a science fiction scenario, but researchers are exploring this field and are constantly working to achieve it.

Because of scientists' painstaking efforts, the dream that has come into reality is neuroprosthetics, an area of neuroscience that uses artificial microdevices to replace the function of impaired nervous systems or sensory organs. Different biomedical devices implanted in the CNS—so-called *neural interfaces*—have already been developed to control motor disorders and to translate willful brain processes into specific actions with the control of external devices. These implants could help increase the independence of people with disabilities by allowing them to control various devices with their thoughts. SWCNTs have received great attention because of their unique physical and chemical features, which allow the development of devices with outstanding electrical properties.

Towards a new generation of future neuroprosthetic devices, Andrea et al. (2007), developed a SWCNT-neuron hybrid system and demonstrated that CNTs can directly stimulate brain circuit activity. Examples of existing brain implants include brain pacemakers to ease the symptoms of diseases such as epilepsy, Parkinson's disease, dystonia, and depression. Other examples are retinal implants consisting of an array of electrodes implanted on the back of the retina, a digital camera worn on the user's body, and a transmitter and image processor that converts the image to electrical signals sent to the brain. Most recently, cyberkinetics devices, such as the BrainGate™ Neural Interface system, have been used successfully by quadriplegic patients to control a computer with their thoughts alone.

The application of deep brain stimulation (DBS) has been proven to be an effective clinical treatment for a host of neurologic disorders such as Parkinson's disease. Thus, carbon nanomaterials can provide greater insight into the complexity of the brain, thereby enabling greater specificity and enhancement to the current DBS technology. To enable this interface, a novel method has been developed to fabricate vertically aligned CNFs into nanoelectrode arrays. Such vertically aligned CNFs can be precisely grown on underlying electronic circuits using techniques compatible with Si microfabrication. Consequently, the CNFs can be directly integrated into nanoelectrode arrays on a multiplex microchip for neural electrophysiology. The chip design is composed of multiple types of arrays for both electrical stimulation and electrochemical monitoring of neurotransmitters, thus providing real-time feedback of neurologic processes.

The forest-like structure of the as-grown CNFs has demonstrated attractive electrical properties. Because of the large surface area of the three-dimensional array structure, the electrode exhibits a high specific capacitance of 0.4 mF/cm^2 and very low impedance. This structure is further enhanced with the electrochemical deposition of polypyrrole (PPy), an electronic-conducting polymer. The capacitance has been found to increase by 100 times to 40 mF/cm^2 with a PPy film of only 24 nm. Correspondingly, the impedance decreased and is negligible compared with the ohmic resistance of the solution and the impedance of the Pt counter electrode.

In addition, PPy coating is conformal to the individual CNF and increases its mechanical strength. PPy-coated CNFs retain their vertical alignment after submersion in solution and are strong enough to withstand cell culture. Both features result in an extremely efficient electrode that can be used for more specific and effective stimulation of the neural cells.

The embedded CNFs that form the nanodisk electrode array have been shown to be an extremely sensitive electrochemical detector. Such an array has demonstrated a detection limit in the nanomolar region with dynamic range of eight orders of magnitude (Campuzano et al., 1996). The nanoscale feature of CNFs exhibits ideal nanoelectrode behavior, which makes them perfectly suited as an ultra-sensitive detector of low-concentration electroactive molecules such as catecholamine neurotransmitters. The nanodisk array is able to harness the unique signal at the individual CNFs while enhancing the amplitude through the summation of the temporal signals corresponding to number of CNFs exposed in the array. It has been demonstrated that this electrochemical method can be used to measure dopamine at the 60-nM level. Furthermore, the physical dimension of the CNFs also makes them attractive for high temporal resolution detection, which is necessary to capture the transient signals of neurotransmitter release (Thuy-Duong et al., 2005).

29.8 Summary

CNTs possess a high tensile strength, are ultra lightweight, and have excellent chemical and electronic properties and thermal stability plus it also possesses semimetallic and metallic conductive properties. The distinct and outstanding electronic properties make it an excellent candidate that can be used as an artificial microdevice to replace the function of impaired nervous systems or sensory organs. Such neural interface implants could help increase the independence of people with disabilities by allowing them to control various devices with their thoughts. Hybrid SWCNT-Neuron which can directly stimulate brain circuit activity, are found to be demonstrated as a future neuroprosthetic device for patients suffering from neurodegenerative disorders.

References

Cited References

Adrian P., *Sens. Bus. Dig.* (2003) [Online] http://www.sensorsmag.com/resources/businessdigest/sbd0703.shtml/

Afre R. A., Soga T., Jimbo T., Kumar M., Ando Y., and Sharon M., *Chem. Phys. Lett.*, 414(1-3):6–10 (2005).

Afre R. A., Soga T., Jimbo T., Kumar M., Ando Y., Sharon M., Somani P. R., and Umeno M., *Micropor. Mesopor. Mater.*, 96(1-3):184–190 (2006).

Ago H., Komatsu T., Ohshima S., Kuriki Y., and Yumura M., *Appl. Phys. Lett.*, 77:79 (2000).

Ago H., Murata K., Yumura M., Yotani J., and Uemura S., *Appl. Phys. Lett.*, 82:811 (2003).

Ago H., Shaffer M. S. P., Ginger D. S., Windle A. H., and Friend R. H., *Phys. Rev. B*, 61:2286 (2000).

Ago H., Uchimura E., Saito T., Ohshima S., and Ishigami N., *Mater. Lett.*, 60:3851 (2006).

Aitken A. J., Chaudhry M. Q., Boxall A. B. A., and Hull M., *Occ. Med.*, 56(5):300 (2005).

Ajayan P. M., Schadler L. S., Giannaris C., Rubio A., Schwoebel R. L., *J. Appl. Phys.* 40:614 (1969).

Ajayan P. M., Ebeen T. W., Ichihashi T., Ijima S., Tangaki K., and Hiura H., *Nature (Lond)*, 362:522–525 (1993).

Ajayan P. M., Nanotubes from carbon. *Chem. Rev.*, 99:1787–1800 (1999).

Ali Z., Watson G., Shun-quin H., Lupu D., Biris A. R., Olenic L., and Mihailescu G., *Partic. Sci. Technol.*, 24(3):311 (2006).

Allemann J., Dillon A., Gennett T., Jones K., and Parilla P., *Carbon Nanotube Materials for Hydrogen Storage.* Proceedings of the 2000 DOE/NREL (2002).

Allen M., Law F., and Rushton N., *Clin. Mater.*, 17(1):1–10 (1994).

Allen M., Myer B., and Rushton N., *J. Biomed. Mater. Res.*, 58B(3):319–328 (2001).

Althues H., *Langmuir*, 18:7428 (2002).

Alvarez W. E., Pompeo F., Herrera J. E., Balzano L., and Resasco D. E., *Chem. Mater.*, 14(4):1853 (2002).

Amelinckx S., Zhang X. B., Barnaerts D., Zhang X. F., Ivanov V., and Nagy J. B., *Science*, 265:635–639 (1994).

Ammar S., Helfen A., Jouini N., Francoise Villain F. F., Molinié P., and Danot M., *J. Mater Chem.* (11):186–192 (2001).

Anderson J. D., "Aircraft Performance and Design," McGraw-Hill (1st ed.) (1999).

Ando Y., *Fullerene Sci. & Tech.*, 2:173 (1994).

Ando Y. and Iijima S., *Jpn. J. Appl. Phys.*, 32:L107 (1993).

Ando Y. and Ohkohchi M., *J. Crystal Growth*, 60:147 (1982).

Ando Y., Zhao X., and Ohkohchi M., *Carbon*, 35:153–158 (1997).

Ando Y., Zhao X., and Ohkohchi M., *Jpn. J. Appl. Phys.*, 37:L61 (1998).

Ando Y., Zhao X., Hirahara K., Suenaga K., Bandow S., and Iijima S., *Chem. Phys. Lett.*, 323:580–585 (2000).

Ando Y., Zhao X., Inoue S., Suzuki T., and Kadoya T., *Diam. Rel. Mater.*, 14:729 (2005).

Andrea Mazzatenta, Michele Giugliano, Stephane Campidelli, Luca Gambazzi, Luca Businaro, Henry Markram, Maurizio Prato, and Laura Ballerini, *Jour. Neurosci.*, 27:6931 (2007).

Andrew L., *J. Power Sources*, 156(2):128 (2006).

Andrews R., Jacques D, Rao A. M., and Derbyshire F., *Chem. Phys. Lett.*, 303:467–474 (1999).

Andrews, R., Jacques, D., Rantell, T., *Account of Chemical Research*, 35:1008-1017 (2002).

Annadurai P., Mallick A. K., and Tripathy D. K., *J. Appl. Polym. Sci.*, 83:145 (2002).

Arena U. Mastelline M. L., Proceedings of IFSA 2005 Industrial luidization, South Africa, (15-17 Nov, 2005).

Arnal P., Corriu R. J. P., Leclercq D., Mutin P. H., and Vioux A., *Chem. Mater.*, 9:694 (1997).

Aronson M. K., Ooi W. L., Geva D. L., Masur D., Blau A., and Frishman W., *Arch. Intern. Med.*, 151:989–992 (1991).

Arzum E., Kerman K., Meric B., and Ozsoz M., *Electroanalysis*, 13(3):219 (2001).

Atala A., *J. Endourol.*, 14(1):49 (2000).

Avouris P., Hertel T., Martel R., Schmidt T., Shea H., and Walkup R., *Surface Sci.*, 141:201 (1999).

Badzian A., Badzian T., Breval E., and Piotrowski A., *Thin Solid Films*, 398:170 (2001).

Bae D., Bok J., Choi Y. C., Choi Y. G., Frauenheim T., Kim N., Lee S., Lee Y., Nahm K., Park K., Park Y., and Yu S., *Synthetic Metals*, 113:209 (2000).

Bai X. D., Zhong D. Y., Zhang G. Y., Ma X. C., Liu S., Wang E. G., Chen Y., and Shaw D. T., *Appl. Phys. Lett.*, 79:1552 (2001).

Baker R., Hidalgo R., Park C., Rodriguez N., and Tan C., Proceedings of the 1998 U.S. DOE Hydrogen Program Review (1998).

Baker R. T. K., *Carbon*, 27(33):315–323 (1989).

Balakos M. W. and Chuang S. S. C., *React. Kinet. Catal. Lett.*, 49(1):7 (1993).

Ball M., O'Brian A., Dolan F., Abbas G., and McLaughlin J. A., *J. Biomed. Mater. Res.*, 70A(3):380 (2004).

Ballou B., Lagerholm B. C., Ernst L. A., Bruchez M. P., and Waggoner A. S., *Bioconjugate Chem.*, 15:79 (2004).

Ban V. S., *J. Electrochem. Soc.* 125:317 (1978).

Barisci J. N., Tahhan M., Wallace G. G., Badaire S., Vaugien T., and Maugey M., *Adv. Funct. Mater.*, 14(2):133 (2004).

Barlow P. G., Donaldson K., MacCallum J., Clouter A., and Stone V., *Toxicol. Lett.*, 155(3):397 (2005).

Baughman R. H., Cui C., Zakhidov A. A., Iqbal Z., Barisci J. N., Spinks G. M., Wallace G. G., et al., *Science*, 284:1340–1344 (1999).

Baughman R. H., Zakhidov A. A., and de Heer W. A., 297:787–792 (2002).

Bellobono I. R., Morazzoni F., Bianchi R., Mangone E. S., Stanescu R., and Costache C., *Int. J. Photoenergy*, 7:87 (2005).

Bender A., Krishnan K. J., Morris C. M., Taylor G. A., Reeve A. K., Perry R. H., Jaros E., Hersheson J. S., Betts J., and Klopstock T., *Nat. Genet.* 38:515–517 (2006).

Bentzon M. D., Wonterghem J. V., Morup S., Tholen A., and Koch C. J. W., *Philos. Mag. B*, 60:169 (1989).

Berg U. E., Kreuter J., Spieser P. P., and Soliva M., *Pharm. Ind.*, 48:75 (1986).

Besteman K., Lee J. O., Wiertz F. G., Heering H. A., and Dekker C., *Nano Lett.*, 3:727 (2003).

Bethune D. S., Kiang C. H., de Vries M. S., Gorman G., Savoy R., Vazquez J., and Beyers R., *Nature*, 363:605 (1993).

Bhardwaj S., Jaybhaye S. V., Sharon M., Sathiyamoorthy D., Dasgupta K., Jagadale P., Gupta A., et al., *Asian J. Exp. Sci.*, 22(2): 89 (2008).

Bhardwaj S., Sharon M., and Ishihara T., *Curr. Appl. Phys.*, 8(1):71 (2008).

Bhardwaj S., Sharon M., Ishihara T., Jayabhaye S., Afre R., Soga T., and Sharon M., *Carbon Lett.*, 8(4):1 (2007).

Bhardwaj S., Jaybhaye S., Sharon M., Sathiyamoorthy D., Dasgupta K., Jagdale P., Gupta A., Patil B., Ozha G., Pandey S., Soga T., Afre R., Kolita G., and Sharon M., *Asian J Exp. Sci.*, 22(2):89–93 (2008).

Bhargava A., [Online].Available: http://www.ewh.ieee.org/r10/Bombay/news3/page4.html (1999).

Bhowmick R., Bruce M., Clemens B., and Cruden A., *Carbon*, 46(6):907–922 (2008).

Bi H., Wu X., and Li M., *Yuhang Cailiao Gongyi*, 35(2):34–37 (2005).

Bianco A., Kostarelos K., and Prato M., *Curr. Opin. Chem. Biol.*, 9(6):674–679 (2005).

Biehl S., Lüthje H., Bandorf R., and Sick J. -H., *Thin Solid Films*, 515:1171 (2006).

Birrenbch G. and Speiser P. P., *J. Pharm. Sci.*, 65:1763–1766 (1976).

Blacker D., Wilcox M. A., Laird N. M., Rodes L., Horvath S. M., Go R. C., Perry R., Watson Jr. B., Bassett S. S., McInnis M. G., Albert M. S., Hyman B. T., and Tanzi R. E., *Nat. Genet.* 19:357–360 (1998).

Bockrath M., Cobden D. H., McEuen P. L., Chopra N. G., Zettle A., Thess A., and Smalley R. E., *Science*, 275(5308):1922–1925 (1997).

Bolskar R. D., Benedetto A. F., Husebo L. O., Price R. E., Jackson E. F., and Wallace S., *J. Am. Chem. Soc.*, 125:5471–5478 (2003).

Bonard J. M., Salvetat J. P., Stöckli T., Forro L., and Châtelain S., *Appl. Phys. A*, 69:245 (1999).

Bowling, A. C., Schulz, J. B., Brown, Jr, R. H., and Beal, M. F., *J. Neurochem.*, 61:2322–2325 (1993).

Brabec C. J., Maiti A., Roland C., and Bernholc J., *Chem. Phys. Lett.*, 236:150–155 (1995).

Brodie I., *Int. J. Electron*, 38:541 (1975).

Brodie I. and Schwoebel P. R., *Proc. IEEE*, 82(7):1006 (1994).

Brodie I. and Spindt C. A., *Adv. Electron. Electron Phys.*, 83:1 (1992).

Bronikowski M. J., Willis P. A., Colbert T. D., Smith K. A., and Smalley R. E., *J. Vac. Sci. Technol. A*, 19(4):1800 (2001).

Browning D. J., Gerrard M. L., Lakeman J. B., Mellor I. M., Mortimer R. J., and Turpin M. C., *Nano Lett.*, 2:201 (2002).

Buitelaar M. R., Bachtold A., Nussbaumer T., Iqbal M., and Schonenberger C., *Phys. Rev. Lett.*, 88(15):156801 (2002).

Bruijn L. I., Miller T. M., and Cleveland D. W., *Annu. Rev. Neurosci.*, 27:723–749 (2004).

Bueken, A. and Huang, H., *Journal of Hazardous Materials*, (62):1–33 (1998).

Bulteau A. L., O'Neill H. A., Kennedy M. C., Ikeda-Saito M., Isaya G., and Szweda L. I., *Science*, 305:242–245 (2004).

Cagle D. W., Kennel S. J., Mirzadeh S., Alford J. M., and Wilson L. J., *Proc. Natl. Acad. Sci. U. S. A.*, 96:5182 (1998).

Cai H., Xu Y., He P. G., and Fang Y. Z., *Electroanalysis*, 15:1864 (2003).

Campuzano V., Montermini L., Molto M. D., Pianese L., Cossee M., Cavalcanti F., Monros E., Rodius F., Duclos F., and Monticelli A., *Science*, 271:1423–1427 (1996).

Cao A., Ajayan P. M., Ramanath G., and Baskaran R.I, *Appl. Phys. Lett.*, 84:109 (2004).

Capano M. A., Safriet A. J., Donley M. S., Zabinski J. S., *Appl. Phys. Lett.*, 69:188 (1996).

Cassel A. M., Raymarkers J. A., Kong J., and Dai H., *J. Phys. Chem. B*, 103:6484 (1999).

Cattanach K., Kulkarni R., Kozlov M., and Manohar S., *Nanotechnology*, 17:4123 (2006).

Cavadini P., O'Neill H. A., Benada O., and Isaya G., *Hum. Mol. Genet.* 11:217–227 (2002).

Chahine R., Bénard P., Poirier E, Cossement D., Lafi L., Mélançon E., Bose T. K., and Désilets S., *Appl. Phys. A: Materials Science & Processing*, 78(7):961–967 (2004).

Chalamala B.R., Wei Y., and Gnade B. E., *IEEE Spectrum*, 42 (1998).

Chambers A., Park C., Baker R. T. K., and Rodriguez N. M., *J. Phys. Chem. B*, 102:4253 (1998).

Chandrasekhar P., *Conducting Polymers, Fundamentals and Applications: A Practical Approach*. London: Kluwer Academic Publishers, 1999.

Chantrel-Groussard K., Geromel V., Puccio H., Koenig M., Munnich A., Rötig A., and Rustin P., *Hum. Mol. Genet.* 10:2061–2067 (2001).

Charlier J.C., De Vita A., Blasé X., *R. Car. Science.* 275:646 (1997).

Charlier J.C., Blasé X., and De Vitta A., *R Car. Appl. Phys.A.* 68:276 (1999).

Chatterjee A. K., Sharon M., Banerjee R., and Neumann-Spallart M., *Electrochimica Acta.*, 48(23):3439 (2003).

Chen B. X., Wilson S. R., Das M., Coughlin D. J., and Erlanger B. F., *Immunology*, 98:10809 (1999b).

Chen P., Zhang H. B., Lin G. D., Hong Q., and Tsai K. R., *Carbon*, 35:1495–1501 (1997).

Chen P., Lin J., Tan K., and Wu X., *Science*, 285(5424):91–93 (1999a).

Chen R. J., Choi H. C., Bangsaruntip S., Yenilmez E., Tang X., and Wang Q., *J.Am. Chem. Soc.*, 126:1563 (2004).

Cheng Y. and Zhou O., *C. R. Physique.* 4:1021 (2003).

Cheng H., Cong H., Fan Y., Liu C., and Liu M., *Science*, 286 (1999).

Cheng H. M., Li F., Su G., Pan H. Y., He L. L., Sun X., and Dresselhans M. S., *Appl. Phys. Lett.*, 72:3282 (1998a).

Cheng H. M., Li F., Sun X., and Brown S. D. M., *Chem. Phys. Lett.*, 289:602–606 (1998b).

Cherukuri P., Bachilo S. M., Litovsky S. H., and Weisman R. B., *J. Am. Chem. Soc.*, 126:15638–15639 (2004).

Cheung C. L., Kurtz A., Park H., and Lieber C. M., *J. Phys. Chem. B*, 106:2429 (2002).

Chhowalla M. M., Ducati C., Rupesinghe N. L., Teo K. B. K., and Amaratunga G. A. J., *Appl. Phys. Lett.*, 79:2079 (2001a).

Chhowalla M. M., Teo K. B. K., Ducati C., Rupesinghe N. L., Amaratunga G. A. J., Ferrari A. C., Roy D., Robertson J., and Milne W. I., *J. Appl. Phys.*, 90:5308 (2001b).

Chiang I. W., Brinson B. E., Huang A. Y., Willis P. A., Bronikowski M. J., Margrave J. L., Smalley R. E., and Hauge R. H., *J. Phys. Chem. B*, 105(35):8297 (2001a).

Chiang I. W., Brinson B. E., Smalley R. E., Margrave J. L., and Hauge R. H., *J. Phys. Chem. B*, 105(6):1157 (2001b).

Chi-Chang H. and Wang C., *Electroche. Commun.*, 4:554 (2002).

Choi H. C., Kim W., Wang D., and Dai H., *J. Phys. Chem. B*, 106(48):12361 (2002).

Choi W. B., Chung D. S., Kang J. H., Kim H. Y., Jin Y. W., Han I. T., Lee Y. H., Jung J. E., Lee N. S., Park G. S., and Kim J. M., *Appl. Phys. Lett.*, 75:3129 (1999).

Choi W. B., Jin Y. W., Kim H. Y., Lee S. J., Yun M. J., Kang J. H., Choi Y. S., Park N. S., Lee N. S., and Kim J. M., *Appl. Phys. Lett.*, 78:1547 (2001).

Chopra S., mc Gruire K., Gothard N., Rao A. M., *Appl. Phys. Letts.*, 80(24):4632 (2002).

Cima L. G., Vacanti J. P., Vacanti C., Ingber D., Mooney D, and Langer R., *J. Biomech. Eng.*, 113:143 (1991).

Cochet M., Maser W. K., Benito A. M., Callejas M. A., Martinez M. T., Benoit J. M., Schreiber J., and Chauvet O., *Chem. Commun.*, 16:1450 (2001).

Coleman J. N., Curran S., Dalton A. B., Davey A. P., McCarthy B., Blau W., and Barklie R. C., *Phys. Rev. B.*, 58:7492 (1998).

Collins P. G. and Avouris P., *Sci. Am.*, 283:62 (2000).

Collins P. G., Arnold M. S., Avouris P., *Science*, 292:706–709 (2001).

Collins P. G., Bradley K., Ishigami M., and Zettl A., *Science*, 287:1801–1804 (2000).

Colvin V., *Nat. Biotechnol.*, 21:1166 (2003).

Corrias M., Caussat B., Ayral, A., Durand, J., Kihn, Y., Kalck P., and Serp. P., *Chem. Eng. Sci.*, 58:4475 (2003).

Correa-Duarte M. A., Wagner N., Rojas-Chapana J., Morszeck C., Thie M., and Giersig M., *Nano Lett.*, 4(11):2233 (2004).

Couteau E., Hernadi K., Seo J. W., Thiên-Nga L., Mikó C., Gaál R., and Forró L., *Chem. Phys. Lett.*, 378:9–17 (2003).

Coquay, P., Vandenberghe, R. E., De Grave, E., Fonseca, A., Piedigrosso, P., and Nagy, J. B., *J. Appl. Phys.*, 92(3):1286 (2002).

Crichton M., *'PREY'*, Harper Collins (2003).

Cruts, M., van Duijn, C. M., Backhovens, H., Van den Broeck, M., Wehnert, A., Serneels, S., Sherrington, R., Hutton, M., Hardy, J., St. George-Hyslop, P. H., Hofman, A., and Van Broeckhoven, C., *Hum. Mol. Genet.* 7:43–51 (1998).

Cui D., Tian F., Ozkan C. S., Wang M.,and Gao H., *Toxicol. Lett.*, 155(1):73 (2005).

Cui F. Z. and Li D. J., *Surface Coatings Technol.*, 131(1–3):481 (2000).

Cui J. B., Ristein J., and Ley L., *Phys. Rev. Lett.*, 81:429 (1998).

Cui J. B., Stammler M., Ristein J., and Ley L., *J. Appl. Phys.*, 88:3667 (2000).

Cui H., Eres G,. Howe J. Y., Puretkzy A., Varela M., Geohegan D. B. and Lowndes D. H., *Chem. Phys. Lett.*, 374:222 (2003).

Cui Y., Wei Q., Park H., and Lieber C. M., *Science*, 293:1289–1292 (2001).

Cullis C. F. and Hirschler M. M., *In the Combustion of Organic Polmers*, Clarendon Press, (Oxford). 240–241 (1981).

Cumings J. and Zettle A., *Science*, 289:602–604 (2000).

Daenen M. J. M., The nanotube site. http://www.pa.msu.edu/cmp/csc/nanotube.html.

Dahiyat M., Cumming A., Harrington C., Wischik C., Xuereb J., Corrigan F., Breen G., Shaw D., and St. Clair D., *Ann. Neurol.*, 46:664–667 (1999).

Dai H., *Acc. Chem. Res.*, 35:1035 (2002).

Dai H. and Flesher. wrow.medicalreport.com. August (2005).

Dai H., G. P., Liu, C., Liu, M., Wang, M. Z., and Cheng H. M., *Nano Lett.*, 2:503 (2002).

Dai H., Hafner J. H., Rinzler A. G. et. al., *Nature*, 384:147 (1996).

Dai H., Rinzler A., Nikolaev P., and Thess A., *Chem. Phys. Lett.*, 260:471–475 (1996).

Dai J., *Appl. Phys.Lett.*, 77:2840 (2000).

Da silva L. B., Fagan S. B., Mota R., and Fazzio A., *Nanotechnology*, 17:4088 (2006).

Deck C. P. and Vecchio K., *Carbon*, 44(2):267–275 (2006).

De la Casa-Lillo M. A., Lamari-Darkrim F., Cazorla-Amoros D., and Linares-Solano A., *Journal of Physical Chemistry B*, 106:10930–10934 (2002).

De Vlta A., Charlier J. C., Blasé X., and Car R., *Appl. Phys.* 68:283 (1999).

Demirbas Ayhan J., *Anal. Appl. Pyrolysis*, 72:97–102 (2004).

Denis M. C., Mahmood U., Benoist C., Mathis D., and Weissleder R., *Proc. Natl. Acad. Sci. USA*, 101(34):12634–12639 (2004).

Deo R. P., Wang J., Block I., Mulchandanic A., Kanchan A., Trojanowick M., Scholz F., Chen W., Lin J., *Analytical Chemical Acta* 530(2):185–189 (2005).

Deshpande K., Mukasyan A., and Varma A., *Chem. Mater.*, 16(24):4896 (2004).

Dhas N. A., Raj C. P., and Gedanken A., *Chem. Mater*, 10:1446 (1998).

Dicks, Andrew L. *J. Power Sources*, 156(2):128–141 (2006).

Dickson D. P. E. *Bionanomaterials* Chapter 18 in *Nanomaterials* by Edelstein A. S. and Robert C. C., Boca Raton, FL: CRC Press.

Dillon A. C., Jones K. M., Bekkedahl T. A., Kiang C. H., Bethune D. S., and Heben M. J., *Nature*, 386:377 (1997).

Dillon. A. C., Gennet, T., Alleman, J. L., Jones, K. M., Parilla P. A., Heben M. J., (1999) *Proceedings of the U. S. DOE Hydrogen Program Review*, (http://www.eren.doe. gov/hydrogen/docs/26938toc.html) (1999).

Ding L., Stilwell J., Zhang T., Elboudwarej O., Jiang H., Selegue J. P., Cooke P.A., Gray J. W., and Chen F. F., *Nano Lett.*, 5(12):2448 (2005).

Dodziuk H., Ejchart A., Anczewski W., Ueda H., Krinichnaya E., and Dolgonos G., *Chem. Commun.*, 8:986 (2003).

Donaldson K., Li X. Y., Dogra S., Miller B. G., and Brown G. M., *J. Pathol.*, 168(2):243 (1992).

Donaldson K. and Tran C. L., *Inhal. Toxicol.*, 14(1):5 (2002).

Donaldson K., Stone V., Tran C. L., Kreyling W., and Borm P., *Nanotoxicol. Occup. Environ. Med.*, 61(9):727 (2004).

Donaldson K. and Tran C. L., *Mutation Res.*, 553(1–2):5 (2004).

Donaldson K., Tran C. L, and MacNee W., *Eur. Respir. Monogr.*, 7:77 (2002).

Doss E. D., Kumar R., Ahluwalia R. K., and Krumpelt M., *J. Power Sources*, 102(15):1 (2001).

Douglas B. C. and Hubler G. K., *Pulsed Laser Deposition of Thin Films*, New York: Wiley (1994).

Douglas H. L., D. B. Geohegan, A. A. Puretzky, D. P. Norton, and C. M. Rouleau *Science*, 273(5277):898–903 (1995).

Downs C., Nugent J., Ajayan P. M., Duquette D. J., and Santhanam S. V., *Adv. Mater*, 11:1028 (1999).

Dresselhaus M. S., *Science*, 292:650–651 (2001).

Dresselhaus M. S., Dresselhaus G., and Eklund P. C., *Science of Fullerenes and Carbon Nanotubes*, San Diego: Academic Press (1996).

Dresselhaus M. S., Dresselhaus G., and Eklund P. C., eds. *Carbon Nanotubes: Synthesis, Structure, Properties, & Applications*. New York: Springer-Verlag (2000).

Drexler, K. E., *Engines of Creation* (1986).

Ebbesen T. W., *Ann. Rev. Mater. Sci.*, 24:235 (1994).

Ebbesen T. W., *J. Phys. Chem. Solids*, 57:951 (1996).

Ebbesen T. W., Ajayan P. M., *Nature*, 358:220 (1992).

Ebbesen T. W. (ed.,) *Carbon Nanotubes—Preparation and Properties*. Boca Raton, FL: CRC Press (1996).

Elias K. L., Price R. L., and Webster T. J., *Biomaterials*, 23(15):3279 (2002).

El Khakani M. A. and Yi J. H., *Nanotechnology*, 15:S534 (2004).

El Khakani M. A., Yi J. H., and Aissa B., *Diam. Rel. Mater.*, 15:1064–1069 (2006).

Emmenegger, C., Bonard, J. M., Mauron, P., Sudan, P., Lepora, A., Groberty, B., Zuttel A., and Schlapbach, L., *Carbon*, 41:539–547 (2003).

Endo M., Takeuchi K., Igarashi S., and Kobori K., *J. Phys. Chem. Solids*, 54:1841 (1993).

Erlanger B. F., Chen B. X., Zhu M., and Brus L., *Nano Lett.*, 1(9):465 (2001).

Ernst S., Fritz M., and Weitkamp J., *Int. J. Hydrogen Energy*, 20(12):967 (1995).

Esler W. P. and Wolfe M. S., *Science*, 293:1449–1454 (2001).

Fan S., Chapline M., Frankline N., Tombler T., Cassel A., and Dai H., *Science*, 283:512–514 (1999).

Fan J., Wan M., Zhu D., B. F., Seta P., Chang B., Pan Z., and Xie S., *Synth. Met.*, 102:1266 (1999).

Fan, Y.Y., Liao, B., Liu, M., Wei, Y. L., Lu, M. Q., Cheng, H. M., *Carbon*, (37):1649–1652 (1999).

Farajian A. A., Ohno K., Esfarjani K., Maruyama Y., and Kawazoe Y. J., *Chem. Phys.*, 111:2164 (1999).

Faravelli L., Rossi R., "Proc. 3rd Int. Workshop on Structural Control," World Scientific, Singapore, 2000:201–213 (2000).

Feigin L. A. and Svergum D. J., *Small angle X-ray and Neutron Scattering*. New York: Plenum Press (1987).

Ferrari A. C. and Robertson D. J., *Phys. Rev. B*, 61:14095 (2000).

Fink R. L., Li Z., and Tolt Z., *Surf. Coating Techn.*, 108:570 (1998).

Flannigan D. and Suslick K., *Nature*, 434:52 (2005).

Foley S., Crowley C., Smaihi M., Bonfils C., and Erlanger Larroque C., *Biochem. Biophys. Res. Commun.*, 294:116 (2002).

Frackowiak E., Gautier S., Bonnamy S., and Beguin F., *Carbon*, 37:61–69 (1999).

Frackowiak E. and Beguin F., *Carbon*, 40:1775–1787 (2002).

Freitas J. R. A., *Nanomed. Nanotechnol. Biol. Med.*, 1:2 (2005).

Fritz J., Baller M. K. H., Lang P., Rothuizen H., Vettiger P., Meyer E., Güntherodt H. J., Gerber C., and Gimzewski J. K., *Science*, 288:316–318 (2000).

Gabay T., Jakobs E., Ben-Jacob E., and Hanein Y., *Physica A*, 250:611 (2005).

Gallagher J., Sams Jr. R., Inmon J., Gelein R., Elder A., and Oberdorster G., *Toxicol. App. Pharmacol*, 190(3):224 (2003).

Gandhi S. C., Mikhail E. I., Niyogi S., Zhao B., and Haddon R. C., *229th ACS National Meeting*, San Diego, IEC-099. Washington DC: American Chemical Society, March 13–17 (2005).

Gao B., *Chem. Phys. Lett.*, 327:69 (2000).

Gao B., Kleinhammes A., Tang X. P., Bower C., Fleming L., Wu Y., and Zhou O., *Chem. Phys. Lett.*, 307:153 (1999).

Gao M., Dai L., Baughman R. H., Spinks G. M., and Wallace G. G., *Proc. SPIE*, 3987 (2000).

Gao X. H., Cui Y. Y., Levenson R. M., Chung L. W. K., and S. M., *Nat. Biotechnol.*, 22(8):969 (2004).

Geis W., Twichell J. C., Macaulay J., and Okano K., *Appl. Phys. Lett.*, 67:1328 (1996).

Gensterblum G., *Physicalia Mag.*, 17(1):3 (1995).

Georgakilas V., Tagmatarchis N., Pantarotto D., Bianco A., Briand J. P., and Prato M., *Chem. Commun.*, 24:3050 (2002).

Ghosh S., Sood A. K., and Kumar N., *Science*, 299(5609):1042–1044 (2003).

Gibson C. P. and Putzer K. J., *Science*, 267:1338–1340 (1995).

Goate A., Chartier-Harlin M. C., Mullan M., Brown J., Crawford F., Fidani L.,Giuffra L., Haynes A., Irving N., and James L., *Nature* 349:704–706 (1991).

Goia D. V. and Matijevic E., *New J. Chem.*, 1203 (1998).

Goto, H., Furuta, T., Fujiwara, Y., and Ohashi, T., *Honda Giken Kogyo Kabushiki Kaisha, Japan*, 166386 (2003007924) (2002).

Govindaraj A., Satishkumar B. C., Nath M., and Rao C. N. R., *Chem. Mater.*, 12:202 (2001).

Golap K., Jagadale P., Sharon M., Sharon M., *Int. J. Synthesis and Reactivity in Inorganic, Metal-Organic, and Nano-Metal Chemistry*, 37:467–471, (2007).

Golap K., S. Adhikari, Aryal H. R., Umeno M., Afre R., Soga T., and Sharon M. *Appl. Phys. Lett.*, 92:508 (2008).

Gref R., Minamitake Y., Peracchia M. T., Trubetskoy V., Torchilin V., and Ranger R., *Science*, 263:1600 (1994).

Grill A., *Diam. Relat. Mater.*, 12(2):166 (2003).

Guillard T., Cetout S., Flamant G., and Laplaz. D., *J. Mater. Science.*, 35:419 (2000).

Guo T. Nikolaev, P., Thess A., Colbert D. T., and Smalley R.E., *Chem. Phys. Letts.*, 243:49–54 (1995).

Gupta A. K. and Gupta M., *Biomaterials* 26:3995–4021 (2005).

Guiseppi-Elie A., Lei C. H., and Baughman R. H., *Nanotechnology*, 13:559 (2002).

Gupta A. K., Nair P. R., Akin D., Ladisch M. R., Broyles S., Alam M. A., and Bashir R., *Proc. Natl. Acad. Sci., U. S. A.*, 103:36 (2006).

Gutensohn K., Beythein C., Bau J., Fenner T., Grewe P., and Koester R., *Thrombo. Res.*, 99(6):577 (2000).

Hafner J. H., Bronikowski M. J., and Azamian B. R., *Chem. Phys. Lett.*, 296:195–202 (1998).

Hafner J. H., Cheung C. L., and Leiber C. M., *J. Am. Chem. Soc.*, 121:9750 (1999a).

Hafner J. H., Cheung C. L., and Leiber C. M., *Nature*, 398:761 (1999b).

Hardy, J. and Selkoe D. J., *Science*, 297:353–356 (2002).

Harmia T., Spiser P. P., and Kreuter J., *Int. J. Pharm.*, 33:45 (1986a).

Harmia T., Spiser P. P., Kreuter J., and Kubis A., *Int. J. Pharm.*, 33:187 (1986b).

Harris P.J.F, *Biomaterials*, 28:344 (2007).

Haubold A. D., *Blood Carbon Interactions*, 6:88 (1983).

Haubold A. D., Shim H. S., and Bokros J. C., Carbon in biomedical devices. In: Williams D. F. (ed.), *Biocompatibility of Clinical Implants Materials*, vol. 2. Boca Raton, FL: CRC Press (1979).

Haufler R. E., Conceicao J., Chibante L. P. F., Chai Y., Byrne N. E., Flanagan S., Haley M. M., O'Brien S. C., Pan C., Xiao Z., Billups W. E., Ciufolini M. A., Hauge R. H., Margrave J. L., Wilson, Curl R. F., and Smalley R. E., *J. Phys. Chem.*, 94 (1990).

He R., Qian X., Yin J., and Zhu Z., *J. Mater. Chem.*, 12:3783 (2002).

Heer W. A. and Martel R., *Physics World*, 13(6):49 (2004).

Heidenreich R. D., Hess W. M., and Ban L. L., *J. Appl. Crystallography*, 1 (1968).

Height M. J., Howard J. B., and Tester J. W., *Materials Research Society Symposia Proceedings*, 55–61, (2003).

Hellar D. A., Baik S., Eurell T. E., and Strano M. S., *Adv Mater.*, 17:2793 (2005).

Helmer C., Joly P., Letenneur L., Commenges D., and Dartigues J. F., *Am. J. Epidemiol.*, 154:642–648 (2001).

Hernadi K., Fonseca A., Nagy J. B., Bernaerts D., and Lucas A. A., *Carbon*, 34:1249–1257 (1996).

Herrera J., Balzano L., Pompeo F., and Resasco D. E., *J. Nanosci. Nanotech.*, 3:1 (2003).

Hilding J., Grulke E. A., Zhang Z. G., and Lockwood F., *J. Dispers. Sci. Technol.*, 24(1):1 (2003).

Hirahara K., Suenaga K., Badow S., Kato H., and Okazaki T., *Phys. Rev. Lett.*, 85:5384 (2000).

Hiramatsu M., Shiji K., Amano H., and Hori M., *Appl. Phys. Lett.*, 23:4708 (2004).

Hirsch A. and Vostrowsky O., *Top. Curr. Chem.*, 245:193–237 (2005).

Hirscher M., Becher M., Haluska M., Quintel A., Skakalova V., Choi Y. M., Dettlaff-Weglikowska U., Roth S., Stepanek I., Bernier P., Leonhardt A., and Fink J., *J. Alloys Comp.*, 330:654 (2002).

508   References

Hoffmann F., Cinatl J., Kabickova Jr. H., Cinatl J., Kreuter J., and Stieneker F., *Int. J. Pharm.*, 157:189 (1997).

Holmgren J. D., Gibson J. O., and Sheer C., *J. Electrochem. Soc.*, 111:362 (1964).

Hong E. H., Lee K., Oh S. O., and Park C., *Advanced Functional Materials*, 13(12):961 (2003).

Hovel H., Bodecker M., Grimm B., and Rettig C., *J. Appl. Phys.*, 92:771 (2002).

Hu G., Cheng M., Ma D., and Bao X., *Chem. Mater.*, 15:1470 (2003a).

Hu H., Ni Y., Montana V., Haddon R. C., Parpura V., *Nano Lett.*, 4(3):507 (2004).

Hu Y., Xiqun J., Ding Y., Zhan L., Yang C. Z., Zhang J., Chen J., and Yang Y., *Biomaterials*, 24:2395 (2003b).

Huczko A., Lange H., Calko E., Grubek-Jaworska H., and Droszcz P., *Fullerene Sci. Tech.*, 9(2):251 (2001).

Hutton M., Perez-Tur J., and Hardy J., Genetics of Alzheimer's disease. *Essays Biochem.*, 33:117–131 (1998).

Hwang K. C., *J. Chem. Soc., Chem. Comm.*, 173:2 (1995).

Ibarra J., Koski A., and Warren R. F., *Orthop. Clin. North Am.*, 31:411 (2000).

Iijima S., *Nature*, 354:56 (1991).

Iijima S. and Ichihashi T., *Nature*, 363:603 (1993).

Iijima S., Yudasaka M., Yamada R., Bandow S., Suenaga K., Kokai F., and Takahashi K., *Chem. Phys. Lett.*, 309:165–170 (1999).

Inganas O. and Lundstrum I., *Science*, 284(5418):1281 (1999).

Inoue K., Takano H., Yanagisawa1 R., Sakurai M., Ichinose T., Sadakane K., and Yoshikawa T., *Resp. Res.*, 6:106 (2005).

Islam M. Z., Alam M., Mominuzzaman S. M., Rusop M., Soga T., Jimbo T., and Umeno M., *J. Crystal Growth*, 288(1):195 (2006).

Iwatsubo T., Odaka A., Suzuki N., Mizusawa H., Nukina N., and Ihara Y., *Neuron* 13:45–53 (1994).

Jagadale P., Sharon M., Sharon M., and Golap K., *Synthesis and Reactivity in Inorganic, Metal-Organic, and Nano-Metal Chemistry*, 37(6):467–471 (2007).

Jang J. and Bae J., *Sensors and Actuators B: Chemical*, 122(1):7 (2007).

Javey A., Wang Q., Qi P., and Dai H., *Proc. Natl. Acad. Sci. U. S. A.*, 101(37) (2004), 13408–13410.

Jaybhaye S., Sharon S., Sharon S., and L. N. Singh, *Int. J. of Synthesis and Reactivity in Inorganic, Metal-Organic, and Nano-Metal Chemistry*, 36(2):37–42 (2006).

Jaybhaye S., Sharon S., Sharon S., Sathiyamoorthy D., and Dasgupta K., *Int. J. Synthesis and Reactivity in Inorganic, Metal-Organic, and Nano-Metal Chemistry*, 37(6):473–476 (2007).

Jenson A. W., Wilson S. R., and Schuster D. I., *Bioorg. Med. Chem.*, 4(6):767 (1996).

Jeong S. H., Hwang H. Y., Lee K. H., Jeong Y., *Appl. Phys. Lett.*, 78:2052 (2001).

Jeong, S. H., Lee, O. J., and Lee, K. H., Presentation at AIChE Annual Meeting (2003).

Jia G., Wang H., Yan L., Wang X., Pei R., Yan T., et al., *Environ. Sci. Technol.*, 39(5):1378 (2005).

Jia X., Xie Q., Zhang Y., and Yao S., *Analytical Science*, 23:689–696 (2007).

Jia Z. B., Wei Y., and Wang H. M., *J. Inorg. Mater*, 15:926 (2000).

Jiang K., Schadler L. S., Siegel R. W., Zhang Z., Zhang H., and Terrones M., *J. Mater. Chem.*, 14(1):37 (2004).

Jia-Yaw C., Ghule A., Chang J.-J., Tzing S.-H., and Lin Y.L., *Chem. Phys. Lett.*, 363:583–590 (2002).

Jin M.H. and Dai L., *Organic photovoltaic mechanisms, materials and devices,* S. Sun and N.S. Sacricftci (eds.), New York: CRC Press (2005).

Jinquan W., Zhu H., Wu B., Wei B., *Appl. Phys. Lett.*, 84(24):4869–4871 (2004).

Johanson M. P., Suenaga K., Hellgren N., Colliex C., Sundgren J. E., and Hultman L., *Appl. Phys. Lett.*, 76:825 (2000).

Jorgensen W. L., *Science*, 303:1813–1818 (2004).

J. P. Shim, Y. S. Park, H. K. Lee, J. S. Lee, *J. Power Sources*, 74(1): 151–154 (1998).

Jorio A., Santos A. P., Ribeiro H. B., Fantini C., Souza M., Vieira J. P. M., Furtado C. A., Jiang J., Saito R., Balzano L., Resasco D. E., and Pimenta M. A., *Phys. Rev. B*, 72:075207 (2005).

Journet C., Maser W. K., Bernier P., Loiseau A., de LA Chapelle M. L., Lefrant S., Deniard P., Lee R., and Fisher J. E., *Nature*, 388:756 (1997).

Joseph S., Mashl R. J., Jakobsson E., and Aluru N. R., *Nano Lett.*, 3(10):1399 (2003).

Joseph H., Swafford B., and Terry S., *Sens. Mag.*, 14:47 (1997).

Ju Y., Yi L. F., and Zhong W. R., *J. Serb. Chem. Soc.*, 70(2):277 (2005).

Jung H., Kettunen M. I., Davletov B., and Brindle K. M., *Bioconjugate Chem.*, 15:983 (2004).

Jung S. H., Kim M. R., Jeong S. H., Kim S. U., Lee O. J., Lee K. H., Suh J. H., and Park C. K., *Appl. Phys. A Materials Science & Processing*, 76 (2):285 (2003).

Kam N. W. S., Jessop T. C., Wender P. A., and Dai H., *J. Am. Chem. Soc.*, 126, 6850 (2004).

Kam N. W. S., Liu Z., and Dai H., *.J. Am. Chem. Soc.*, 127:12492–12493 (2005a).

Kam N. W., O'Connell M., Wisdom J. A., and Dai H., *Proc. Natl. Acad. Sci. U. S. A.*, 102(33):11600–11605 (2005b).

Kam N. W. S. and Dai H., *J. Am. Chem. Soc.*, 127(16):6021–6026 (2005c).

Kam N. W., Liu Z., and Dai H., *Angew Chem. Int. Ed. Engl.*, 45(4):577 (2006).

Kannan R. K., Balasubramanian, and Marko B., *Anal. Bioanal. Chem.*, 385:452 (2006).

Kellogg E. W., 3rd and Fridowich I., *J. Biol. Chem.*, 250:8812 (1975).

Kerman K., Morita Y., Takamura Y., Ozsoz M., and Tamiya E., *Electroanalysis*, 16:1667 (2004).

Kerman K., Morita Y., Takamura Y., Tamiya E., Maehashi K., and Matsumoto K., *Nanobiotechnology*, 1:65 (2005).

Kerman K., Morita Y., Takamura Y., and Tamiya E., *Anal. Bioanal. Chem.*, 381:1114 (2005).

Khairnar V., Jaybhaye S., Hu C., Afre R., Soga T., Sharon M., and Sharon M., *Carbon Science Lett.*, 9(3):1–100 (2008).

Kiang, C. H., Goddard, W. A., Beyers R., Salem J. R., Bethune, D., *J. Phys. Chem. Of Solids*, 57(35) (1996).

Kichambre P. D., Sharon M., Kumar M., Avery N. R., and Black K. J., *Molecular Crystal and Liquid Crystal*, 340:523 (2000).

Kimata K., Mariuchi K., Hosoya K., Arai T., and Tanka N., *Anal. Chem.*, 65:3717 (1993).

Kiran N., Ekinci E., and Snape C. E., *Resour. Conserv. Recycle*, 29(4):273–283 (2000).

Ko T. and Hwang D. -K., *Mater. Lett.*, 57(16-17):2472 (2003).

Kolmakov A., Lanke U., Karam R., Shin J., Jesse S., and Kalinin S. V., *Nanotechnology*, 17:4014 (2006).

Komarneni S., Roy R., and QH L. I., *Mater. Res. Bull.*, 27, 1393–1405 (1992).

Komatsu T., Inoue H., *Molecular Crystal Liquid Crystal*, 387(337):113, (340):116 (2002).

Kong J., Cassel A. M., and Dai H., *Chem. Phys. Lett.*, 292:567–574 (1998).

Kong J., Franklin N. R., Zhou C., Chapline M. G., Peng S., Cho K., and Dai H., *Science*, 287:622–625 (2000).

Kopf H., Joshi R. K., Soliva M., and Speiser P. P., *Pharm. Ind.*, 38:281 (1976).

Kopf H., Joshi R. K., Soliva M., and Speiser P. P., *Pharm. Ind.*, 39:993 (1977).

Kormann C., Bahnemann D. W., and Hoffmann M. R., *Environ. Sci. Technol.*, 22:798 (1988).

Kramer P. A., *J. Pharm. Sci.*, 63:1646 (1974).

Krätchmer W., Lamb L. D., Fostiropoulos K., and Huffman D. R., *Nature*, 347:354 (1990).

Kraytsberg Y., Kudryavtseva E., McKee A. C., Geula C., Kowall N. W., and Khrapko K., *Nat. Genet.*, 38:518–520 (2006).

Kreuter J., *Int. J. Pharm.*, 14:43 (1983a).

Kreuter J., *Pharm. Acta. Helv.*, 58:196 (1983b).

Kreuter J., In *Microcapsules and nanoparticles*. Donbrow M. (ed). Boca Raton, FL: CRC Press, p. 125 (1991).

Kreuter J., In *Colloidal Drug Delivery Systems*. New York: Dekker, pp. 219–324 (1994a).

Kreuter J., In *Encyclopedia of Pharmaceutical Technology*, New York: Dekker (1994b).

Kreuter J., Alyautdin R. N., Kahrkevich D. A., and Ivanov A. A., *Brain Res.*, 674:171 (1995).

Krishnan A., Dujardin E., Treacy M. M. J., Hugdahl J., Lynum S., and Ebbesen T. W., *Nature*, 388:451 (1997).

Kroto H. W., Heath J. R., O'Brien S. C., Curl R. F. and Smalley R. E., *Nature*, 318(6042), 162–163 (1985).

Kshirsagar D. E., Puri V., Sharon1 M., and Sharon M., *Carbon Science*, 7(4):245–248, (2006).

Kshirsagar D. E., Puri V., Sharon M., and Sharon M., *Carbon Sci.*, 7(4):245–248 (2006).

Kudin A. P., Bimpong-Buta N. Y., Vielhaber S., Elger C. E., and Kunz W. S., *J. Biol. Chem.* 279:4127–4135 (2004).

Kumaresh S. S., Tejraj M. A., Anandrao R. K. and Walter E. R., *J. Contr. Release*, 70: 1– 20 (2001).

Kumar M., Kichambre P. D., Sharon M., Avery N. R., and Krista B., *Materials Chemistry and Physics*, 66:83 (2000).

Kumar M., Zhao X., and Ando Y., *Int. Symposium on Nanocarbons*. Nagano, Japan, extended abstract, 244–245 (2001).

Kumar M., Zhao X., Ando Y., Sharon M., and Iijima S., *Mol. Cryst. Liquid Cryst.*, 387:117 (2002).

Kumar M. and Ando Y., *Diam. Rel. Mater.*, 12:998 (2003).

Kumar M. and Ando Y., *Chem. Phys. Lett.*, 374:521–529 (2003).

Kumar M. and Ando Y., *Carbon*, 43(3):533–540 (2005).

Kumar R. V., Diamant Y., and Gedanken A., *Chem. Mater*, 12:2301 (2000).

Kunz W. S., Kudin A. P., Vielhaber S., Blumcke I., Zuschratter W., Schramm J., Beck H., and Elger C. E., *Ann. Neurol.*, 48:766–773 (2000).

Kundu A., Upadhyay C., and Verma H. C., *Phys. Lett.*, 311(4-5):410 (2003).

Lam C., James J. T., McCluskey R., and Hunter R., *Toxicol. Sci.*, 77(1):126 (2004).

Langer K., Coester C., Weber C., Von Briesen H., and Kreuter J., *Eur. J. Phar. Biopharm.*, 49:303 (2000a).

Langer K., Coester C., Weber C., Von Briesen H., and Kreuter J., *Int. J. Pharm.*, 196:147 (2000b).

Lanza R. P., Langer R. S., and Vacanti J., *Principles of Tissue Engineering*, (2nd ed.) San Diego: Academic Press (2000).

Lanza G. M., Winter P. M., and Caruthers S. D., *Nanomedicine*, 1:321 (2006).

Lee D. C., Mikulev F. V., and Korgel B. A., *J. Am. Chem. Soc.*, 126:4951–4957 (2004).

Lee E. C., Kim Y. S., Jin Y. G., and Chang K. J., *Phys. Rev. B*, 66 (2002).

Lee N. S., Chung D. S., Han I. T., Kang J. H., Choi Y. S., Kim H. Y., Park S. H., Jin Y. W., Yi W. K., Yun M. J., Jung J. E., Lee C. J., You J. H., Jo S. H., Lee C. G., and Kim J. M., *Diam. Rel. Mater.*, 10:265 (2001).

Lee S. and Lee Y., *Appl. Phys. Lett.*, 76 (20):2877 (2000).

Lee Y. H., Kim S. G., and Tomanek D., *Phys. Rev. Lett.*, 78:2393 (1997).

Levoska J. and Leppavuori S., *Appl. Surface Sci.*, 86:180 (1995).

Levy-Lahad E., Lahad A., Wijsman E. M., Bird T. D., and Schellenberg G. D., *Ann. Neurol.*, 38:678–680. (1995).

Li Y. F., Hatakeyama R., Izumida T., Okada T., and Kato T., *Nanotechnology* 17: 4143–4147 (2006).

Li J., Powell D., Getty S., and Lu Y., *Nanoletters* 3(7):929 (2003).

Lia G., Hub G. G., Zhoua H. D., Fan X. J., and, Lia X. G., *Mater. Chem Phys.*, 75:101 (2002).

Li M., Hu, Wang Z., Wu X., Chen Q., and Tian Y., *Diam. Rel. Mater.*, 13:111 (2004).

Li W. Z., Xie S., Qian L. X., Chang B. H., and Zou B. S., *Science*, 274:1701–1703 (1996).

Li X. and Kale G. M., *J. of Physics: Conference Series*, 26:319 (2006).

Li Y., Kim W., Zhang Y., Rolandi M., Wang D., and Dai H., *J. Phys. Chem. B*, 105:11424. (2001).

Li Y. F., Hatakeyama R., Kaneko T., Izumida T., Okada T., and Kato T., *Nanotechnology*, 17:4143 (2006).

Lii C. Y., Stobinski L., Tomansik P., and Liao C., *Carbohydrate Polymer*, 50(1):93 (2003).

Liju Y., Padmapriya P. B., Mohammad R. C., Kwan S. L., Bhunia A. K., Ladisch M., and Bashir R., *The Royal Society of Chemistry Lab Chip*, 6:896 (2006).

Liao K. J., Wang W. L., Zhang Y., Duan L. H., and Ma Y., *Microfab. Technol.*, 4:57 (2003).

Lim Y. T., Kim S., Nakayama A., Stott N. E., Bawendi M. G., and Frangioni J. V., *Mol. Imaging*, 2, (1):50 (2003).

Lin T., Bajpai V., Ji T., and Dai L., *Aust. J. Chem.*, 56(7):635 (2003).

Lin Y., Taylor S., Li H., Fernando S. K. A., Qu L., and Wang W., *J. Mater. Chem.*, 14(4):527 (2004a).

Lin Y., Yantasee W., Lu F., Wang J., Musameh M., Tu Y., Ren Z., *Encyclopedia of Nanoscience and Nanotechnology*, vol. B:361 Dekker. (2004b).

Linder S., Pinkowski W., and Aepfelbacher M., *Biomaterials*, 23(3):767 (2002).

Little K. and Parkhouse J., *Lancet*, 2(7261):857–861 (1962).

Liu C., Fan Y. Y., Liu M., Cong H. T., Cheng H. M., and Dresselhaus M. S., *Science*, 286:1127–1129 (1999).

Liu J. and Dai H. *Design, Fabrication, and Testing of Piezoresistive Pressure Sensors Using Carbon Nanotubes.* [Online]. Available: http://www.nnf.cornell.edu/2002re u/ Liu.pdf (2002).

Liu J., Rinzler A. G., Dai H., Hafner J. H., Bradley R. K., Boul P. J., Lu A., Iverson T., Shelimov K., Huffman C. B., Rodriguez-Macias F., Shon Y. S., Lee T. R., Colbert D. T., and Smalley R. E., *Science*, 280:1253–1256 (1998).

Liu J., Shao M., Xie Q., Kong L., Y, W., and Qian Y., *Carbon*, 41:2101–2104 (2003).

Liu X., Huang B., and Coville N. J., *Fullerenes, Nanotubes and Nanostructures*, 10(4):339-352 (2002).

Liuzzi F. J. and Tedeschi B., *Neurosurg. Clin. N. Am.* 2:31 (1991).

Loiseau A. *Understanding Carbon Nanotubes: From Basics to Applications.* New York: Springer.

Lolli G., Zhang L., Balzano L., Sakulchaicharoen N., Tan Y., and Resasco D. E., *J. Phys. Chem. B*, 110:2108 (2006).

Lopes C. M. A., Peixoto G. G., and Rezende M. C, *Proceedings of the SBMO/IEEE MTT-S International Microwave and Optoelectronics Conference*, 10th, Foz do Iguacu, Brazil, Sept. 20–23, pp 771–774 (2003).

Lovat L. B., Jamieson N. F., Novelli M. R., Mosse A., Selvasekar C., Thorpe S. and Brown S. G., *Gastrointestinal Endoscopy*, 62:617–623 (2005).

Lu T., Iverson, K. Shelimov, C. B. Huffman, F. Rodriguez-Macias, Y. S. Shon, T. R. Lee, D. T. Colbert, R. E. Smalley, *Science*, 280:1253–1256 (1998).

Lucci M, Reale A., Di Carlo A., Orlanducci S., Tamburri E., Terranova M. L., Davoli L., Di Natale C., D'Amico A., and Paolesse R., *Sensors and Actuators B*, 118:226 (2006).

Lueking, A., Yang, R. T., *J. Catalysis*, 206:165–168 (2002).

Luo T., Chen L., Bao K., Yu W., and Qiyan Y., *Carbon* 44(13), 2844–2848 (2006).

Lutsev L. V., Yakovlev S. V., Zvonareva T. K., Alexeyev A. G., Starostin A. P., and Kozyrev S. V., *J. Appl. Phys.*, 97(10):104327/1–104327/6 (2005).

Lyu S. C., Liu B. C., Lee S. H., Park C. Y., Kang H. K., Yang C. W., and Lee, C. J., *J. Phys. Chem. B*, 108:1613 (2004).

Lyu S. C., Liu B. C., Lee S. H., Park C. Y., Kang H. K., Yang C. W., and Lee C. J., *J. Phys. Chem. B*, 108:2192 (2003).

Ma Y. C., Xia Y. Y., Zhao M. W., Wang R. J., and Mei L. M., *Phys. Rev. B*, 6311, 115422 (2001).

MacDonald R. A., Laurenzi B. F., Viswanathan G., Ajayan P. M., and Stegemann J. P., *J. Biomed. Mater. Res.*, 74A:489 (2005).

Makeiff D. A. and Huber T., *Synthetic Metals*, 156:497 (2006).

Magrez A., Sandor K., Valérie S., Nathalie P., Won J., Celio S., Catsicas S., Schwaller B., and Forró B., *Nano Lett.*, 6(6):1121 (2006).

Maiti A., Brebec C. J., Roland C., and Bernholc J., *Phys. Rev. B.*, 52:14850 (1995).

Maiti C. J. and Brabec Bernholc, *J. Phys. Rev.* B, 55:6097 (1997).

Majumdar A., Thundat T., and Ridge O., *Nat. Biotechnol.*, 19:856 (2001).

Mann C. C., *Tech. Rev.*, 60 : (2004).

Manna S. K., Sarkar S., Barr J., Wise K., Barrera E. V., and Jejelowo O., *Nano Lett.*, 5(9):1676 (2005).

Marlow J., *Bacteria building nanobots*; http://sci.techarchive.net/Archive/Sci.nanotech/2004-08/1003.html.

Martin C. R. and Kohli P., *Nat. Rev. Drug Discov.*, 2:29 (2003).

Marty J. J. and Oppenheim R. C., *Aust. J. Pharm. Sci.*, 6:65 (1977).

Maruyama S., Kojima R., Miyauchi Y., and Chiashi S., *Chem. Phys. Lett.*, 360:229–234 (2002).

Marquardt C. L., Williams R. T., and Nagel D. J., *Mater. Res. Soc. Symp. Proc.*, 38:325 (1985).

Masashi S. and Masafumi A., Work function of carbon nanotubes. *Carbon*, 39(12):1913–1917 (2001).

Maser K. W., Benito A. M., Munoz E., Marta de Val G., Martinez M. T., Larrea A., and Fuente G. F., *Nanotechnology*, 12:147 (2001).

Mathiowitz E., Jacob J. S., Jong Y. S., Carino G. P., Chickering D. E., Chaturvedi P., Santos C. A., Vijayaraghavan K., Montgomery S., Basset M., and Morell C., *Nature*, 386:410 (1997).

Matyshevska O. P., Karlash A. Y., Shtogun Y. V., Benilov A., Kirgizov Y., and Gorchinskyy K. O., *Mater. Sci. Eng. C*, 15(1–2):249 (2001).

Maurin G., Bouhquet Ch., Henn F., Bernier P., Almairac R., and Simon B., *Solid State Ionics*, 1295:136 (2000).

Maurin I., Stepanek P., Bernier J. F., Colomer J. B., and Henn N. F., *Carbon*, 39:1273–1278 (2001).

Mauron P. H., Emmenegger Ch., Sudan P., Wenger P., Rentsch S., and Zuttel A., *Diam. Rel. Mater.*, 12:780 (2003).

Maynard R. L. and Howard B., (eds) *Particulate Matter: Properties and Effects Upon Health*. Oxford: Bios Scientific Publishers (1999).

McCreey R. L., Carbon electrodes: structural effects on electron transfer kinetics. In Bard, A. J. (ed). *Electroanalytical Chemistry*. New York: Marcel Dekker, 221–374 (1991).

McKenzie J. L., Waid M. C., Shi R., and Webster T. J., *Biomaterials*, 25(7-8):1309 (2004).

McKhann G., Drachman D., Folstein M., Katzman R., Price D., and Stadlan E. M., *Neurology*, 34:939–944 (1984).

McKinight T. E., Melechko A. V., Guillorn M. A., Merkulov V. I., Lowndes D. H., and Simpson M. L., In Rosenthal S. J. and Wright D. W. (eds). *Synthetic Nanoscale Elements for Delivery of Material Into Viable Cells. Nanobiotechnology Protocols*. Totowa, New Jersey: Humana Press, pp. 191 (2005).

Megaridis C. M., Yazicioglu A. G., Libera J. A., and Gogotsi Y., *Phys. Fluids*, 14(2): L5 (2002).

Mertens Jr., Jacobs E, Callaerts A., Buekens A., *Makromol. Chem. Rapid Commun.* 3, 349–356 (1982).

Meyyappan M., ed. *Carbon Nanotubes: Science and Applications*. Boca Raton, FL: CRC Press (2004).

Michael W., Topinka M. A., McGehee M. D., Prall Hans-Jurgen, Dennler G., Rowell S., Niyazi S., Hu L., and Gruner G., *Appl. Phys. Lett.*, 88(23):233506 (2006).

Mikos G., Bao Y., Cima L. G., Ingber D. E., Vacanti J. P., and Langer R., *J. Biomed. Mater. Res.*, 27:183 (1993).

Miller M. J. and Patrick C. W. Jr., *Clin. Plast. Surg.*, 30:91 (2003).

Mitra S. and Zafar I., http://foodnavigaitor.com/europe. (2005).

Mohammadizadeh M. R., *Physica status solidi. C*, 3(9):3126–3129 (2006).

Moloni K., Lal A., and Lagally M., *Proc. SPIE*, 4098:76 (2000).

Mominuzzaman S. M., Rusop M., Xuemin T., Soga T., and Jimbo T., *ICECE* (2002).

Monteiro-Riviere N. A., Nemanich R. J., Inman A. O., Wang Y. Y., and Riviere J. E., *Toxicol Lett.*, 155(3):377 (2005).

Monthioux M., Mith B. W., Burteaux B., Ciaye A., Fischer J. E., and Luzzide D. E., *Carbon*, 39:1251–1272 (2001).

Moon J. M., An K. H., Lee Y. H., Park Y. S., Bae D. J., and Park G. S., *J. Phys. Chem. B*, 105(24):5677 (2001).

Moore, D. J., West, A. B., Dawson, V. L., and Dawson T. M., *Annu. Rev. Neurosci.*, 28:57–87 (2005).

Mornet S., Vasseur S., Grasset F., and Duguet E., *J. Mater. Chem.*, 14:2161 (2004).

Motiei M., Hacohen Y. R., Calderon-Moreno J., and Gedanken A., *J. Am. Chem. Soc.*, 123:8624–8625 (2001).

Motojima S., Nagahara D., Kuzuya T., and Hishikawa Y., *Transactions of the Materials Research Society of Japan*, 29(2):461 (2004).

Moumen N. and Pileni M. P., *Chem. Mater.*, 8:1128 (1996).

Moumen N. and Pileni M. P., *J. Phys. Chem.*, 100:1867 (1996).

Moumen N., Bonnville M., and Pileni P., *J. Phys. Chem.*, 100:14410 (1996).

Mukhopadhyay K., Krishna K. M., and Sharon M., *Phys. Rev. Lett.*, 72(20):3184 (1994).

Mukhopadhyay K., Krishna K. M., and Sharon M., *Curr. Sci.*, 67(8):602 (1994).

Mukhopadhyay K., Krishna K. M., and Sharon M., *Carbon*, 34(2):251–264 (1996).

Mukhopadhyay K. and Sharon M., *Materials and Manufacturing Processes*, 12(3):541 (1997).

Mukhopadhyay K. and Sharon M., *Materials Chemistry and Physics*, 49:105 (1997).

Muller R. H., *Colloidal Carriers for Controlled Drug Delivery And Targetting, Wissenschaft*. Stuttgart: Verlagsges GmbH (1990).

Muller J., Huaux F., Moreau N., Misson P., Heiler J. F., and Delos M., *Toxicol Appl, Pharmacol.*, 207(3).221 (2005).

Munoz, E., Maser, W. K., Benito, A. M., Fuente, G. F., Righi, A., Sauvajol, J. L., Anglaret, E., and Maniette, Y., *Appl. Phys. A*, 70:145 (2000).

Murakami Y., Chiashi S., Miyauchi Y., and Hu M., *Chem. Phys. Lett.*, 385:298–303 (2003).

Maruyama S., Kojima R., Miyauchi Y., Chiashi S., *Chem. Phys. Lett.*, 360:229 (2002).

Maruyama, S., Marukami, Y., Miyauchi, Y., and Chashi, S., *Catalytic CVD generation and optical characterization of single–walled carbon nanotubes from alcohol*. Presentation at AIChE Annual Meeting (2003).

Musa I., Baxendale M., Amaratunga G. A. J., and Eccleston W., *Synth. Met.*, 102:1250 (1999).

Marquardt C. L., R. T., Williams, and D. J., Nagel, *Mater. Res. Soc. Proc.*, 38:325 (1985).

Naguib N. N., Mueller Y. M., Bojuczuk P. M., Rossi M. P., Katsikis P. D., and Gogotsi Y., *Nanotech.*, 16(4):567 (2005).

Narayan R. J., *Mater. Res. Soc. Symp. Proc.*, vol. 785. Materials Research Society (2004).

Nerushev O. A., Dittmar, S., Morjan, R. E., Rohmund, F., and Campbell, E. E. B., *J. Appl. Phys.*, 93(7):4185 (2003).

Nguyen H. and Huh J., *Sensors and Actuators B*, 117:426 (2006).

Nguyen C. V., So C., Stevens R. M. D., Li Y., Delzeit L., Sarrazin P., and Meyyappan M., *J. Phys. Chem. B*, 108(9):2816 (2004).

Nikolaev P., Bronikowski M. J., and Bradley R.K,. *Chem. Phys. Lett.*, 313:91–97 (1999).

Nikula K. J., Snipes M. B., Barr E. B., Griffith W. C., Henderson R. F., and Mauderly J. L., *Fundamen. App. Toxicol.*, 25:80 (1995).

Nojeh A., Lakatos G. W., Peng S., Cho K., and Pease R. F. W., *Nano Lett.*, 3(9):1187 (2003).

Numata M., Asai M., Kaneko K., Hasegawa T., Fujita N., and Kitada Y., *Chem. Lett.*, 33(3):232 (2004).

Oberdörster G., Maynard A., Donaldson K., Castranova V., Fitzpatrik J., Ausman K., Carter J., Karn B., Kreyling W., Lai D., Olin S., Monteiro-Riviere N., Warheit D. and Yang H., *Particle & Fiber Toxicology*, 2:1–35(2005).

Odom T. W., Huang J. L., Kim P., and Lieber C. M., *Nature*, 391:62 (1998).

Ohkohchi M., *Jpn. J. Appl. Phys.*, 38 4158 (1999).

Ohkohchi M., Ando Y., Bandow S., and Saito Y., *Jpn. J. Appl. Phys.*, 32: L1248 (1993).

Ohkohchi M., Zhao X., Wang M., and Ando Y., *Fullerene Sci. & Tech.*, 4:97 (1996).

Okitsu K., Mizukoshi Y., Bandow H., Maeda Y., Yamamoto T., and Nagata Y., *Ultrason. Sonochem.*, 3:S249 (1996).

Okuno J., Maehashi K., Kerman K., Takamura Y., and Tamiya M. K., *Biosens. Bioelectron.*, 14 (2006).

O'Loughlin J. L., Kiang C. H., Wallace C. H., Reynolds T. K., Rao L., and Kaner R. B., *J. Phys. Chem. B*, 105:1921 (2001).

Ong K. G., Zheng K., Grimes C.A., *IEEE Sensor* 2:2 82 (2002)

Osofsky M. S., Lubitz P., Harford M. Z., Singh A. K., and Qadri S. B., *Appl. Phys. Lett.*, 53:1663 (1998).

Ouyang M., Huang J. L., Cheung C. L., and Lieber C. M., *Science*, 292:702–705 (2001).

Paganelli G., Guzennec Y., Rizzoni G., and Moran M. J.J., *Energy Resources Technology-Trans. ASME*, 124:20 (2002).

Palmer A. M., Stratmann G. C., and Procter A. W., *Ann Neurol.*, 23:610–620 (1988).

Pan Z. W., Xie S. S., Chang B. H., Sun L. F., Zhou W. Y., and Wang G., *Nature*, 394:631 (1998).

Pantarotto D., Partidos C., Hoebeke D., Brown J., Kramer F., and Bianco A., *Chem. Biol.*, 10:961 (2003a).

Pantarotto D., Partidos C. D., Graff R., Hoebeke J., Briand J. P., and Prato M., *J. Am. Chem. Soc.*, 125(20):6160 (2003b).

Pantarotto D., Briand J. P., Prato M., and Alberto B., *Chem. Commun.*, 1:16 (2004).

Parihar S., Sharon M., Sharon M., *Synthesis and Reactivity in Inorganic, Metal-Organic, and Nano-Metal Chemistry*, 36(1):107 (2006).

Park J. B., Choi G. S., Cho Y. S, Hong S. Y., Kim D., Choi S. Y., Lee J. H., and Cho K. I., *J. Crystal Growth*, 244(2):211 (2002).

Park J. W., *Breast Cancer Res.*, 4:95 (2002).

Park K. H., Chhowalla M., Iqbal Z., and Sesti F., *J. Biol. Chem.*, 278(50):50212 (2003).

Partlow K. C., Chen J., and Brant J. A., *FASEB J.*, 21:1647 (2007).

Patolsky F., Zheng G., Hayden O., Lakadamyali M., Zhuang X., and Lieber C. M., *Proc. Natl. Acad. Sci. U. S. A.*, 101:14017–14022 (2004).

Pavel N., Bronikowski M. J., Kelley R. Bradley, Rohmund F., Colbert D. T., Smith K. A., and Smalley R. E., *Chem. Phys. Lett.*, 313(1-2):91 (1999).

Peng H., Alemany L. B., Margrave J. L., and Khabashesku V. N., *J. Am. Chem. Soc.*, 125(49):15174–15182 (2003).

Penza M., Antolini F., and Antisari M. V., *Sensors & Actuators B-Chemical*, B 100(1–2):47 (2004).

Penza M., Cassano G., Aversa P., Antolini F., Cusano A., Cutolo A., Giordano M., and Nicolais L., *Appl. Phys. Lett.*, 85:2379 (2004).

Penza M., Cassano G., Aversa P., Cusano A., Consales M., Giordano M., and Nicolais L., *Sensors Journal IEEE*, 6(4):867 (2006).

Peracchia M. T., Desmaele D., Couvreur P., and d'Angelo J., *Macromolecules*, 30:846 (1997).

Peracchia M. T., Desmaele D., Vauthier C., Gulik A., Dedieu J. C., Demoy M., d'Angelo J., and Couvreur P., *Pharm. Res.*, 15:550–556 (1998).

Perez-Cabero M., Rodriguez–Ramos I., and Guerrero–Ruiz A., *J. Catalysis*, 215:305 (2003).

Pinho M. S., Gregori M. L., Nunes R. C. R., and Soares B. G., *Eur. Polym. J.*, 38:2321 (2002).

Porro S., Musso S., Teo K. B. K., and Milne W. I., *J. Non-Crystlline Solids*, 352:1310 (2006).

Pradhan B. K., Harutyunyan A., and Eklund P., *Abstracts of Papers of the Am. Chem. Soc.*, 224:024 (2002).

Pradhan D., Sharon M., Kumar M., and Ando Y., *J. Nanosci. Nanotechnol.*, 3(3):215 (2003).

Pramanik P., *Bulletin of Materials Science*, 19(6):957 (1996).

Price R. L., Waid M. C., Haberstroh K. M., and Webster T. J., *Biomaterials*, 24(11):1877 (2003).

Puppels G. F. and Jovin T. M., *Nature*, 347:301 (1990).

Qin L. C., Zhao X., Hirahara K., Miyamoto Y., Ando Y. and Iijima S., *Nature*, 408:50 (2000).

Qingrun H., Gao G., *J. Phys. Condensed Matter*, 9:10333 (1997).

Radomski A., Jurasz P., and Alonso-Escolano D., *Br. J. Pharmacol.*, 146:882–893 (2005).

Rajamathi M. and Seshadri R., *Curr. Opin. Solid State. Mater. Sci.*, 6:337 (2002).

Ramachandran N., Hainsworth E., Bhullar B., Eisenstein S., Rosen B., Lau A. Y., Walter J. C., and La Baer J., *Science*, 305:86–90 (2004).

Ramadan A. A., Abdel-Hady S., Abdel-Ghany S., and Soltan S. E., *Egypt J. Sol.*, 23(1):59 (2000).

Ranadeep B., Bruce M. C., and Brett A. C., *Carbon*, 46(6):907–922 (2008).

Rao A. M., Richter E., Bandow S., Chase B., Eklund P. C., Williams K. A., Fang S., Subbaswamy K. R., Menon M., Thess A., Smalley R. E., Dresselhaus G., and Dresselhaus M. S., *Science*, 275:187–191 (1997).

Reich S., Thomsen C., and Maultzsch J., *Carbon Nanotubes: Basic Concepts and Physical Properties*. New York: John Wiley & Sons (2004).

Resasco et al., Resasco, D. E., Alvarez, W.E., Pompeo, F., Balzano, L., Herrera, J. E., Kitiyanan, B., Borgna, A., *J. Nanoparticle Research*, 00:1–6 (2001).

Ristow M., Pfister M. F., Yee A. J., Schubert M., Michael L., Zhang C. Y., Ueki K., Michael II M. D., Lowell B. B., and Kahn C. R., *Proc. Natl. Acad. Sci. U. S. A.* 97:12239–12243 (2000).

Rivas G. A., Miscoria S. A., Desbrieres J. and Barrera G. D., *Talanta* 71(1):270–5 (2007).

Robertson D. H., Brenner D. H., and Mintmire J. W., *Phys. Rev. B*, 45:12592 (1992).

Rockenberger J., Scher R. C., and Alivisatos P., *J. Am. Chem. Soc.*, 121:11595 (1999).

Roland M. B., *Energy*, 28(497):198 (2002).

Romero M., Figueroa R., and Madden C., *Med. Dev. Diag. Ind. Mag.*, http://www.devicelink.com/mddi/archive/00/10/004.html (2000).

Rong H., Qian X., Yin J., and Zhu Z., *J. Mater. Chem.*, 12:3783 (2002).

Rosen D. R., Siddique T., Patterson D., Figlewicz D. A., Sapp P., Hentati A., Donaldson D., Goto J., O'Regan J. P., and Deng H. X. *Nature* 362:59–62 (1993).

Rosen R., Simendinger W., Debbault C., Shimoda H., Fleming L., Stoner B., and Zhou O., *Appl. Phys. Lett.*, 76:668 (2000).

Rotig A., de Lonlay P., Chretien D., Foury F., Koenig M., Sidi D., Munnich A., and Rustin P., *Nat. Genet.*, 17:215–217 (1997).

Ruoff R. S., Tse D. S., Malhotra R., and Lorente D. C., *J. Phys. Chem.*, 97:3379 (1993).

Rusop M., Soga T., Jimbo T., Umeno M., and Sharon M., *Surface Rev. Lett.*, 12(4):579 (2005).

Rusop M., Soga T., and Jimbo T., *Solar Energy Materials & Solar Cells*, 90(3):291 (2006).

Saido, T. C., Overview—Aβ metabolism: from Alzheimer research to brain aging control. In: Saido, T. C. (ed.), *Aβ Metabolism and Alzheimer's Disease*. Georgetown, TX: Landes Bioscience, pp. 1–16 (2003).

Saito Y., Hamaguchi K., Hata K., Uchida K., Tasaka Y., Ikazaki F., Yumura M., Kasuya A., and Nishina Y., *Nature*, 389:554 (1997).

Saito Y. and Uemura S., *Carbon*, 38:169–182 (2000).

Sakai K., *J. Chem. Eng. Jap.*, 30(4):587 (1997).

Sakamoto N., Ohshima K., Montermini L., Pandolfo M., and Wells R. D., *J. Biol. Chem.* 276:27171–27177 (2001).

Salgado J., Ricardo C., Antolini E., and Gonzalez E. R., *Appl. Catalysis B: Environmental*, 57(4):283 (2005).

Sandrock G. D. and Huston E. L., *Chemtech.*, 11:754 (1981).

Sasaki K., *Appl. Phys. A*, 69:115 (1999).

Satishkumar B. C., Govindaraj A., Sen R., and Rao C. N. R., *Chem. Phys. Lett.*, 293:47–52 (1998).

Satiskumar B. C., Govindaraj A., and Rao C. N. R., *Chem. Phys. Lett.*, 307:158–162 (1999).

Sato T., Shingeo F., Satoshi I., and Mitsugu H., *Jpn. J. Appl. Phys.*, 26:L1487 (1987).

Satoru T., Takayoshi K., Seiji A., and Yoshikazu N., *J. Appl. Phys.*, 40:4314 (2001).

Savage N. and Diallo M. S., *J Nanoparticle Res.*, 7:331 (2006).

Scheffel U., Rhodes B. A., Natarajan T. K., and Wagner H. N. Jr., *J. Med.*, 13:498. (1972).

Scheibe H. J., Drescher D., Schultrich B., Falz M., Leonhardt G., and Wilberg R., *Surface coat. Technol.*, 85:209 (1996).

Schellenberger E. A., Reynolds F., Weissleder R., and Josephson L., *Chem Bio Chem.*, 5:275 (2004).

Schmieder A. H., Winter P. M., Caruthers S. D., Harris T. D., Williams T. A., and Allen J. S., *Magnet. Reson. Med.*, 53(3):621 (2005).

Schoeters J. G., Buekens A. G., Recycling Berlin '79, Verlag fur Umwelttechnik: Berlin, (1979).

Scott C. D., Arepalli S., Nikolaev P., and Smalley R. E., *Appl. Phys. A: Mater. Sci. Process.*, 72:573 (2001).

Sen R., Zhao B., Perea D., Itkis M. E., Hu H., and Love J., *Nano Lett.*, 4(3):459–464 (2004).

Senda S., Tanemura M., Sakai Y., Ichikawa Y., Kita S., Otsuka T., Haga A., and Okuyama F., *Rev. Sci. Ins.*, 75(5):1366 (2004).

Seraphin S., Zhou D., Jiao J., and Loufty R., *Nature*, 363:503 (1993).

Seraphin S., Zhou D., Jio J., Withers J. C., and Loufty R., *Appl. Phys Lett.*, 63:2073 (1993).

Service R. F., *Science*, 281:940 (1998).

Shah N., Wang Y., Panjala D., and Huffman, G. P., *Energy and Fuels*, A–I (2004).

Shand H. and Wetter K., Shrinking science: An introduction to nanotechnology. In *State of the World 2006: Special Focus: China and India. The Worldwatch Institute.* New York: WW Norton & Company (2006).

Shandas R. and Lanning C., *Med. Biol. Eng. Comp.*, 41(4):416 (2003).

Shao, M., Wang, D., Yu, G., Hu, B., Yu, W., and Qian, Y., *Carbon*, 42:183–185 (2004).

Sharma B., Preethi D. K., and Agnihotri O. P., *Proc. SPIE, 3316, Physics of Semiconductor Devices*, Kumar V. and Agrawal S. K. (eds). 825 (1998).

Sharon M., Mukhopadhyay I., and Mukhopadhyay K.; *Sol. Energy Mat. Sol. Cells*; 45:35 (1997a).

Sharon M., Mukhopadhyay K., Mukhopadhyay I. , Soga T. and Umeno M., *Carbon*; 35:863–864 (1997b).

Sharon M., Mukhopadhyay K., Kalaga K. M., *Materials Chem. and Phys.* 49:252–257 (1997c).

Sharon M., Krishna K.M., Soga T., Mukhopadhyay K. and Umeno M., *Solar Energy Materials and Solar Cells*, 48(1–4):25 (1997d).

Sharon, M., Sharon M., Jain S., Kichambre P. D., and Kumar M., *Mater. Chem. Phys.*, 8:331 (1998).

Sharon M., Soga T., Afre E., Sathiyamoorthy D, Dasgupta K., Bhardwaj S., Sharon M., and Jaybhay S., *International Journal of Hydrogen Energy*; 32:4238 (2007).

Sharon M., Khairnar V., Jaybhaye S., Hu C., Afre R., Soga T. and Sharon M., *Carbon Science*, 9(3):188–194 (2008a).

Sharon M., Rusop M., T. Soga and Afre R., *Carbon Science*, 9(1):17–22, (2008b).

Sharon M., *Encylopedia of Nanoscience and Nanotechnology*, 1:517 (2004).

Sharon M. and Pal B., *Bull. Electrochem.*, 12(3–4):219 (1996).

Sharon M., Banerjee R., and Chatterjee A. K., *J. Power Sources*, 117:39 (2003a).

Sharon M., Banerjee R., and Chatterjee A. K., *J. Power Sources*, 5251:1 (2003b).

Sharon M., Datta S., Shah S., Sharon M., Soga T., and Afre R., *Carbon Lett.*, 8(3):189 (2007).

Sharon M., Hsu W. K., Kroto H. W., Walton D. R. M., Kawahara A., Ishihara T., and Takita Y., *J. Power Sources*, 45:1 (2001).

Sharon M., Hsu W. K., Kroto H. W., Walton D. R. M., Kawahara A., Ishihara T., and Takita Y., *J. Power Sources*, 104:148 (2002).

Sharon M., Kumar M., Kichambre P., and Neumann-Spallart M., *J. Appl. Electrochem.*, 28:1399 (1998).

Sharon M., Kumar M., Kichambre P. D., Avery N. R., and Black K. J., *Mol. Cryst. Liq. Cryst.*, 340:523 (2000).

Sharon M., Mukhopadhyay K., Mukhopadhyay I., and Krishna K. M., *Carbon*, 33(3):331–333 (1995).

Sharon M., Mukhopadhyay K., and Krishna K., *Carbon*, 34(2):251–264 (1996).

Sharon M., Mukhopadhyay K., Yase K., Ijima S., Ando Y., and Zhao X., *Carbon*, 36(5-6):507–511 (1998).

Sharon M., Pal. B., and Kamat. D. V., *J. Biomed. Technol.* 1:365 (2005).

Sharon M., Pradhan D., Ando Y., and Xinluo Z., *Curr. Appl. Phys.*, 2:445 (2002).

Sharon M., Pradhan D., Zacharia R., and Puri V., *J. Nanosci. Nanotechnol.*, 5(12):2117 (2005).

Sharon M., Soga T., Afre R., Sharon M., and Gupta A., *Abstract International Conference on Nanoscience and Techology—ICONSAT* (2006).

Sharon M. and Sharon M., *Int. J. Synthesis and Reactivity in Inorganic, Metal-Organic, and Nano-Metal Chemistry* 36:1–15, (2006).

Sharon M., Soga T., Afre E., Sathiyamoorthy D., Dasgupta K., Bhardwaj S., Sharon M. and Jaybhay S., *Int. J. Hydrogen Energy*, 32:4238 (2007).

Sharon M., Sundarakoteeswaran N., Kichambre P. D., Kumar M., Ando Y., and Xinluo Zhao, *Diam. Rel. Mater.*, 8:485 (1999).

Sheeja D., Tay B. K., and Nung L. N., *Diam. Relat. Mater,* 13(1):184 (2004).

Shinohara H., Sato H., Saito Y., Ohkohchi M., and Ando Y., *J. Phys. Chem.*, 96:3571 (1992).

Shinohara H.,Yamaguchi H., Hayashi N., Sato H., Ohkohchi M., Y. Ando Y., Saito Y., *J. Phys. Chem.*, 97:4259 (1993).

Shinohara H., Sato H., Ohkohchi M., Ando Y., Kodama T., Shida T., Kato T., and Saito Y., *Nature*, 357:52 (1992).

Shinohara H., Yamaguchi H., Hayashi N., Sato H., Ohkohchi M., Ando Y., and Saito Y., *J. Phys. Chem.*, 97:4259 (1992).

Shiraishi Masashi and Masafumi Alta, *Carbon* 39(12) 1913–1917 (2001).

Shults M. C., Rhodes R. K., Updike S. J., Gilligan B. J., and Reining W. N., *IEEE Trans. Biomed. Eng.*, 41 (10):937 (1994).

Shvedova A. A., and Castranova V., *J. Toxicol. Environ. Health A*, 66(20):1909 (2003).

Shvedova A. A., Vincent C., Kisin E. R., Schwegler-Berry D., Murray A. R., Gandelsman V. Z., Baron M. A., P. *J. Toxicology and Environmental Health Part A.*, 66(20):1909–26 (2004).

Sigma-Aldrich. *Handbook of Fine Chemicals and Laboratory Equipment.* (2003).

Singh R., Pantarotto D., McCarthy D., Chaloin O., Hoebeke J., and Partidos C. D., *J. Am. Chem. Soc.*, 127(12):4388 (2005).

Singh R., Pantarotto D., McCarthy D., Chaloin O., Hoebeke J., Partidos C. D., Briand J. P., Prato M., Bianco A., and Kostarelos K., *J. Am. Chem. Soc.* 127:4388–4396. (2005).

Singh R., Pantarotto D., and Lacerda L., *PNAS*, 103:3357 (2006).

Sinha N. and Yeow J. T. W., *IEEE Trans. Nanobiosci.*, 4(2):180 (2005).

Sirdeshmukh R., Teker K., and Panchapekesan B., *Proceedings of the 2004 International Conference on MEMS, NANO and Smart Systems* (2004).

Sisodia M. L. and Raghuwanshi G. S., *Basic Microwave Techniques And Laboratory Manual.* New Age International Publishers.

Sivakumar R., Ramaprabhu S., RamaRao K. V. S., Anton H., and Schmidt P. C., *J. Alloys Compounds*, 285:143 (1999).

Sloan J., Hammer J., Zwiefta-Sibley M., and Green M. L. H., *Chem. Commun.*, 347 (1998).

Slooter A. J., Cruts M., Kalmijn S., Hofman A., Breteler M. M., Van Broeckhoven C., and van Duijn C. M., *Arch. Neurol.*, 55:964–968 (1998).

Smalley R. E., G·io T., Nikolaev, P., Thess A., Colbert D. T., *Chem. Phys. Lett.*, 243: 49–54 (1995).

Souad A., Arnaud H., Noureddine J., Fernand F., Izio R., Françoise V., Philippe M., and Michel D., *J. Mater. Chem.*, 11:186 (2001).

Soppimath K. S., Aminabhavi T. M., Kulkarni A., and Rudzinski, *J. Controlled Rel.*, 70:1 (2001).

Sotiropoulou S. and Chaniotakis N. A., *Anal. Bioanal. Chem.*, 375:103 (2003).

Spencer P. S., Nunn P. B., Hugon J., Ludolph A. C., Ross S. M., Roy D. N., and Robertson, R. C., *Science*, 237:517–522 (1987).

Staii C., Johnson A. T., and Chen M., *Nano Lett.*, 5:1774 (2005).

Stampfer C., Jungen, A., and Hierold C., *Sensors,. Proceedings of IEEE* 24 (27):1056 (2004).

Standler R. B., *Protection of Electronic Circuits from Overvoltages.* New York: Wiley (1989).

Star A., Tu E., Niemann J., Gabriel J. C., and Joiner C. S., *Proc. Natl. Acad. Sci. U. S. A.*, 103(4):321 (2006).

Starke T. K. H., and Coles G. S. V. *IEEE Sensors J.* 2:14 (2002).

Stehling O., Elsässer H. P., Brückel B., Muhlenhoff U., and Lill R., *Hum. Mol. Genet.*, 13(23):3007–3015 (2004).

Stepanova A. N., Givargizov E. I., Bormatova L. V., ZhirnovV. V., Mashkova E. S., and Molchanov A. V., *J. Vac. Sc. Technol. B*, 16:678 (1998).

Stevefelt J. and Collins C. B., *J. Phys. D: Appl. Phys.*, 24:2149 (1991).

Stevens R. M. D., Frederick N. A., Smith B. L., Morse D. E., Stucky G. D., and Hansma P. K., *Nanotechnology*, 11(1):1 (2001).

Stobinski L., Tomasik P., Lii C. Y., Chan H. H., Lin H. M., and Liu H. L., *Carbohydr. Polym.*, 51(3):311 (2003).

Stone V., Shaw J., Brown D. M., Macnee W., Faux S. P., and Donaldson K., *Toxicol. In Vitro*, 12(6):649 (1998).

Strobel R., Jorissen L., Schliermann T., Trapp V., Schutz W., Bohmhammel K.,Wolf G., and Garche J., *J. Power Sources*, 84:221 (1999).

Sugie H., Tanemura M., Filip V., Iwata K., Takahashi K., and Okuyama F., *Appl. Phys. Lett.*, 78(17):2578 (2001).

Sun Z., *Diam. Rel. Mater.*, 9:1979, (2000).

Supronowicz P. R., Ajayan P. M., Ullman K. R., Arulanandam B. P., Metzger D. W., and Bizios R., *J. Biomed. Mater. Res.*, 59A(3):499 (2002).

Tagmatarchis N. and Prato M., *J Mater. Chem*, 14(4):437 (2004).

Tai Y., Inukai K., Osaki T., Tazawa M., Murakami J., Tanemura S., and Ando Y., *Chem. Phys. Lett.*, 224:118–122 (1994).

Takezawa T., *Biomaterials*, 24:2267 (2003).

Tamir S. and Drezner Y., *Appl. Surface Sci.*, 252:4819 (2006).

Tamura K., Takashi N., Akasaka T., Roska I. D., Uo M., and Totsuka Y., *Key Eng. Mater.*, 254:919 (2004).

Tanemura M., Tanaka J., Itoh K., Fujimoto Y., Agawa Y., Miao L., and Tanemura S., *App. Phys. Lett.*, 86:113107 (2005).

Tans S. J., Devoret M. H., Dai H., Thess A., Smalley R. E., Geerligs L. J., and Dekker C., *Nature*, 386:474 (1997).

Tao L., Luyang C., Keyan B., Yu W., and Yitai Q., *Carbon*, 44(13):2844–2848 (2006).

Teo K. B. K., Lee S. B., Chhowalla M., Semet V., Binh V. T., Groening O., Castignolles M., Loiseau A., Pirio G., Legagneux P., Pribat D., Hasko D. G., Ahmed H., Amaratunga G. A. J., and Milne W. I., *Nanotechnology*, 14:204 (2003).

Terrones M., Hsu W. K., Kroto H. W., and Walton D. R. M., In Hirsch A. (ed). *Fullerenes and Related Structures.* New York: Springer, p. 189 (1999).

Texier-Mandoki N., Dentzer J., Piquero T., Saadallah S., David P., and Vix-Guterl C., *Carbon*, 42(12–13):2744–2747 (2004).

Thess A., Lee R., Nikolaev P., Dai H., Petit P., Robert J., Xu C., Lee Y. H., Kim S. G., Rinzler A. G., Colbert D. T., Scuseria G. E., Tomanek D., Fischer J. E., and Smalley R. E., *Science*, 273:483–487 (1996).

Thews G., Mutschler E., and Vaupel P., *Anatomic, Physiologie, Pathophysiologie des Menschen.* Stuttgart, p. 217 (1999).

Thien-Nga L., Hernadi K., Ljubovic E., Garaj S., and Forro L., *Nano Lett.*, 2(12):1349 (2002).

Thomson L. A., Law F. C., and Franks R. N., *J. Biomater.*, 12(1):37 (1991).

Thostenson E. T., Ren Z., and Chou T. W., *Compos. Sci. Technol.*, 61:1899 (2001).

Thuy-Duong, Nguyen-Vu B., Chen H., Cassell A., Koehne J., Purewal H., Meyyappan M., Andrews R., and Li J., 10th Annual Conference of the International FES Society, Montreal, Canada (2005).

Tian Y., Hu Z., Yang Y., Wang X., and Chen X., *J. Am. Chem. Soc.*, 126:1180 (2004).

Tibbetts G. G., in *Carbon Fibers Filaments and Composites*. Figueiredo et al (eds). Netherlands: Kluwer Academic Publishers, pp. 73–94 (1990).

Tibbetts G. G., *J. Cryst. Growth*, 66:632 (1984).

Tibbetts G. G., Devour MG., and Rodda E. G., *Carbon*, 25:367–375 (1987).

Timea K., Zoltan K., and Imre K., *Langmuir*, 20:1656 (2004).

Toth E., Bolskar R. D., Borel A., Gonzalez G., Helm L., and Merbach A. E., *J. Am. Chem. Soc.*, 127(2):799–805 (2005).

Tran C. L., Buchanan D., Cullen R. T., Searl A., Jones A. D., and Donaldson K., *Inhal. Toxicol.*, 12(12):1113 (2000).

Tsang S. C., Chen Y. K., Harris P. J. F., and Green M. L. H., *Nature (Lond)*, 372:159 (1994).

Tsang S. C., Harris P. J. F., and Green M. L. H., *Nature (Lond)*, 362:520 (1993).

Tu W. and Liu H., *Chem. Mater.*, 12:564 (2000).

Ugarte D., *Nature*, 359:707–709 (1992).

Ugarte D., Stockli T., Bonard J. M., Chatelian A., and de Heer W. A., *Appl. Phys. A*, 67:101 (1998).

Umeno M. and Adhikary S., *Diam. Rel. Mater.*, 14(11-12):1973 (2005).

Utsumi T., *IEEE Trans. Electron Dev.*, 38:2276 (1991).

Vats A., Tolley N. S., Polak J. M., and Gough J. H., *Clin Otolaryngol.*, 28:165 (2003).

Venkatesan N., Yoshimitsu J., Ito Y., Shibata N., and Takada K., *Biomaterials*, 26(34):7154 (2005).

Vinoy K. J. and Jha R. M., *Radar Absorbing Materials: From Theory to Design and Characterization*. Boston: Kluwer Academic Publishers (1996).

Vivekchand S. R. C., Sudheendra L., Sandeep M., Govindaraj A., and Rao C. N. R., *J. Nanosci. Nanotechnol.*, 2:631 (2002).

Voevodin A., Capano M. A., Gatriel A. J., Donley M. 3., Zabiński J. S., *Appl. Phys. Lett.*, 69:188 (1996).

Vohrer U., Kolaric I., Haque M. H., Roth S., and Detlaff-Weglikowska U., *Carbon*, 42:1159–1164 (2004).

Vora J., Bapat N., and Broujerdi M., *Drug Deliv. Ind. Pharm.*, 19:759 (1993).

Walker J. P. L., Rakszawski J. F., and Imperial G. R., *J. Phys. Chem*, 63:133 (1959).

Walt A. de Heer, Châtelain A., and Ugarte D., *Science*, 270(5239):1179-1180 (1995).

Walter P., Eléonore W., Philippe H., Nestor J. , Zaluzec C. D., Jacques C., Veyssie`re P., Bréniaux R., Lévêque J. L., and Tsoucaris G., *Nano Lett.*, 6(10) (2006).

Wang J., *Electroanalysis*, 17:7 (2004).

Wang J., Musameh M., and Lin Y., *J. Am. Chem. Soc.*, 125:2408 (2003).

Wang J., Kawde A. N., and Jan M. R., *Biosens. Bioelectron.*, 20(5):995 (2004).

Wang M., Zhao X., Ohkohchi M., and Ando Y., *Fullerene Sci. Tech.*, 4:1027 (1996).

Wang S. F., Shen L., Zhang W. D., and Tong Y. J., *Biomacromolecules*, 6:3067 (2005).

Wang Y., Wei F., Luo G., Yu H., and Gu G., *Chem. Phys. Lett.*, 364:568–572 (2002).

Wang Y., Wang X., Ni X., Wu H., *Computational Materials Science*, 32(2):141–146 (2005).

Wang Z., Swinney K., and Bornhop D. J., *Electrophoresis*, 24:865 (2003).

Warheit D. B., Laurence B. R., Reed K. L., Roach D. H., Reynolds G. A. M., and Webb T. R., *Toxicol. Sci.*, 77(1):117 (2004).

Wartena R., Curtight A. E., Craig B. A., Aberto P., and Swider-Lyons K. E., *J. Power Sources*, 126:193 (2004).

Weber C., Reiss S., and Langer K., *Int. J. Pharm.*, 211:67 (2000).

Webster T. J., Waid M. C., McKenzie J. L., Price R. L., and Ejiofor J. U., *Nanotechnology*, 15(1):48–54 (2004).

Wei B. Q., Vajtai R., and Ajayan P. M., *Appl. Phys. Lett.*, 79:1172 (2001).

Wei J., Zhu H., and Wu D., *Appl. Phys. Lett.*, 84:24 (2004).

Wei X. and Cui T., *Sensors and Actuators A: Physical*, 16:510 (2007).

Wei Z., Wan M., Lin T., and Dai L., *Adv. Mater.*, 15:136 (2003).

Weia C. and Srivastava D., *Appl. Phys. Lett.*, 85(12) (2004).

Weisman C. V., *Nat. Biotechnol.*, 21:1166 (2003).

Weizhong Q., Fei W., Zhanwen W., Tang L., Hao Y., Guohua L., Lan X., and Xiangyi D., *AICHE J.*, 49(3):619 (2003).

Wen X., Shi D., and Zhang N., *Applications of Nanotechnology in Tissue Engineering, Handbook of Nanostructured Biomaterials and Their Applications in Nanobiotechnology*, H. S. Nalwa (ed.), 1:1–23, (2005).

Widder K. J., Senyei A., And Czerliski E., *J. Appl. Phys.*, 49:3578 (1978a).

Widder K. J., Senyei A. E., and Sarpelli D. G., *Proc. Soc. Exp. Biol. Med.*, 58:141 (1978b).

Widder K. J., Flouret G., and Senyei A. E., *J. Pharm. Sci.*, 68:79 (1979).

Widder K. J., Morris R. M., Poore G. A., Howard D. P., and Senyei A. E., *Eur. J. Cancer*, 19:135–139 (1983a).

Widder K. J., Marino P. A., Moris R. M., Howard D. P., Poore G. A., and Senyei A. E., *Eur. J.Cancer*, 19:141–147 (1983b).

Wildoer J. W. G., Venema L. C., Rinzler A. G., Smalley R. E., and Dekker C., *Nature*, 391:59 (1998).

Williams D. F., *The Williams Dictionary of Biomaterials*. Liverpool, UK: Liverpool University Press (1999).

Wilson M., Chapman and Hall "*Nanotechnology: Basic Science and Emerging Technologies*" Academic Press (2002).

Wilson M., Kannangara K., Smith G., Simmons M., and Raguse B. *Nanotechnology-Basic Science and Emerging Technologies*, Boca Raton, FL: CRC Press, (2000).

Wong S. S., Harper J. D., Lansbury P. T., and Lieber C. M., *J. Am. Chem. Soc.*, 120:603 (1998).

Wong S. S., Joselevich E., Woolley A. T., Cheung C. L., and Lieber C. M., *Nature*, 394:52 (1998).

Wu Y. H., Qiao P. W., Chong T. C., and Shen Z. X., *Adv. Mater.*, 14:64 (2002).

Xie S., Li W. Z., Qian L, Chang B. H., Zou B. S., Zou W. Y., Zau R. A., and Wang G, *Science*, 274:1701 (1996).

Xing N. W. and Yan, Z. F., *Nanotechnology in Mesostructured Materials*, 146:771 (2003).

Xue H. G., Sun W. L., He B. J., and Shen Chin Z. Q., *Chem. Lett.*, 13:799 (2002).

Xueen J. I. A., Qingji X. I. E., Zang Y., and Shouzhuo Y. A. O., *Analytic. Sci.*, 23:689 (2007).

Yacaman M. J., Yoshida M. M., and Rendon L., *Appl. Phys. Lett.*, 62:657 (1993).

Yakobson B. I. and Smalley R. E., *Am. Sci.*, 84(4):324 (1997).

Yamago S., Tokuyama H., Nakamura E., Kikuchi K., Kananishi S., and Sueki K., *Chem. Biol.*, 2(6):385 (1995).

Yang D., Lee, Woo-Sung C., Seung-I. L. M., Yun-Hi L., Kim J. K., Sahn N., and Byeong-Kwon J., *Chem. Phys. Lett.*, 433:105–109 (2006).

Yang J., Shen Z. M., and Hao Z. B., *Carbon*, 42(8-9):1882–1885 (2004).

Yao Z., Postma H. W. C., Balents L., and Dekker C., *Nature*, 402:273 (1999).

Yasuda A., Kawase N., and Mizutani W., *J. Phys. Chem. B*, 106(51) (2002).

Ye H. F., Wang J. B., and Zhang H. W., *Comp. Mater. Sci.*, 44(4) (2009), 1089–1097.

Ye J., Shi X., Jones W., Rojanasakul N., Cheng N., and Cheng D., *Am. J. Physiol.*, 276(3):426 (1999).

Yin Y., Lu Y., Gates B., and Xia Y., *J. Am. Chem. Soc.*, 123:8718–8729 (2001).

Ying L., Wang M., Zhao F., Xu Z., and Dong S., *Biosensors and Bioelectronics*, 21(6):984 (2005).

Yinghuai Z., Peng A. T., Carpenter K., Maguire J. A., Hosmane N. S., and Takagaki M., *J. Am. Chem. Soc.*, 127(27):9875–9880 (2005).

Yongsheng Z., Ke Y., Rongli X., Desheng J., Laiqiang L., and Ziqiang Z. 120(1):142 (2005).

Yongqiang T. and Resasco D. E., *J. Phys. Chem. B*, 109:14454 (2005).

Yoon T. and Cowan J. A., *J. Biol. Chem.* 279:25943–25946 (2004).

Yu X., Munge B., Patel V., Jensen G., Bhirde A., Gong J. D., Kim S. N., Gillespie J., Gutkind J. S., Papadimitrakopoulos F., and Rusling J. F., *J. Am. Chem. Soc.*, 128(34):11199 (2006).

Yudasaka M., Ichihashi T., Kasuya D., Kataura H., and Ijima S., *Carbon*, 41:1273–1280 (2003).

Yue G. Z., Qui Q., Gao B., Cheng Y., Zhang J., Shimoda H., Chang S., Lu J. P., and Zhou O., *Appld. Phys. Lett.*, 81(2):355 (2002).

Yuji M., Kei K., Kazumi M., and Hirofumi Y., *Nanotechnology*, 18:5 (2007).

Yung-Tai W. and Hu C.-C., *J. Electrochem Soc.*, 151(12):2060 (2004).

Yutaka M., Makoto K., Masahiro H., Tadashi H., Shin-ichi K., Yongfu L., Takatsugu W., Takeshi A., Said K., Nobutsugu M., Toshiya O., Yuhei H. Kenji H., Lu J., and Shigeru N., *J. Am. Chem. Soc.*, 128:12239–12242 (2006).

Zanello L. P., Zhao B., Hu H., and Haddon R. C., *Nano Lett.*, 6:562 (2006).

Zhang J., Li J., Cao J., Qian Y., *Mater. Lett.*, 62:1839–1842 (2008).

Zhang L., Yongqiang T., and Resasco D. E., *Chem. Phys. Lett.*, 422:198–203 (2006).

Zhang Y., Li J., Shen Y., Wang M., and Li J., *J. Phys. Chem. B*, 108(39):5343 (2004).

Zhang Z., Kleinstreuer C., Donohoe J. F., and Kim C. S., *J. Aerosol. Sci*, 36(2):211 (2005).

Zhao M., Beauregard D. A., Loizou L., Davletov B., and Brindle K. M., *Nat. Med.*, 11:1241–1244 (2001).

Zhao N., Cao Ting, Shi C., Li J., and Guo W., *Fuhe Cailiao Xuebao*, 20(5):63 (2003).

Zhao X., Ohkohchi M., Wang M., Iijima S., Ichihashi T., and Ando Y., *Carbon*, 35:775–781 (1997).

Zhao X., Guo T. L., Ohkohchi M., Okazaki T, Iijima S., and Ando Y., *Acta. Metal. Sinica*, 12:345 (1999).

Zhao X., Inoue S., Jinno M., Suzuki T., and Ando Y., *Chem. Phys. Lett.*, 373:266–271 (2003).

Zhao X., Liu Y., Inoue S., Suzuki T., Jones R. O., and Ando Y., *Phys. Rev. Lett.*, 92:125502 (2004).

Zheng X., Chen G. H., Li Z., Deng S., and Xu N., *Phys. Rev. Lett.*, 92:106803 (2004).

Zhenhui K., Enbo W., Baodong M., Zhongmin S., Lei C., and Lin X., *Nanotechnology*, 16:1192 (2005).

Zhu W., Bower C., Kochanski G. P., and Jin S., *Solid State Electronics*, 45.921 (2001).

Zhu W., Kochanski G. P., Jin S., and Seibles L., *J. Vac. Sci. Technol. B* 14:2011 (1996).

Zhu H. W., Li X. S., Ci L. J., Xu C. L., Wu D. H., and Mao Z. Q., *Mater. Chem. Phys.*, 78:670 (2003).

Ziegler K. J., *Trends Biotechnol.*, 23(9):440 (2005).

Zolle I., Hosain F., Rhodas B. A., and Wagner H. N. Jr., *J. Nucl. Med.*, 11:379 (1970).

Zorbas V., Ortiz-Acevedo A., Dalton A. B., Yoshida M. M., Dieckmann G. R., and Draper R. K., *J. Am. Chem. Soc.*, 126(23):7222–7227 (2004).

Additional Resources

http://people.nas.nasa.gov/~cwei/Publication/swcnt_strainrate_T.pdf.

http://people.nas.nasa.gov/~cwei/on/mwcnt_strainrate.pdf.

http://en.wikipedia.org/wiki/Carbon_nanotube#cite_note-carbon-0 (Accessed on 2-7-08).

http://people.nas.nasa.gov/~cwei/Publication/nanofiber.pdf.

http://www.iris.ethz.ch/NT0206p12_21.pdf.

http://en.wikipedia.org/wiki/Carbon_nanotube#cite_note-carbon-0 (Accessed on 2-7-08).

http://www.cict.nasa.gov/assets/pdf/014_CICT_ITSR_CNT_Biosensors_A_web.pdf.

http://en.wikipedia.org/wiki/Adsorption.

http://www.iupac.org/publications/pac/2001/pdf/7308x1381.pdf.

http://www.mdpi.org/sensors/papers/s10100013.pdf.

http://nano.eecs.berkeley.edu/publications/NanoLett_2003_NO2.pdf.

INDEX

523